Finite Element Method for Engineers

From Theory to Practice

Finite Element Method for Engineers

From Theory to Practice

C.V. Girija Vallabhan
Mehmet Zülfü Aşık

Alpha Science International Ltd.
Oxford, U.K.

Finite Element Method for Engineers
From Theory to Practice
292 pgs. | 151 figs. | 13 tbls.

C.V. Girija Vallabhan
Emeritus Professor of Civil Engineering
Texas Tech University
Lubbock, Texas, USA

Mehmet Zülfü Aşık
Professor of Engineering Sciences
Middle East Technical University (METU)
06800 Ankara, Turkey

ALPHA SCIENCE INTERNATIONAL LTD.
7200 The Quorum, Oxford Business Park North
Garsington Road, Oxford OX4 2JZ, U.K.

www.alphasci.com

ISBN 978-1-84265-553-5

Printed in India

Dedication

This book is dedicated to my dear parents Sri. C.R. Velayudhan and Smy. M.K. Devaky their tireless efforts and sacrifices throughout their lives to fulfill their dreams to inspire all the ten children to achieve the highest level of education possible

C.V. Girija Vallabhan

Preface

While there are more than a dozen text books on finite element methods, one might ask why add one more. My response is that I started research on finite element method in early sixties, when there was no text book on finite element method. Since then, during my several years of teaching and research on finite element methods, I developed my own class notes with a beginning student in mind. I accepted a challenge to continuously push myself to present finite element method in a clear and simple format. With that challenge in mind, I updated my class notes in the form of a text book, that would help even students who aspire to learn finite element methods on their own. I synthesized the years of feedback from colleagues and more importantly, from students, to present a beginner's guide to finite element method. This is the purpose of the book.

Most students in civil, mechanical or aeronautical engineering who studied fundamentals of solid mechanics are familiar with the minimum potential energy theorem which is essentially another form of the principle of virtual work or the Castigliano's theorem used in solving structural mechanics problems. It is interesting to note that in the early days of the development of finite element methods, this energy principle was adopted.

Finite element methods in solid mechanics can be introduced in two different approaches: one is by the use of an energy principle and the other is to use a variational technique called the "weak form of the Galerkin principle." Some texts employ the Galerkin principle which is more general and is necessary for solving fluid mechanics problems using finite element method. Since solid mechanics is emphasized here, the main thrust of the methodology adopted is the energy principle. The weak form of the Galerkin principle is discussed at the end of the book to provide a brief review of this principle.

I am extremely thankful to my dear friend and former graduate student, Dr. Mehmet Asik, Professor of Engineering Sciences, Middle East Technical University, Ankara, Turkey, who agreed to be a co-author for the book and helped me to complete the manuscript.

Above all, I am deeply indebted to my brother Dr. Shanmukhan Chiyyarath who pains takingly read the entire manuscript and made valuable suggestions and corrections.

C.V. Girija Vallabhan

Abstract

Organization of the Book

The finite element method can be considered as the application of the classical Ritz method, which is based on the minimum potential energy theorem. The first chapter of the book explains the application of the Ritz method to solve simple (Euler) beam problems in solid mechanics. It is shown that convergence to exact solutions is improved as the number of degrees of freedom in the problem is increased. It is further shown that the Ritz principle can lead to the well known Galerkin principle as well.

Chapter two describes the essential equations in solid mechanics. Also the student is introduced to two-dimensional plane elasticity, axisymmetric and three-dimensional problems in elasticity. An emphasis is made to describe what are known as $\mathbf{C}_0$ and $\mathbf{C}_1$ continuities in functions since there are two different types of finite elements commonly used; one type for analysis of solid continuua and the other for beams, plates and shells.

Chapter three deals with one dimensional simple bar elements (of $\mathbf{C}_0$ type) where all the simple fundamental techniques in finite element method are introduced. The main four steps in the derivation of the stiffness matrix of an element are introduced and emphasized as the same steps can be used for obtaining stiffness matrices of all finite elements. The principle and technique used to assemble elements to form the global stiffness matrix equation, applying boundary conditions by modifying the equations and formation and solution of these equations in a banded form are also presented here. Obtaining equivalent nodal forces using energy principles is also presented here. Higher order bar elements are also introduced. Here the bar elements are extended to analyze two-and three-dimensional trusses. This chapter provides a complete perception of the overall finite element methodology for all other types of finite elements.

Chapter four is dedicated to plane stress/plane strain problems (using $\mathbf{C}_0$ type elements). Here the student gets a first hand feeling for solving a 2-D plane elasticity problems using triangular elements. Higher order triangular elements are introduced in this chapter. Chapter five is for rectangular and quadrilateral finite elements used to solve 2-D plane elasticity problems. Numerical integration schemes using Gaussian quadrature which stimulated the finite element menthodology into developing newer and more complicated finite elements are introduced. Chapter six and seven are devoted for finite element analysis of axisymmetric and 3-D problems in elasticity.

In Chapter eight, beam elements which are of the type with C_1 continuity are discussed. First the familiar Euler beam elements are introduced. Here it is shown that be combining simple bar and beam elements, two-and three-dimensional rigid frames and arches can be analyzed. Further, for the benefit of the beginners in solid mechanics, the Timoshenko beam theory is introduced so that a new beam element can be developed that includes the shear deformation in the beam. Even though this new beam element is not used much for the analysis of beams or frames, the concept used here provides the basis of the plate element that includes the bending and shear deformations in plates. Chapter nine starts with the development of the Classical Kirchhoff plate elements and then later is used to introduce the Mindlin plate elements that are very popular to create quadrilateral elements. Shell elements are introduced in this chapter.

Chapter ten is used to introduce the application of finite elements to solve some two-dimensional steady state flow problems. Towards the end of the chapter the weak form of Galerkin method is presented for those interested in that methodology. Some similarities of the finite method with the classical finite difference method is briefly introduced here.

An Appendix is provided to shown different types of equation solvers that are used for solving finite element equations.

Another important aspect of the book is its emphasis on sign conventions. Since the finite element method extends to many complex three-dimensional problems, there is no doubt that a consistent sign convention is essential. The distinction between the sign convention for bending moments and applied moments is emphasized.

Features of the Book

One of the many features of this book is that it is specifically tailored for self-learning of the finite element method from fundamentals. Basic principles in assembling elements, applying boundary conditions etc. are explained in detail. In addition, it is designed for a one semester course on finite element method. I regularly taught the entire contents of this book in one full semester. This book is not intended to be a references book on finite element method; only a minimum number of references is provided.

In order to understand and feel the thrill of the finite element methodology completely, one has to know how the theory is carried onto the computer through coding. Due to this reason, I am not in favor of introducing commercial finite element codes to beginners. To make students understand the transfer of theory to develop computer code, several simple subroutines are presented that helps them to learn the different computer schemes. Subroutines to compute element stiffness matrix with numerical integration if necessary, to assemble element stiffness matrices to form global equations in a half-banded form, to modify the global equations to satisfy the boundary conditions and to solve for the banded equations are presented. Additionally, the book encourages the student to have a hands-on experience on finite element programming using any one computer language of his/her choice. Fortran language is used in this book. A simple 2-D plane elasticity program is given at the end of this book. By using the methodology that is presented here, one can write a finite element program in any other preferred computer language. Writing one full simple finite element program is emphasized because the student gets a very clear picture of the organization of all finite element codes that he/she may use in future. Many of my graduate students wrote their own new finite element codes with their own choice of computer language using the information provided here. This experience is extremely valuable to students who comprehend the basic philosophy used in commercial codes. Experience in coding a finite element program helps those who desire to continue research in the area of finite element methods.

C.V. Girija Vallabhan
Mehmet Zülfü Aşık

A Special Acknowledgement

I consider a great honor and privilege to work as the co-author with my former advisor and friend Dr. C. V. Girija Vallabhan who was a distinguished professor with immense knowledge and experience on finite element method. Even though there are many text books on the subject, I agreed to be a coauthor because of the organization and contents of the book which immensely appealed to me. This book is different from others: the methodology followed in the book is logical, descriptive and clarifies the fundamentals of the topic, relates theory, concepts or ideas towards an easy learning of the finite element procedure; it skillfully integrates the knowledge on mechanics, mathematics and numerical computation in engineering. It is a great chance for me to make a contribution for the completion of such a promising book. Again I would like to acknowledge my sincere thanks for allowing me to be a coauthor.

Mehmet Zülfü Aşık

Contents

1

CHAPTER

Introduction

1.1 WHAT IS FINITE ELEMENT METHOD?

Often beginners learning engineering and science are curious 'what is this finite element method?' A simple concept can be obtained by imagining a structure of a building or a bridge as composed of columns, beams and plates, which can be individually called 'elements' or 'finite elements', the properties of these elements when assembled together using principles of mechanics and matrix algebra can represent the whole structure. The overall property such as the deformation and stresses in the structure due to the prescribed forces is computed by solving the assembled equations of its individual elements. This procedure in short is called the finite element method.

In the finite element method, the major emphasis is to study systematically the properties of the individual elements in their own local coordinates, and assembling these element properties mathematically in a global coordinate system to simulate or approximate the behaviour of the whole structure. One can then perform comprehensive studies on the various characteristics of the structure, such as displacements, strains and stresses as it is subjected to different combinations of forces. The finite element method is not restricted to the analysis of trusses or framed structures; it is used to analyze plates, shells and other solid continuum problems, where the entire solid is discretized into finite elements. Here one develops algebraic equations to characterize the behavior of these finite pieces of the body and these equations are assembled to characterize the original structure. The structure can be an airplane, building, machine part, dam or even a human bone or bones, etc. The only criterion is that the finite part of the overall body should have some definite known material behavior that can be represented mathematically.

The word 'finite element' was coined by Clough[2] in 1960, but Turner et al.[10] developed some fundamental ideas in 1956. In fact, Courant[4] had published fundamental concepts on finite elements as early as in 1943, which got attention of engineers only after a few years later. Ever since, during the early years the finite element concept grew in the world along with the development of high speed computers, many pioneers like Zienkiewicz[12] and many others popularized, Oden[8], Washizu[11] and others connected the finite element theory with the many existing mathematical theories. The application of the finite element method is not limited to structural mechanics problems; engineers[12] have extended the concept to analyze nonlinear solids, complex behavior of fluids, heat and mass transfer phenomena, electromagnetic fields and even to solve complicated differential equations.

The basis of finite element formulation is similar to that of the differential equations in mechanics. In order to derive governing differential equations in mechanics, a small differential element having sides dx and dy in a two dimensional, or dx, dy and dz in a three dimensional continuum is assumed, and the gradients of the pertinent parameters on different sides of this differential element are formulated. Using these gradients and other properties, governing differential equations are developed based on some general established theorems of mechanics such as equilibrium. These governing differential equations are applicable to the entire continuum. In the finite element method, instead of using an infinitesimal differential element, e.g., for a two dimensional problem in a continuum much bigger elements of arbitrary shapes such as triangular, rectangular, quadrilateral with straight or even curved boundaries are used. A set of discrete algebraic equations are established that relate pertinent characteristic parameters of the element. All these individual algebraic equations of each element are assembled to form a set of simultaneous equations. To obtain detailed and accurate solutions, one may have to employ a large number of small finite elements resulting in a large number of algebraic equations. The advances in computers in their speed and capacity aiding the solution of large number of equations have made the finite element method a popular tool among engineers and scientists.

For those interested in learning the finite element method, one of the best ways is to start with the classical energy methods such as Ritz and Galerkin used in earlier times for solving problems in solid mechanics.[1,5,7,9,11] This chapter is devoted to an introduction to these methods, especially to solve simple problems for which the results are known from elementary principles of solid mechanics. The classical beam problem is selected as the example problem.

1.2 CLASSICAL EULER BEAM PROBLEM

One of the differential equations[4] that engineers are normally exposed to during their early studies of solid mechanics is related to the lateral deflection v of a prismatic beam subjected to a prescribed distributed load $\tilde{q}$. A brief discussion on the elementary beam theory usually referred to as Euler beam theory[6] is presented in Chapter 4. As an example, a simply supported beam having a uniform stiffness EI carrying a distributed load $\tilde{q}$ as shown in Fig. 1.1 is considered here.

Using all the assumptions of the Euler beam theory, the differential equation for the lateral deflection of the beam can be obtained as

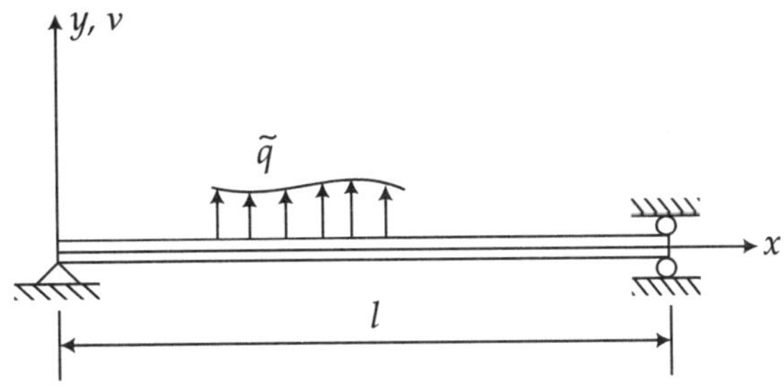

Fig. 1.1 A Simple Beam Carrying a Distributed Load

$$EI \frac{d^4 v}{dx^4} = \tilde{q}, \text{ for } 0 < x < l \tag{1.1}$$

Equation (1.1) is sometimes called the field or domain equation for the beam. This equation being a fourth order differential equation, requires four boundary conditions. For the simply supported case, the boundary conditions are:

$$@\, x = 0,\; v = 0 \text{ and the applied moment } \tilde{M}_0 = -EI \left. \frac{d^2 v}{dx^2} \right|_0 = 0$$

$$@\, x = l,\; v = 0 \text{ and the applied moment } \tilde{M}_l = EI \left. \frac{d^2 v}{dx^2} \right|_l = 0 \tag{1.2}$$

If $\tilde{q}$ is constant, by simple integration and applying the boundary conditions, the solution of Eq. (1.1) is obtained as

$$v = \frac{\tilde{q}}{EI}\left(\frac{x^4}{24} - \frac{x^3 l}{12} + \frac{x l^3}{24} \right) \tag{1.3}$$

The problem represented by Eq. (1.1) and its boundary conditions, is a relatively simple boundary value problem in mathematics, which has a unique solution given by Eq. (1.3). While it is easy to solve Eq. (1.1) directly, there are many boundary value problems in solid mechanics where exact solutions are indeed too complex to obtain and perhaps one has to rely on some approximate methods. Let us try to solve this simple equation by an approximate method so as to build some confidence in these methods of approximation and compare the results obtained with the exact solution.

1.3 APPROXIMATE METHODS OF SOLUTION

Mathematicians and engineers in the past have contemplated on obtaining approximate solutions for differential equations that are acceptable for practical purposes.[5,7,9] Earlier, they tried finite difference methods[3] to solve differential equations, and these methods became handy only when computers became available for long and tedious calculations. Another very popular and classical approximate method used to solve problems in solid mechanics comes from the use of the minimum potential energy theorem. The basic principle of the finite element method, which is in fact an approximate method of solution, is rooted in this energy theorem. The fundamental aspect of this theorem is given in chapter two; however an attempt is made to make use of the theorem here, so that one can get a feel of this technique for use in the finite element method. According to the theorem, an energy function[3,5,6,7,9]

$$\Phi = \mathbf{U} + \mathbf{V} \tag{1.4}$$

is a minimum at the equilibrium configuration of the beam, where **U** is the elastic strain energy stored in the beam and **V** is equal to the inner product of the prescribed loads and the corresponding displacements of the beam with a negative sign in the front. Some call this function **V** as 'work potential of the prescribed loads' or 'potential energy of the prescribed loads', but whatever way it is called, one has to pay close attention to its definition as given above. The function Φ is called 'the total potential energy function' and will be refered in this text on several occasions. For the beam in Fig. 1.1, the energy function

$$\Phi = \frac{1}{2}\int_0^l EI\left(\frac{d^2v}{dx^2}\right)^2 dx - \int_0^l \tilde{q}v\,dx \tag{1.5}$$

is a minimum at equilibrium position. The mathematical equivalence of the above statement and the equilibrium equation given by Eq. (1.1) is shown later in this chapter. Ritz in 1909, suggested that by assuming an approximate function for the displacement of the beam that will a priori satisfy the boundary conditions, a satisfactory solution can be obtained by minimizing the function Φ. For example, assume an approximate solution,

$$\bar{v} = c\sin\frac{\pi x}{l} \tag{1.6}$$

Upon examination, the function $\bar{v}$ satisfies all the boundary conditions but will not satisfy the field equation. Here the function $\sin(\pi x/l)$ defines approximately the shape of the lateral displacement of the beam, while the term c is an adjustable parameter that controls the amplitude of the displacement. Substituting $\bar{v}$ in Eq. (1.5),

$$\Phi = \frac{EI}{2}c^2\int_0^l\left(-\frac{\pi^2}{l^2}\sin\frac{\pi x}{l}\right)^2 dx - \int_0^l \tilde{q}\,c\sin\frac{\pi x}{l}dx$$

$$= \frac{EI}{4}\frac{\pi^4}{l^3}c^2 - \frac{2\tilde{q}\,l}{\pi}c \tag{1.7}$$

is obtained as a quadratic in c, and has to be a minimum at the equilibrium position. Since c is an adjustable parameter and Φ is a function of c, it is minimized by differentiating with respect to c and set it equal to zero, i.e.,

$$\frac{d\Phi}{dc} = \frac{EI\,\pi^4}{2l^3}c - \frac{2\tilde{q}\,l}{\pi} = 0,$$

which leads to

$$c = \frac{4\tilde{q}\,l^4}{EI\,\pi^5}$$

i.e.
$$\bar{v} = \frac{4\tilde{q}\,l^4}{EI\,\pi^5}\sin\frac{\pi x}{l} \tag{1.8}$$

and,
$$\bar{v}_{\max}\Big|_{x=\frac{l}{2}} = \frac{4\tilde{q}\,l^4}{EI\,\pi^5} = 0.01307\,\frac{\tilde{q}\,l^4}{EI},$$

while using the exact solution from Eq. (1.3), the maximum deflection of the beam in the middle is,

$$v_{\max}\Big|_{x=\frac{l}{2}} = \frac{5}{384}\frac{\tilde{q}\,l^4}{EI} = 0.01302\,\frac{\tilde{q}\,l^4}{EI}$$

It is observed from Table 1.1 that the error is small for the maximum displacement, but the error increases for the maximum value of the slope, bending moment and shear force in the beam, progressively. Also, it can be observed that

$$EI\frac{d^4\bar{v}}{dx^4} = EI\frac{\pi^4}{l^4}\frac{4\tilde{q}\,l^4}{EI\,\pi^5}\sin\frac{\pi x}{l}$$

$$= \frac{4}{\pi}\tilde{q}\,\sin\frac{\pi x}{l},$$

and not equal to $\tilde{q}$, which is a constant.

Table 1.1 Comparison of Results

	Exact Solution Eq. (1.3)	*Approximate Solution Eq. (1.8)*	*% of Error*
Max. Displacement × $\frac{EI}{\tilde{q}l^4}$	0.01302	0.01307	- 0.384
Max. Slope × $\frac{EI}{\tilde{q}l^3}$	0.0417	0.0411	1.44
Max. Bending Moment × $\frac{1}{\tilde{q}l^2}$	0.125	0.129	- 3.20
Max. Shear Force × $\frac{1}{\tilde{q}l}$	0.500	0.4053	18.94

The comparison between the distribution of the actual and the computed bending moment, shear force and prescribed load are shown in Fig. 1.2(a). Naturally, now the question will be, 'how to improve the solution?'. A new approximate solution with two variables is tried as,

$$\bar{v} = c_1\sin\frac{\pi x}{l} + c_2\sin\frac{3\pi x}{l} \tag{1.9}$$

with two adjustable parameters that will satisfy all the boundary conditions. It should be noted that the function $\sin\frac{2\pi x}{l}$, being antisymmetric with respect to the center of the beam, is not a possible deflection shape for this loading condition, hence not considered in this example, and, even if it is included, its coefficient will be found to be equal to zero. Substituting Eq. (1.9) in Eq. (1.5),

$$\Phi = \frac{EI}{2}\left(\frac{\pi^4}{l^4}c_1{}^2 + \frac{81\pi^4}{l^4}c_2{}^2\right)\frac{1}{2} - \frac{2\tilde{q}l}{\pi}\left(c_1 + \frac{c_2}{3}\right)$$

which is a quadratic in c_1 and c_2. Minimizing Φ with respect to c_1 and c_2, i.e.,

$$\frac{d\Phi}{dc_1} = \frac{EI\pi^4}{2l^3}c_1 - \frac{2\tilde{q}l}{\pi} = 0; \qquad \text{which gives } c_1 = \frac{4\tilde{q}l^4}{EI\,\pi^5}$$

and

$$\frac{d\Phi}{dc_2} = \frac{81EI\pi^4}{2l^3}c_2 - \frac{2\tilde{q}l}{3\pi} = 0; \qquad \text{which gives } c_2 = \frac{4\tilde{q}l^4}{243EI\,\pi^5}$$

which results,

$$\bar{v} = \frac{4\tilde{q}l^4}{EI\pi^5}\left(\sin\frac{\pi x}{l} + \frac{1}{243}\sin\frac{3\pi x}{l}\right)$$

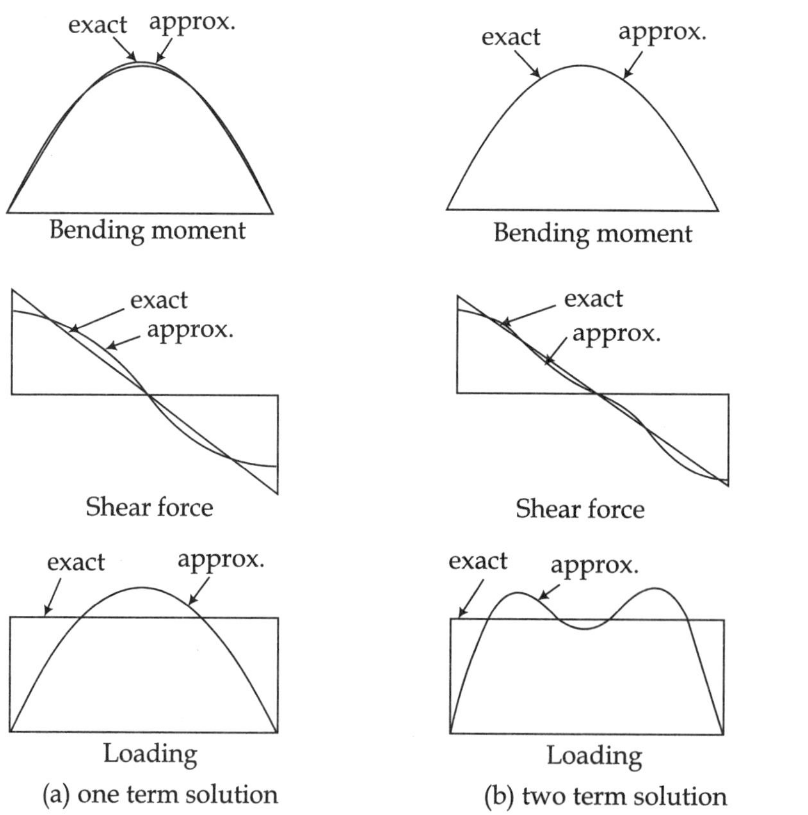

Fig. 1.2 Comparison of Exact and Approximate Solutions

The term c_2 is considerably smaller than c_1; however, the approximate shear forces and loading that result from this solution are shown in Fig. 1.2(b). Evidently, one can conclude that by adding more terms, c_i, $i = 1, 2,\ldots,n$, for $\bar{v}$ such that

$$\bar{v} = \sum_{i=1}^{n} c_i \sin\frac{(2i-1)\pi x}{l} \tag{1.10}$$

the solution can be improved. Substituting Eq. (1.10) in Eq. (1.5) and differentiating with respect to c_i and setting equal to 0, the solution is obtained as:

$$\bar{v} = \sum_{i=1}^{n} \frac{4\tilde{q}l^4}{\pi^5(2i-1)^5} \sin\frac{(2i-1)\pi x}{l}$$

In other words, the approximate solution, converges to the exact solution as n tends to ∞. The convergence for displacement is rather very fast, since the values of $c_1 = 0.01307\ \tilde{q}l^4$, $c_2 = -0.0000537\ \tilde{q}l^4$ and for higher values of i the coefficients c_i becomes very small. However, values of the slope, bending moment and shear force converge slightly slower and the fourth derivative of the displacement will slowly converge to a uniformly distributed load.

1.3.1 A Beam Problem with a Concentrated Load

Consider another example of a simply supported beam carrying a concentrated load P at the center, as shown in Fig. 1.3.

Again, let $\bar{v} = c \sin\frac{\pi x}{l}$,

that will satisfy all the boundary conditions, then the potential energy function is given by

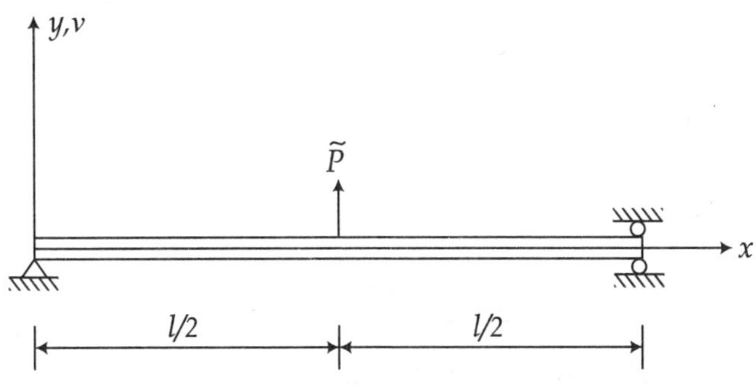

Fig. 1.3 A Beam Carrying a Concentrated Load

$$\Phi = U + V = \frac{1}{2}\int_0^l EI\left(\frac{d^2v}{dx^2}\right)^2 dx - \tilde{P}v\Big|_{x=\frac{l}{2}}$$

$$= \frac{EI}{4}\frac{\pi^4}{l^3}c^2 - \tilde{P}c\sin\frac{\pi x}{l}\Big|_{x=\frac{l}{2}}$$

$$= \frac{EI}{4}\frac{\pi^4}{l^3}c^2 - \tilde{P}c$$

Setting
$$\frac{d\Phi}{dc} = \frac{EI\pi^4}{2l^3}c - \tilde{P} = 0$$

$$c = \frac{2\tilde{P}l^3}{\pi^4 EI}, \quad \text{which leads to the solution,}$$

$$\bar{v} = \frac{2\tilde{P}l^3}{\pi^4 EI}\sin\frac{\pi x}{l} \tag{1.11}$$

and the exact solution is

$$v = \frac{\tilde{P}l^3}{48EI}\left(\frac{3x}{l} - \frac{4x^3}{l^3}\right), \text{ for } 0 < x < \frac{l}{2} \tag{1.12}$$

Table 1.2 Analysis of the Results

Displacement $\frac{EI\,v}{\tilde{P}l^4}$	*Exact*	*Approximate*	*% of error*
@ $x = l/2$	0.02083	0.02053	+ 1.44
@ $x = l/4$	0ty.01432	0.01452	- 1.40

Comparison of the distribution of the bending moment, shear force and loading between the approximate solution and the exact solutions are shown in Fig. 1.4(a) for a one term expansion of the displacement function. Now, if the approximate solution is expanded as a linear combination of several appropriate sine functions, like a Fourier series, i.e.,

$$\tilde{v} = \sum_{i=1}^{n} c_i \sin\frac{(2i-1)\pi x}{l}$$

then, the solution is obtained as

$$\tilde{v} = \frac{2\tilde{P}l^3}{\pi^4 EI}\sum_{i=1}^{n}\frac{(-1)^{i+1}}{(2i-1)^4}\sin\frac{(2i-1)\pi x}{l} \tag{1.13}$$

and n becomes large the solution tends to the exact one.

Fig. 1.4(b) and (c) illustrate the distribution of bending moment, shear force and loading for a two and three term expansions of the displacement function. From these figures, it is now apparent that by increasing the number of the adjustable parameters c_i with suitable shape functions, one can make the approximate solution to converge to the exact solution. In other words, the fourth derivative of the displacement can be made to converge to a concentrated load.

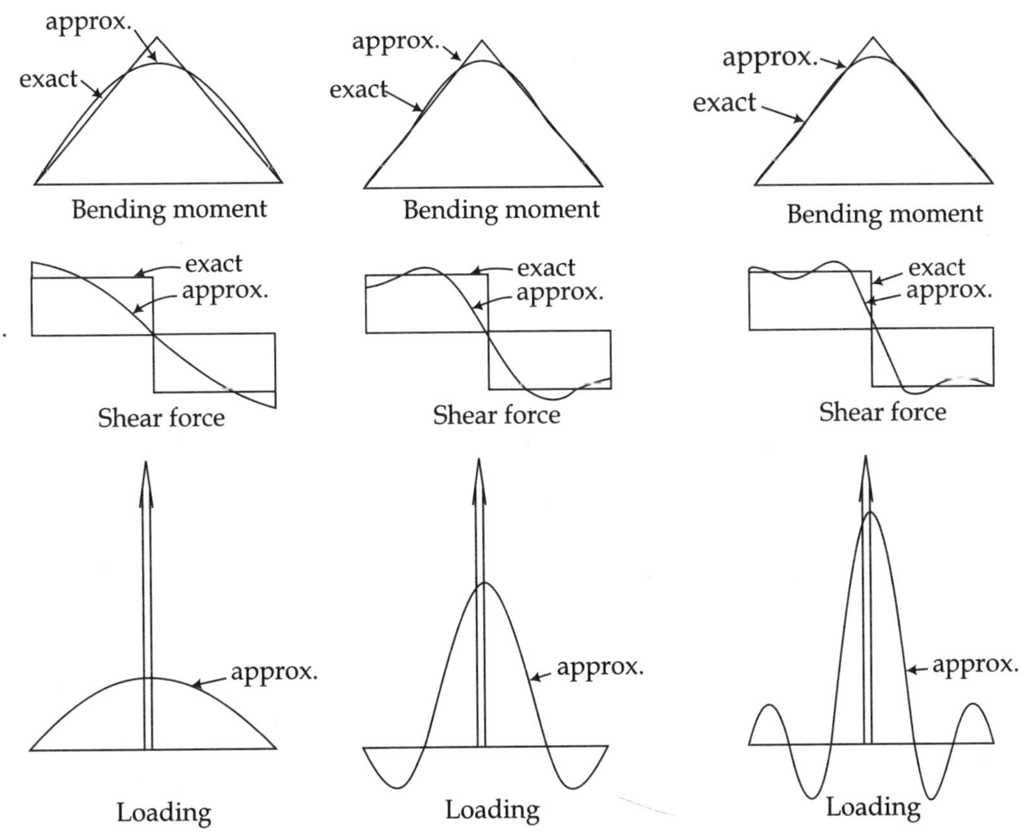

Fig. 1.4 Comparison of Exact and Approximate Solutions

1.4 GENERALIZATION OF THE RITZ METHOD

The Ritz method[4,5,6,8] applied to solve solid mechanics problems can be stated as follows:

If the displacement of a linearly elastic deformable body, subjected to externally applied forces, is approximated by the linear combination of a set of complete admissible functions, ψ_i that satisfy the prescribed boundary conditions, i.e.,

$$\bar{v} = \sum_{i=1}^{n} c_i \psi_i , \tag{1.14}$$

then the potential energy function Φ becomes a quadratic function of c_i, and minimizing Φ with respect to c_i, for $i = 1,2,\ldots,n$, n equations and with n unknown parameters are obtained, i.e.,

$$\frac{\partial \Phi}{\partial c_i} = \frac{\partial \mathbf{U}}{\partial c_i} + \frac{\partial \mathbf{V}}{\partial c_i} = 0 \tag{1.15}$$

$$\frac{\partial U}{\partial c_i} = \sum_{j=1}^{n} \frac{\partial^2 U}{\partial c_i \cdot \partial c_j} c_j = -\frac{\partial V}{\partial c_i} = f_i$$

and

$$\begin{bmatrix} & & \\ & A & \\ & & \end{bmatrix} \begin{Bmatrix} c_1 \\ c_2 \\ \cdot \\ \cdot \\ c_n \end{Bmatrix} = \begin{Bmatrix} f_1 \\ f_2 \\ \cdot \\ \cdot \\ f_n \end{Bmatrix}$$

where

$$a_{ij} = \frac{\partial^2 U}{\partial c_i \partial c_j},\ i = 1, 2, ..., n \text{ and } j = 1, 2, ..., n \text{ and } f_i = -\frac{\partial V}{\partial c_i},\ i = 1, 2, ..., n. \text{ equations.}$$

These n become the final system equations. Solution of these equations will yield the parameters c_i and an approximate solution of the system can be obtained. The terms c_i, $i = 1,2,...,n$ are called 'generalized coordinates'; the corresponding set of functions ψ_i are called the 'basis functions'. The selection of the basis functions is not completely arbitrary; it should include all the possible displacement shapes before any new shape is added to the set.

The Ritz method can be extended to solve multiple variables such as u, v and w displacement functions needed in three dimensional solid mechanics problems. The method consists of expressing these variables using combinations of admissible functions in the domain, i.e., functions that satisfy the prescribed boundary conditions, and minimizing the potential energy function with respect to the respective generalized coordinates.

1.5 DERIVATION OF THE BEAM EQUATION FROM THE ENERGY THEOREM

The differential equation for the beam can be derived by minimizing the energy function Φ. This derivation reinforces the concept behind the Ritz method for arriving at some approximate solutions.

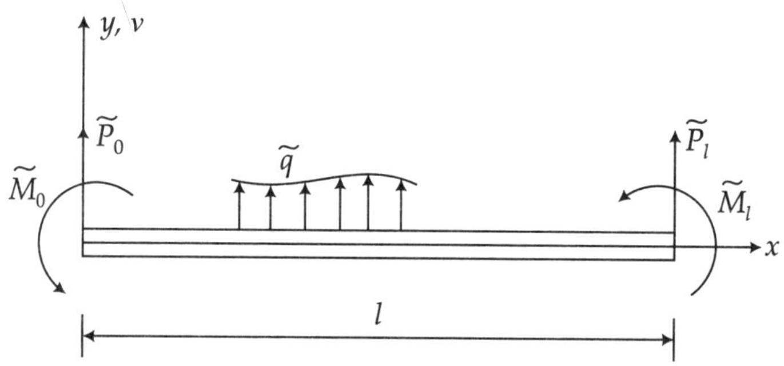

Fig. 1.5 An Euler Beam Problem

Let the beam be subjected to prescribed forces and moments at the ends in addition to a prescribed lateral load $\tilde{q}$, as shown in Fig. 1.5. All the prescribed forces shown in the diagram are positive. Now, it is a good idea to differentiate the prescribed forces at the ends with the internal bending moments and shear forces in a beam. The internal bending moments and shear forces are the stress resultants, defined as integrals of internal stresses in the beam (see Figs. 1.6a and 1.6b) and therefore have to follow the usual sign conventions used for the stresses. Since the prescribed forces (and moments) are vectors, sign conventions for vectors that are consistent with the coordinate axes have to be employed. *This fact will help engineers to differentiate between internal bending moment, denoted by m_x at any x from the prescribed (applied) moment $\tilde{M}_a$* at $x = a$. (It should be more appropriate to designate the prescribed moment as $\tilde{M}_z$ at $x = a$, since it is a vector in the direction of z-axis).

The sign conventions for the internal bending moment and shear force where the y-axis is taken positive upwards are shown in Fig. 1.6a. Many structural engineers are used to a convention that a sagging bending moment caused by downward forces is positive. This convention is consistent with the sign conventions given here, if the y-axis is taken as positive downwards as shown in Fig. 1.6b. Using a consistent sign convention for internal moments and shear forces and differentiating them from external moments and forces will help to evaluate results in a three dimensional frame, plate and shell analyses.

If the prescribed forces at the ends are $\tilde{P}_0$, $\tilde{M}_0$ at $x = 0$ and $\tilde{P}_l$, $\tilde{M}_l$ at $x = l$ (see Fig. 1.5), and if the resulting bending moments and the shear forces at the ends of the beam are m_0, and V_0, and m_l and V_l respectively, then:

$$\tilde{M}_0 = m_0 \text{ and } \tilde{M}_l = m_l, \qquad (1.16)$$

$$\tilde{P}_0 = -V_0 \text{ and } \tilde{P}_l = V_l$$

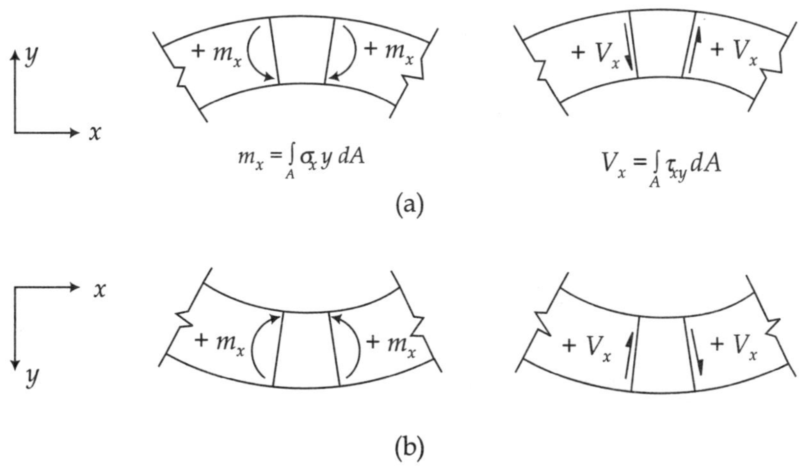

Fig. 1.6 Sign Conventions for Bending Moments and Shear Forces

Now, by definition,

$$\Phi = \mathbf{U} + \mathbf{V} = \frac{1}{2}\int_0^l EI\left(\frac{d^2v}{dx^2}\right)^2 dx - \left[\underbrace{\int_0^l \tilde{q}\,v\,dx}_{\text{due to prescribed distributed forces}} + \underbrace{\tilde{P}_0 v_0 + \tilde{M}_0\theta_0 + \tilde{P}_l v_l + \tilde{M}_l\theta_l}_{\text{due to prescribed discrete forces at the ends}}\right] \tag{1.17}$$

For Φ to be a minimum, the first variation of Φ, i.e., $\delta\Phi$ for a small variation in δv, is set equal to zero, i.e.,

$$\delta\Phi = \int_0^l EI\,\frac{d^2v}{dx^2}\,\delta\left(\frac{d^2v}{dx^2}\right) dx$$

$$- \left[\int_0^l \tilde{q}\,\delta v\,dx + \tilde{P}_0\,\delta v_0 + \tilde{M}_0\,\delta\theta_0 + \tilde{P}_l\,\delta v_l + \tilde{M}_l\,\delta\theta_l\right] = 0$$

Integrating the first integral by parts,

$$\delta\Phi = EI\left(\frac{d^2v}{dx^2}\right)\delta\left(\frac{dv}{dx}\right)\Bigg]_0^l - \int_0^l \frac{d}{dx}\left[EI\left(\frac{d^2v}{dx^2}\right)\right]\delta\left(\frac{dv}{dx}\right)dx$$

$$- \left[\int_0^l \tilde{q}\,\delta v\,dx + \tilde{P}_0\,\delta v_0 + \tilde{M}_0\,\delta\theta_0 + \tilde{P}_l\,\delta v_l + \tilde{M}_l\,\delta\theta_l\right]$$

Integrating again the first integral above by parts,

$$\delta\Phi = EI\,\frac{d^2v}{dx^2}\,\delta\theta\Bigg]_0^l - EI\,\frac{d^3v}{dx^3}\,\delta\theta\Bigg]_0^l + \int_0^l \frac{d^2}{dx^2}\left(EI\frac{d^2v}{dx^2}\right)\delta v\,dx -$$

$$\left[\int_0^l \tilde{q}\,\delta v\,dx + \tilde{P}_0\,\delta v_0 + \tilde{M}_0\,\delta\theta_0 + \tilde{P}_l\,\delta v_l + \tilde{M}_l\,\delta\theta_l\right]$$

Rearranging the equations,

$$\delta\Phi = \int_0^l \left[\frac{d^2}{dx^2}\left(EI\frac{d^2v}{dx^2}\right) - \tilde{q}\right]\delta v\,dx + \left[\left(EI\frac{d^2v}{dx^2} - \tilde{M}_l\right)\delta\theta_l\right.$$

$$- \left(EI\frac{d^2v}{dx^2} + \tilde{M}_0\right)\delta\theta_l$$

$$\left. - \left(EI\frac{d^3v}{dx^3} + \tilde{P}_l\right)\delta v_l + \left(EI\frac{d^3v}{dx^3} - \tilde{P}_0\right)\delta v_0\right] = 0 \tag{1.18}$$

There are two parts of this variation of $\delta\Phi$ that vanish; the first part is an integral in the domain, while the second part is at the boundaries of the beam. For the variation $\delta\Phi$ to vanish for all values of δv, both parts have to vanish independently, and the products of the terms on the boundaries also have to vanish independently. From the first part, for the integral to be zero for all values of δv, the integrand must also be zero which becomes the field equation as,

$$\frac{d^2}{dx^2}\left(EI\frac{d^2 v}{dx^2}\right) - \tilde{q} = 0 \text{ for } 0 < x < l \tag{1.19}$$

and from the second part, the boundary conditions at the ends of the beam can be obtained as:

either $EI\left.\frac{d^2 v}{dx^2}\right|_l - \tilde{M}_l = 0$ or $\delta\theta_l = 0$ i.e. $\tilde{\theta}_l$ is prescribed

either $EI\left.\frac{d^2 v}{dx^2}\right|_0 + \tilde{M}_0 = 0$ or $\tilde{\theta}_0$ is prescribed

either $EI\left.\frac{d^3 v}{dx^3}\right|_l + \tilde{P}_l = 0$ or $\tilde{v}_l$ is prescribed

either $EI\left.\frac{d^3 v}{dx^3}\right|_0 - \tilde{P}_0 = 0$ or $\tilde{v}_0$ is prescribed (1.20)

Natural Boundary Conditions *Essential Boundary Conditions*

It is important to realize that in the above equations, the terms $\tilde{P}_0$, $\tilde{P}_l$, $\tilde{M}_0$ and $\tilde{M}_l$ are prescribed forces and not shear forces or bending moments at the ends. These equations lead to the conventional boundary conditions given in most textbooks on the elementary theory of bending of beams. The reason for giving these equations in this form is to familiarize the reader with the use of consistent sign convention for the prescribed forces in the total potential energy function Φ. In finite element formulations, these conventions have to be consistent and therefore will help to avoid confusion in more complicated derivations.

It is clear that four boundary conditions are necessary to solve the above beam problem. The boundary conditions on the right hand side are called 'essential boundary conditions'; at least two of them are necessary to be prescribed, and at least one of two must be either $\tilde{v}_0$ or $\tilde{v}_l$. The idea is that the rigid body motions of the beam must be restrained to obtain a unique solution and therefore the boundary conditions on the right-hand side are called "essential boundary conditions" while on left-hand side are called "natural boundary conditions" as they are derived using all the assumptions of the classical Euler beam theory. These conditions yield a means of relating the prescribed forces with the characteristics of the beam. The distinction between essential and natural boundary conditions is necessary for further study of the approximate methods of analysis using a displacement approach[10].

For solving a beam problem subjected to prescribed loads and moments using Ritz method, the following function Φ has to be minimized with respect to all the adjustable parameters.

$$\Phi = \frac{1}{2}\int_0^l EI(\frac{d^2v}{dx^2})^2\, dx - \int_0^l \tilde{q}v\, dx - \sum_i \tilde{P}_i v\big|_{x=x_i} - \sum_i \tilde{M}_i \theta\big|_{x=x_i} \tag{1.17}$$

where the summation is used for all the prescribed forces and moments.

1.6 THE GALERKIN METHOD

If Eq. (1.18) has to vanish, the domain integral in that equation has to vanish independently,

$$\int_0^l \left[\frac{d^2}{dx^2}\left(EI\frac{d^2v}{dx^2}\right) - \tilde{q}\right] \delta v\, dx = 0,\ i = 1,2,.., n \tag{1.21}$$

then it can be assumed that the field equation is orthogonal to the first variation of displacement, δv. Galerkin took advantage of this condition and developed a new method known as 'Galerkin method' where he generalized for all differential equations of the type[7,9,11]

$$L(v) - f = 0 \text{ in the domain } \Omega \tag{1.22}$$

where $L(.)$ is a differential operator. If an approximate solution $\bar{v}$ is assumed as

$$\bar{v} = \sum_{i=0}^{n} c_i\, \psi_i, \tag{1.23}$$

where ψ_i, i = 1, 2,..., n are selected appropriate functions that satisfy all the boundary conditions, then Galerkin postulated that the error in the governing equation is orthogonal to each displacement mode ψ_i,

$$\int_0^l \left[L(\bar{v}) - f\right]\, \psi_i\, d\Omega = 0, \text{ for } i = 1, 2, \ldots, n \tag{1.24}$$

These conditions lead to n equations with n unknowns. It is interesting to note that for the simply supported beam problem shown in Fig. 1.1, the Galerkin method will yield the same result as the Ritz method. For example, using an approximate solution

$$\bar{v} = c_1 \sin\frac{\pi x}{l} + c_2 \sin\frac{3\pi x}{l}$$

as used in Eq.(1.9) which satisfies all the boundary conditions, the Galerkin method yields,

$$\int_0^l \left[EI\left(c_1\frac{\pi^4}{l^4}\sin\frac{\pi x}{l} + c_2\frac{81\pi^4}{l^4}\sin\frac{3\pi x}{l}\right) - \tilde{q}\right]\sin\frac{\pi x}{l}\, dx = 0$$

and

$$\int_0^l \left[EI\left(c_1\frac{\pi^4}{l^4}\sin\frac{\pi x}{l} + c_2\frac{81\pi^4}{l^4}\sin\frac{3\pi x}{l}\right) - \tilde{q}\right]\sin\frac{3\pi x}{l}\, dx = 0$$

Upon integrations and simplifications,

$$c_1 = \frac{4\tilde{q}l^4}{\pi^5 EI} \text{ and } c_2 = \frac{4\tilde{q}l^4}{243\pi^5 EI}$$

These are the same expressions that resulted from the previous Ritz solution. There are special conditions where both Galerkin and Ritz methods would yield exactly the same solutions. The Ritz method has an added advantage that it can accommodate concentrated loads, while it is rather difficult to use the Galerkin method in such instances.

In finite element methodology, the Galerkin method is used in an integrated form of the Eq. (1.24), and this will be explained later in Chapter 10. The purpose of these exercises is to illustrate the use of Ritz and Galerkin methods to create an approximate solution of equilibrium problems in solid mechanics. The correspondence of the equilibrium equation and the minimum potential energy function for the case of beam problem is established here. The concept can be extended to solve more sophisticated solid mechanics problems, as the fundamental aim in this chapter is to create confidence in the reader for the use of energy techniques. In general, it can be seen that the finite element method is nothing but Ritz or Galerkin method applied in a slightly different form.

1.7 SUMMARY

Finite element method is one of the methods of approximation that has roots in the energy principles and variational calculus. Problems on beams with known solutions are solved by minimizing a potential energy function and compared. This technique is known as the Ritz method. It is shown that in the approximate solution, while the displacement is good, the errors associated with the slope, bending moment and shear force increase as the order for the derivatives increase. The solution need not even satisfy the governing differential equation. But as the number of parameters increases, the solution converges to the exact solution.

One of the important concepts introduced here is the distinction between the internal bending moment and applied moment and their respective sign conventions. Similar statements hold for shear force and applied force. Both applied moment and applied force are characterized as vectors and therefore have a sign convention consistent with the coordinates used for the beam. Since internal bending moments and shear forces are integrals of the stress more precisely refered to as second order tensors, they have to follow the sign conventions used for the stress at a point which are also tensors. The discussion is repeated here, because many engineers are often confused with these sign conventions, and a consistent sign convention is extremely important in finite element formulations especially when they deal with plate and shell analysis.

By minimizing the potential energy function for a beam, the equilibrium equation and the associated boundary conditions are derived. Galerkin method is introduced as another method of approximation.

PROBLEMS

1.1 A simply supported beam carries a concentrated load $\tilde{P}$ at point C as shown in Fig. P1.1. Using a three term expansion of sine series, compute the displacement function for the beam, using the Ritz method. Explain your reasons for the selection of the functions. Compare the displacement, bending moment and shear force with the corresponding theoretical values.

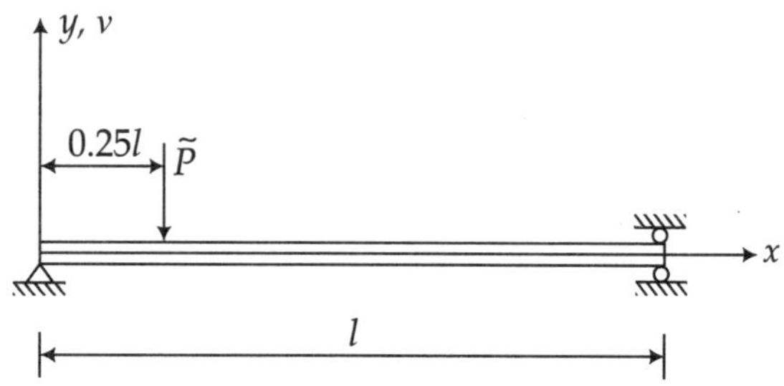

Fig. P1.1

1.2 A simply supported beam carries a uniformly distributed load for the first half span of the beam. See Fig. P1.2. Using a three term expansion of the sine series, determine the displacement function using the Ritz method, and compare the displacement, bending moment and shear force with the corresponding theoretical solutions

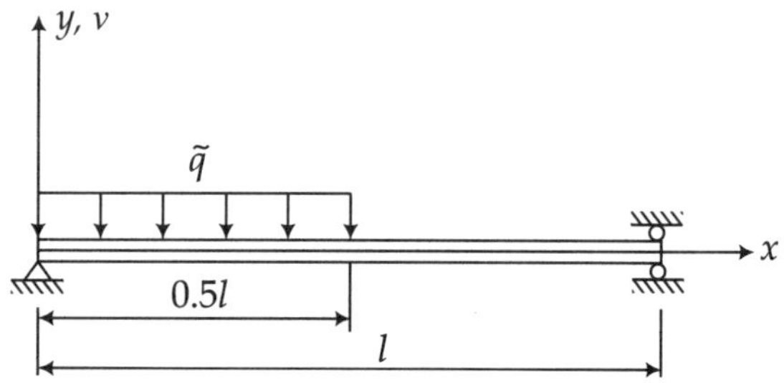

Fig P1.2

1.3 A simple supported beam carries a uniformly distributed load $\tilde{q}$ for the entire span. Assuming that an approximate displacement of the beam can be represented by a simple quadratic satisfying the essential boundary conditions such as $\bar{v}(x) = cx(l - x)$, determine the value of c, using the Ritz method. Compute the displacement, bending moment and shear force and compare with theoretical solutions.

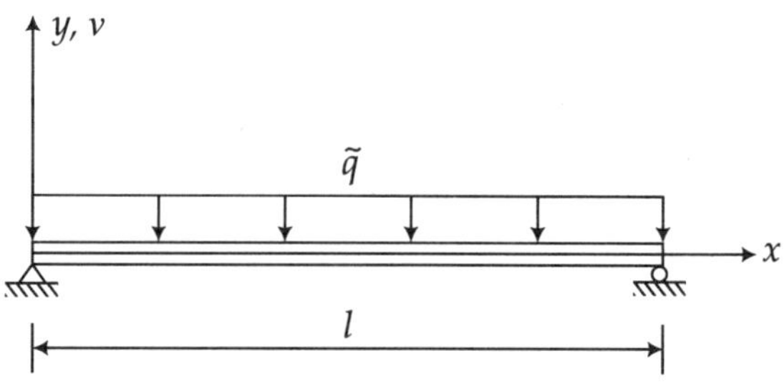

Fig. P1.3

1.4 Compute the displacement of a uniform cantilever beam, carrying a uniformly distributed load q, assuming an approximate solution as

(a) $v(x) = c\left(1 - \cos\dfrac{\pi x}{2l}\right)$

(b) $v(x) = cx^2$

(c) $v(x) = c_1x^2 + c_2x^3$

(d) $v(x) = c_1x^2 + c_2x^3 + c_3x^4$

Compare the displacement, bending moment and shear force obtained from the approximate analysis with the corresponding theoretical solutions.

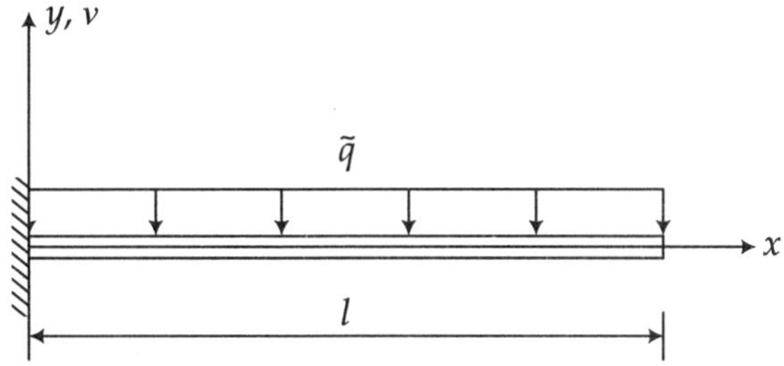

Fig. P1.4

1.5 Repeat the Problem No. 4., where the cantilever is carrying a concentrated load at the free end, instead of a uniformly distributed load.

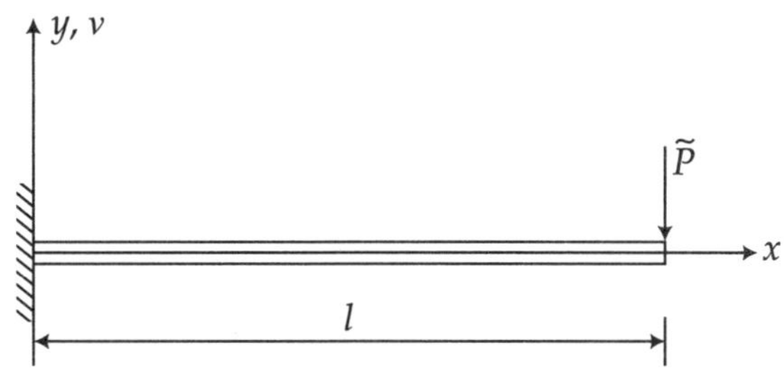

Fig. P1.5

REFERENCES

1. Argyris, J., and Kelsey, S., "Energy Theorems and Structural Analysis", Butterworh & Company (Publishers) Ltd., London, 1960. Reprinted by Plenum Publishing Corporation, 1968.
2. Clough, R.W., "The Finite Element Method in Plane Stress Analysis", Proceedings of the Second Conference on Electronic Computation, American Society of Civil Engineers, New York, 1960, pp. 345-377.
3. Crandall, S., "Engineering Analysis", McGraw Hill Book Company, 1956.
4. Courant, R.,"Variational Methods for Solution of Problems of Equilibrium and Vibration", Bull. Am. Math. Soc. 49 (1943), 1-43.

5. Dym, C.L. and Shames, I.H., "Solid Mechanics: A Variational Approach", McGraw Hill Book Company, New York, 1973.
6. Fung, Y.C., "Foundations of Solid Mechanics", Prentice-Hall, Englewood Cliffs, New Jersey, 1965.
7. Langhaar, H.L., "Energy Methods in Applied Mechanics", John Wiley & Sons, New York, 1962.
8. Oden, J.T., "Finite Elements of Nonlinear Continnua", McGraw Hill Book Company, 1972.
9. Reddy, J.N., "Energy and Variational Methods in Applied Mechanics", John Wiley & Sons, New York, 1984.
10. Turner, M., Clough, R., Martin, H. and Topp, L., "Stiffness and Deflection Analysis of Complex Structures", J. Aeronaut. Sci. 23, No. 9 (1956), 805 - 823.
11. Washizu, K., "Variational Methods in Elasticity and Plasticity", Pergamon Press, Oxford, England, 1982.
12. Zienkiewicz, O.C. and Taylor, R.L., "The Finite Element Method", Vols. I and II, McGraw Hill Book Company, 1991.

2

CHAPTER

Equations in Solid Mechanics

2.1 DISPLACEMENTS AND FORCES

An arbitrary solid body is shown in Fig. 2.1 in a three dimensional Cartesian coordinate system represented by x, y and z axes. The displacement of an arbitrary point P in the interior of the body as the body is subjected to external forces, is represented by three components (u, v, w) along the respective coordinate axes (x, y, z). The applied external forces, usually called prescribed forces can be categorized into three different types[2,5,7]. First type is a concentrated force with components F_x, F_y and F_z in the respective (x, y, z) axes. The second type is called surface traction, which is a distributed force per unit area of the surface and is represented by

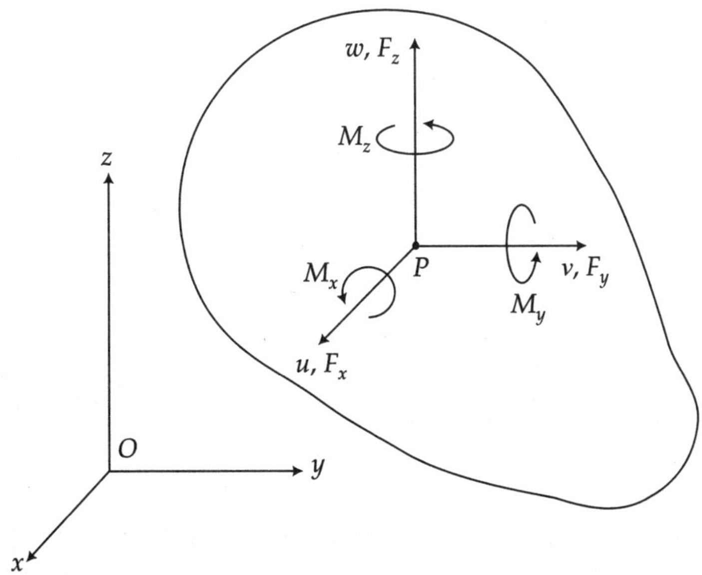

Fig. 2.1 Displacements, Forces and Moments at a Point

its components T_x, T_y and T_z respectively. The third type is called body force per unit volume of the body usually representing the body self weight, magnetic force and inertia force and its components are represented by b_x, b_y and b_z respectively. In addition to the forces described above sometimes, a concentrated external moment can be applied on the body, with components M_x, M_y and M_z along the x, y, z axes.

Other external forces include those produced by local volume changes due to temperature variations or due to some change in the solid crystal structure, these effects are incorporated into the analysis as an equivalent body force vector.

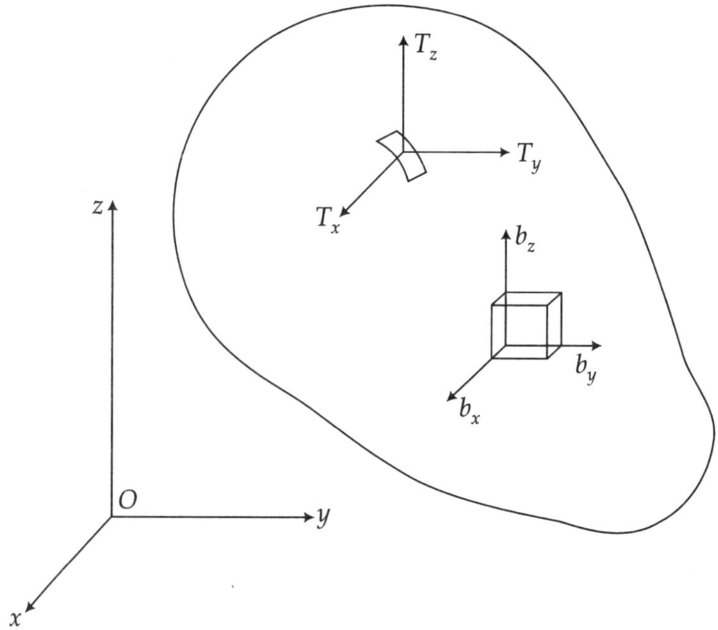

Fig. 2.2 Body Forces and Surface Tractions

Table 2.1 Notations Used for Displacements and Prescribed Forces

Variables	*Components in*			*Vector Notation*
	x-axis	*y-axis*	*z-axis*	
Displacement at a point	u	v	w	$\{u\}$
Concentrated Force at a point	F_x	F_y	F_z	{F}
Surface Traction/unit area	T_x	T_y	T_z	$\{T\}$
Body Force/unit volume	b_x	b_y	b_z	$\{b\}$
Applied Moment at a point	M_x	M_y	M_z	$\{M\}$

All components of forces shown in Figs. 2.1 and 2.2 are positive in the positive directions of the coordinate axes, with a right-handed system where the components of the moments can be easily represented with a simple sign convention as shown in the figures. It is very important to keep a proper sign convention for the displacements, forces and moments consistent with that used for the coordinate system.

The study of solid mechanics involves determination of deformations of a body caused by the prescribed forces. To accomplish this, two other parameters, strain and stress at an interior point are introduced in the theory.

2.2 STRAIN AT A POINT

The strain at a point is represented by six independent components derived from the gradients of the displacement vector (u v w) along the (x, y, z) coordinate axes. These six components are divided into two types; called normal strain and shear strain, respectively. For a linear small deformation problem, the normal strains ε_x, ε_y and ε_z represent the small deformations of a small rectangular block element in the (x, y, z) axes without any angular distortions. The shear strains γ_{xy}, γ_{yz} and γ_{zx} represent the angular distortions of a similar element in the (x-y), (y-z) and (x-z) planes respectively. Deformation of a differential element with sides dx, dy in the (x, y) plane is shown in Fig. 2.3. The six components of the strain at a point are represented by a vector

$$\{\varepsilon\} = \{\varepsilon_x \;\; \varepsilon_y \;\; \varepsilon_z \;\; \gamma_{xy} \;\; \gamma_{yz} \;\; \gamma_{xz}\}^t \tag{2.1}$$

The components of the strain at a point are written in terms of the derivatives of the displacement vector with respect to the coordinate axes as given below.

Normal Strains / Shear Strains

$$\varepsilon_x = \frac{\partial u}{\partial x} \qquad \gamma_{xy} = \frac{\partial u}{\partial y} + \frac{\partial v}{\partial x}$$

$$\varepsilon_y = \frac{\partial v}{\partial y} \qquad \gamma_{yz} = \frac{\partial v}{\partial z} + \frac{\partial w}{\partial y}$$

$$\varepsilon_z = \frac{\partial w}{\partial z} \qquad \gamma_{zx} = \frac{\partial w}{\partial x} + \frac{\partial u}{\partial z} \tag{2.2}$$

These equations are called strain-displacement equations in the linear theory of solid mechanics[3,5,7]. For the two-dimensional case, the above equations can also be derived geometrically from the Fig. 2.3. The element $PQRS$ in the undeformed body transforms into $P'Q'R'S'$ in the deformed body. The change of length PQ to $P'Q'$ along the x-axis, is represented by the normal strain $\varepsilon_x = \frac{\partial u}{\partial x}$.

Similarly the change of length PS to $P'S'$ along the y-axis is the normal strain $\varepsilon_y = \frac{\partial v}{\partial y}$. The angular distortion of the right angle QPS into $Q'P'S'$ is denoted by the sum of two angles, α and β and hence the total angular distortion $\gamma_{xy} = \alpha + \beta = \frac{\partial u}{\partial y} + \frac{\partial v}{\partial x}$. Similarly, the other equations can be derived for the other strain components [7]. The components of shear strain shown here are called "engineering shear strain".

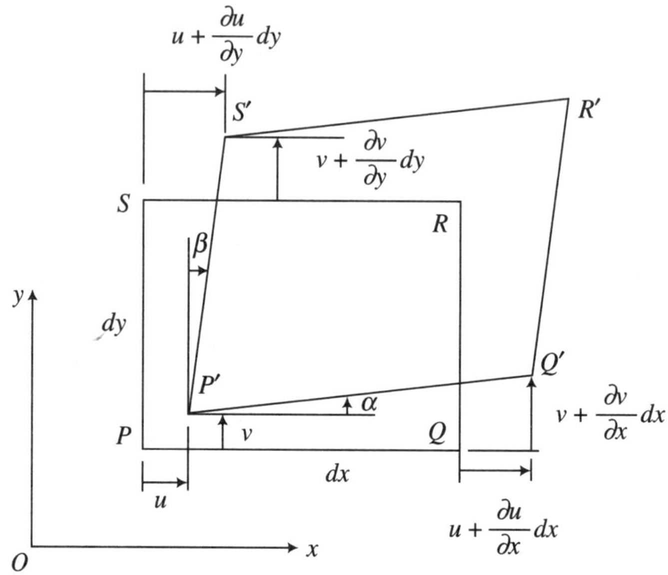

Fig. 2.3 Deformation in (x, y) Plane

2.3 STRESS AT A POINT

Just as there are six independent components for strain at a point, there are six independent components for the stress at a point such as three normal stresses and three shear stresses. The normal stress denoted by σ_x is the internal force along x-axis per unit area normal to the x-axis. The shear stress τ_{xy} is the internal force along the y-axis per unit area normal to the x-axis. All these six stress components on an elemental cube are illustrated in Fig. 2.4.

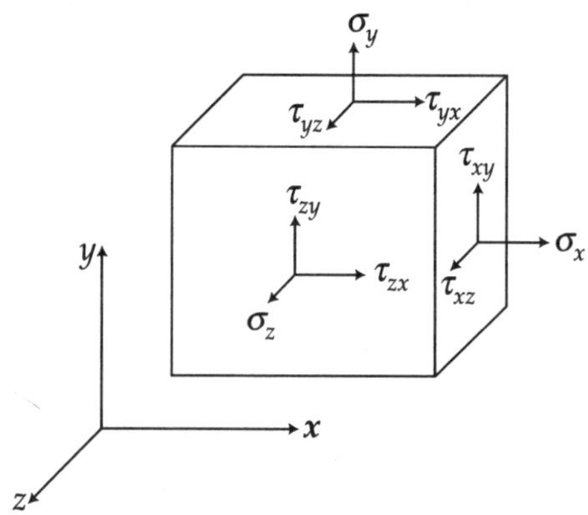

Fig. 2.4 Components of Stress at a Point

It may be noted that the shear stress components $\tau_{xy} = \tau_{yx}$, $\tau_{yz} = \tau_{zy}$, and $\tau_{xz} = \tau_{zx}$, a property that results from the moment-equilibrium of the internal forces. Thus, the stress at a point is also represented by a vector $\{\sigma\}$ having six independent components such that

$$\{\sigma\} = \{\sigma_x \quad \sigma_y \quad \sigma_z \quad \tau_{xy} \quad \tau_{yz} \quad \tau_{xz}\}^t \tag{2.3}$$

It is customary in solid mechanics theory to represent stress at a point as a 3×3 matrix[2] (or as second order tensor), i.e.,

$$[\sigma] = \begin{bmatrix} \sigma_x & \tau_{xy} & \tau_{xz} \\ \tau_{xy} & \sigma_y & \tau_{yz} \\ \tau_{xz} & \tau_{yz} & \sigma_z \end{bmatrix} \tag{2.4}$$

This representation helps to understand the characteristics of the stress at a point in a more elegant way, especially to discuss the principal stresses, failure criteria, etc. However, for the study of finite element method, a vector representation is sufficient and convenient.

2.3.1 Principal Stresses

In stress analysis, often one has to determine the extremum values of the components of stress at a point which are called 'principal stresses'. The principal stresses can be shown to be the eigenvalues of the stress matrix and the directions of the principal stresses are the corresponding eigenvectors[1,7]. The eigenvalues are the roots of the characteristic equation

$$\det\begin{bmatrix} \sigma_x - \lambda & \tau_{xy} & \tau_{xz} \\ \tau_{xy} & \sigma_y - \lambda & \tau_{yz} \\ \tau_{xz} & \tau_{yz} & \sigma_z - \lambda \end{bmatrix} = -\lambda^3 + I_1\lambda^2 - I_2\lambda + I_3 = 0 \tag{2.5}$$

where

$$I_1 = \sigma_x + \sigma_y + \sigma_z = \sigma_1 + \sigma_2 + \sigma_3$$
$$I_2 = \sigma_x\sigma_y + \sigma_y\sigma_z + \sigma_z\sigma_x - \tau_{xy}^2 - \tau_{yz}^2 - \tau_{zx}^2 = \sigma_1\sigma_2 + \sigma_2\sigma_3 + \sigma_3\sigma_1$$
$$I_3 = \det(\text{stress matrix}) = \sigma_1\sigma_2\sigma_3$$

Here σ_1, σ_2 and σ_3 are the principal stresses and they are the roots of Eq. (2.5). The quantities I_1, I_2 and I_3 are called the 'invariants' of the stress matrix. A stress at a point can be transformed into a different set of six independent components in another coordinate axes, but the invariants remain the same[3]. The eigenvectors are computed using the matrix equation

$$\begin{bmatrix} \sigma_x - \lambda & \tau_{xy} & \tau_{xz} \\ \tau_{xy} & \sigma_y - \lambda & \tau_{yz} \\ \tau_{xz} & \tau_{yz} & \sigma_z - \lambda \end{bmatrix} \begin{Bmatrix} x_1 \\ x_2 \\ x_3 \end{Bmatrix} = \begin{Bmatrix} 0 \\ 0 \\ 0 \end{Bmatrix} \tag{2.6}$$

where λ is substituted by each of the principal stresses. For each eigenvector, the equation is solved separately. Since the above matrix is singular to begin with, the eigenvector solution has to be a dependent solution, in other words, one of the component of the eigenvector has to be assumed. If the eigenvectors are normalized, the components of the vector become the direction cosines of the corresponding principal stress. The values of the maximum shear stresses are: $1/2|\sigma_1 - \sigma_2|$, $1/2|\sigma_2 - \sigma_3|$ and $1/2|\sigma_3 - \sigma_1|$. In the case of stress analysis of ductile materials, engineers often use a limiting value for the stress at a point, called 'effective stress' or 'von Mises stress' and this stress is compared to the yield stress in a ductile material. The von Mises stress is defined as

$$\sigma_{vm} = \frac{1}{\sqrt{2}}\sqrt{(\sigma_1 - \sigma_2)^2 + (\sigma_2 - \sigma_3)^2 + (\sigma_3 - \sigma_1)^2} \tag{2.7}$$

$$= \sqrt{I_1^2 - 3I_2}$$

Example Problem 2.1

Determine the principal stresses and the corresponding principal axes for the following stress matrix

$$[\sigma] = \begin{bmatrix} 4 & 2 & 0 \\ 2 & 4 & 2 \\ 0 & 2 & 4 \end{bmatrix}$$

Solution:

The principal stresses are the roots of the characteristic equation equal to

$$\det \begin{bmatrix} 4-\lambda & 2 & 0 \\ 2 & 4-\lambda & 2 \\ 0 & 2 & 4-\lambda \end{bmatrix} = -\lambda^3 + 12\lambda^2 - 40\lambda + 32 = 0$$

and the roots or the principal stresses are:

$$\sigma_1 = 4 + 2\sqrt{2}$$
$$\sigma_2 = 4$$
$$\sigma_3 = 4 - 2\sqrt{2}$$

Solving Eq.(2.6) and using $x_1 = 1.0$, the eigenvector corresponding to σ_1 is obtained as $(1 \; -\sqrt{2} \; 1)^t$. The other two corresponding eigenvectors can be found to be $(1 \; 0 \; -1)^t$ and $(1 \; \sqrt{2} \; 1)^t$ respectively. If one normalizes the eigenvectors, they become the direction cosines of the principal axes.

2.3.2 Principal Stresses for a Two Dimensional Stress

The stress matrix for the two dimensional state of stress is

$$[\sigma] = \begin{bmatrix} \sigma_x & \tau_{xy} \\ \tau_{xy} & \sigma_y \end{bmatrix}$$

and the principal stresses can be found from the characteristic equation

$$\det \begin{bmatrix} \sigma_x - \lambda & \tau_{xy} \\ \tau_{xy} & \sigma_y - \lambda \end{bmatrix} = \lambda^2 - (\sigma_x + \sigma_y)\,\lambda + \sigma_x\sigma_y - \tau_{xy}^2 = 0$$

and the roots or the principal stresses are:

$$\sigma_1 = \frac{1}{2}(\sigma_x + \sigma_y) + \sqrt{\frac{(\sigma_x - \sigma_y)^2}{4} + \tau_{xy}^2}$$

$$\sigma_2 = \frac{1}{2}(\sigma_x + \sigma_y) - \sqrt{\frac{(\sigma_x - \sigma_y)^2}{4} + \tau_{xy}^2} \tag{2.8}$$

The characteristics of two dimensional stress at a point are often taught using the Mohr circle theory. The directions of the principal stresses are obtained from the Mohr circle, but the theory of Mohr circle requires a new sign convention for the shear stresses (used only for the Mohr circle study) which is contradictory to the sign convention used in classical solid mechanics. The directions of the principal stresses are the two roots of the angle θ, given by the equation

$$\tan 2\theta = \frac{2\tau_{xy}}{(\sigma_x - \sigma_y)} \tag{2.9}$$

where θ is the angle made by the principal axis with the x-axis. Using the concept of eigenvectors, Vallabhan[8] (1997) has developed a simple formula for the computing the directions of the principal axes. In other words, the directions of the σ_1 and σ_2 are given by the equations

$$\tan \theta_1 = \frac{\tau_{xy}}{\sigma_1 - \sigma_y} = \frac{\sigma_1 - \sigma_x}{\tau_{xy}}$$

and

$$\tan \theta_2 = \frac{\tau_{xy}}{\sigma_2 - \sigma_y} = \frac{\sigma_2 - \sigma_x}{\tau_{xy}} \tag{2.10}$$

where θ_1 and θ_2 are angles between the x-axis and the principal stress directions. The formulae given in Eq. (2.10) give uniquely the respective principal directions and is convenient for computer applications.

The corresponding von Mises stress is:

$$\sigma_{vm} = \sqrt{\sigma_1^2 + \sigma_2^2 - \sigma_1\sigma_2} \tag{2.11}$$

Example Problem 2.2

Determine the principal stresses and corresponding principal axes for the following two dimensional plane stress case:

$$[\sigma] = \begin{bmatrix} \sigma_x & \tau_{xy} \\ \tau_{xy} & \sigma_y \end{bmatrix} = \begin{bmatrix} -7 & -3 \\ -3 & 1 \end{bmatrix} \text{ MPa.}$$

Solution:

The principal stresses are:

$$\sigma_{1,2} = \frac{1}{2}(-7 + 1) \pm \sqrt{\frac{1}{4}(-7 - 1)^2 + (-3)^2}$$

$$= +\,2.0 \text{ and } -8 \text{ MPa.}$$

The principal axis for σ_1 is given by the Eq. (2.9)

$$\tan \theta_1 = \frac{\tau_{xy}}{\sigma_1 - \sigma_y} = \frac{-3}{2 - 1} = -3$$

and $\theta_1 = -71.57^0$.

Similarly, the principal axis for σ_2 is given by

$$\tan\theta_2 = \frac{\tau_{xy}}{\sigma_2 - \sigma_y} = \frac{-3}{-8-1} = 0.3333$$

and $\theta_2 = +18.43°$.

The directions are illustrated in the Fig. 2.5.

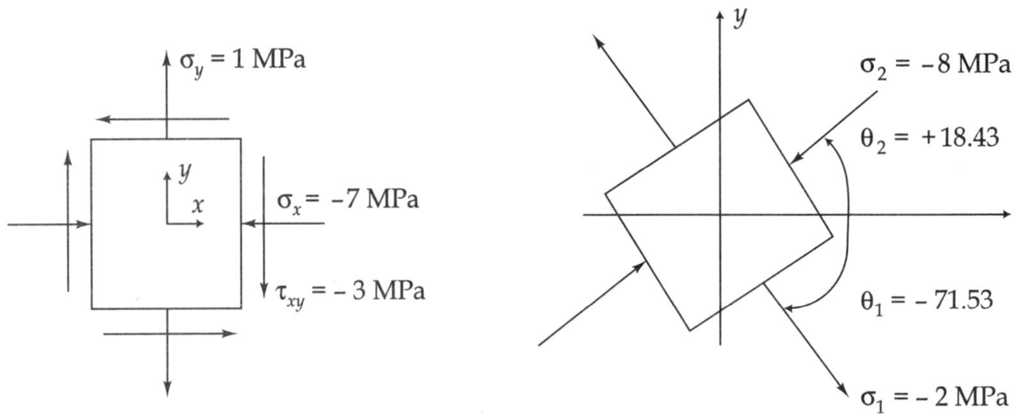

Fig. 2.5 State of Stress and the Principal Axes

2.3.3 Principal Strains

Just like principal stresses for a stress at a point, there are principal strains for a strain at a point. The corresponding matrix to represent the strain at a point is:

$$[\varepsilon] = \begin{bmatrix} \varepsilon_x & \frac{\gamma_{xy}}{2} & \frac{\gamma_{xz}}{2} \\ \frac{\gamma_{xy}}{2} & \varepsilon_y & \frac{\gamma_{yz}}{2} \\ \frac{\gamma_{xz}}{2} & \frac{\gamma_{yz}}{2} & \varepsilon_z \end{bmatrix} \tag{2.12}$$

Here the off diagonal elements of the matrix represent shear strains, but their magnitudes are only equal to one-half of the angular distortions in the respective planes, in order to become a strain tensor[3,5]. The principal strains, principal axes and their invariants can be computed using the similar equations given in Eq. (2.5) through (2.11).

2.4 STRESS STRAIN RELATIONSHIPS

A material is defined as linear, if the components of a strain at a point can be related to the corresponding stress at that point using constants defining the material properties of the body.

If the material is elastic, there are 21 elastic constants[5]. These constants can be reduced to two if the material is homogeneous and isotropic. The commonly used material properties in engineering are Young's modulus of elasticity E and Poisson's ratio μ. The relationship between the strain $\{\varepsilon\}$ and the stress $\{\sigma\}$ for such a material is called Hooke's law [3] and is given in a matrix form as

$$\begin{Bmatrix} \varepsilon_x \\ \varepsilon_y \\ \varepsilon_z \\ \gamma_{xy} \\ \gamma_{yz} \\ \gamma_{zx} \end{Bmatrix} = \frac{1}{E} \begin{bmatrix} 1 & -\mu & -\mu & 0 & 0 & 0 \\ -\mu & 1 & -\mu & 0 & 0 & 0 \\ -\mu & -\mu & 1 & 0 & 0 & 0 \\ 0 & 0 & 0 & 2(1+\mu) & 0 & 0 \\ 0 & 0 & 0 & 0 & 2(1+\mu) & 0 \\ 0 & 0 & 0 & 0 & 0 & 2(1+\mu) \end{bmatrix} \begin{Bmatrix} \sigma_x \\ \sigma_y \\ \sigma_z \\ \tau_{xy} \\ \tau_{yz} \\ \tau_{zx} \end{Bmatrix} \tag{2.13}$$

and commonly denoted by

$$\{\varepsilon\} = [C]\,\{\sigma\}$$

However, in the finite element method, the relationship of stress in terms of strain is required, and this is given as

$$\begin{Bmatrix} \sigma_x \\ \sigma_y \\ \sigma_z \\ \tau_{xy} \\ \tau_{yz} \\ \tau_{zx} \end{Bmatrix} = \frac{E(1-\mu)}{(1+\mu)(1-2\mu)} \begin{bmatrix} 1 & \frac{\mu}{1-\mu} & \frac{\mu}{1-\mu} & 0 & & 0 \\ \frac{\mu}{1-\mu} & 1 & \frac{\mu}{1-\mu} & 0 & 0 & 0 \\ \frac{\mu}{1-\mu} & \frac{\mu}{1-\mu} & 1 & 0 & 0 & 0 \\ 0 & 0 & 0 & \frac{1-2\mu}{2(1-\mu)} & 0 & 0 \\ 0 & 0 & 0 & 0 & \frac{1-2\mu}{2(1-\mu)} & 0 \\ 0 & 0 & 0 & 0 & 0 & \frac{1-2\mu}{2(1-\mu)} \end{bmatrix} \begin{Bmatrix} \varepsilon_x \\ \varepsilon_y \\ \varepsilon_z \\ \gamma_{xy} \\ \gamma_{yz} \\ \gamma_{zx} \end{Bmatrix} \tag{2.14}$$

and in matrix notation, the above equation is written as

$$\{\sigma\} = [D]\,\{\varepsilon\} \tag{2.15}$$

The above linear equation is good only for elastic materials with isotropic and homogeneous properties. Similar equations can be developed for materials with orthotropic or even anisotropic properties by changing the coefficients of the D matrix appropriately[3]. It will be shown that the finite element stiffness matrix equations for all linear materials have the same formula including the D matrix, whether the material is isotropic, orthotropic or anisotropic. The relevant material properties of each finite element can be incorporated into overall finite element analysis using the appropriate D matrices

2.5 STRAIN ENERGY IN AN ELASTIC BODY

When a set of forces is applied on a deformable elastic body, the work done by the forces is stored as potential energy called strain energy in the body. If the forces are applied very slowly, such that the ensuing kinetic energy and the thermal energy can be neglected, then the work done by the external forces can be equated with the elastic strain energy stored in the body, and is completely recoverable if the forces are removed. Theoretically, elastic properties of a material are independent of the linearity of the stress-strain relationship. But, for a linear material, the elastic strain energy **U** is given as [2,3,6],

$$\mathbf{U} = \frac{1}{2}\int_{vol} (\sigma_x \varepsilon_x + \sigma_y \varepsilon_y + \sigma_z \varepsilon_z + \tau_{xy}\gamma_{xy} + \tau_{yz}\gamma_{yz} + \tau_{xz}\gamma_{xz})\ dvol$$

$$= \frac{1}{2}\int_{vol} \{\sigma\}^t \{\varepsilon\}\ dvol \tag{2.16}$$

$= \mathbf{W}$, work done by external prescribed forces

$$= \frac{1}{2}\{\widetilde{F}\}^t \{U\} \tag{2.17}$$

where $\{\widetilde{F}\}$ is the prescribed force vector containing n components such as $\widetilde{f}_1, \widetilde{f}_2, \widetilde{f}_3, \ldots, \widetilde{f}_n$, and $\{U\}$ is the corresponding displacement vector denoted by n components $u_1, u_2, u_3, \ldots, u_n$. The term '*dvol*' at the end of the integration denotes an elemental volume. If there exists a stiffness matrix $[K]$ relating the discrete displacements and the corresponding forces in a linear elastic system, such that

$[K]\{U\} = \{\widetilde{F}\}$, then

$$\mathbf{U} = \frac{1}{2}\{\widetilde{F}\}^t \{U\}$$

$$= \frac{1}{2}\{U\}^t [K]\{U\} \tag{2.18}$$

$$= \frac{1}{2}\sum_{i=1}^{n}\sum_{j=1}^{n} k_{ij}\, u_i\, u_j$$

where k_{ij} is an element of the stiffness matrix $[\boldsymbol{K}]$. Equating the strain energy to the work done by the forces is one technique used for solving problems in solid mechanics, but more powerful principles have been developed based on variational calculus[1,4,9]. These principles are called principles of virtual work; one of them deals with virtual displacements and the other with virtual forces. Using these principles, two major energy theorems such as minimum potential energy and complementary potential energy theorems[1,4,9] are developed. Since users of finite element methods for solving solid mechanics problems predominantly employ a displacement approach, only the displacement approach is discussed here.

2.6 PRINCIPLE OF VIRTUAL WORK

This principle enunciated by Bernoulli in the 1700's is extensively used in the formulation of the finite element equations. In fact, many pioneers in the finite element research have used the principle of virtual work in their derivations. For a complete discussion on this topic, readers are referred to the classic books by Washizu[9] and Reddy[6].

When one refers to the principle of virtual work in finite element method, normally it means the principle of virtual displacements. In this approach, the displacements of the body are unknowns. Of course, the body has to be sufficiently restrained to be free from any rigid body movements, and this condition is achieved by prescribing displacements at the necessary points.

Let a body be subjected to a small variation in displacements, such that the prescribed forces move through a corresponding set of small displacements called virtual displacements, δu_i, $i = 1,2,\ldots,n$, also denoted by $\{\delta U\}$. The virtual displacement field is assumed such that it automatically vanishes at the prescribed displacement boundaries, and these virtual displacements are called kinematically admissible displacements. The work done by the prescribed forces undergoing the virtual displacements is called virtual work, $\delta \mathbf{W}$ and

$$\delta \mathbf{W} = \{\tilde{F}\}^t \{\delta U\} \tag{2.19}$$

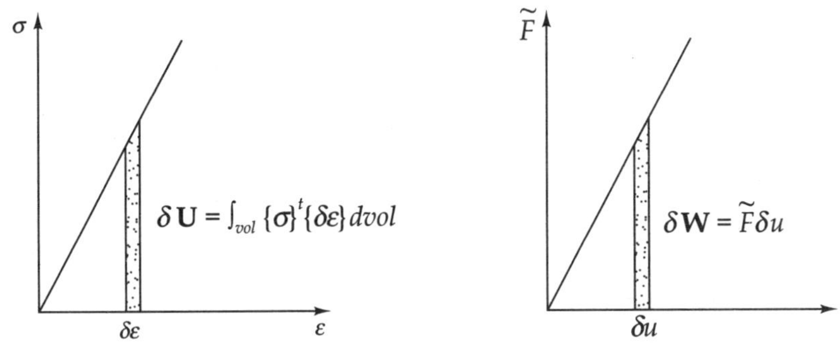

Fig. 2.6

This is shown in Fig. 2.6. For an elastic body, the virtual work is stored as virtual strain energy $\delta \mathbf{U}$ in the body. This virtual strain energy is produced by the virtual strains $\{\delta\varepsilon\}$ as a result of virtual displacements, i.e.,

$$\begin{aligned} \delta \mathbf{U} &= \int_{vol} \{\sigma\}^t \{\varepsilon\}\, dvol \\ &= \delta \mathbf{W} = \{\tilde{F}\}^t \{\delta U\} \end{aligned} \tag{2.20}$$

The advantage of this principle may not be obvious at this level. However from the above equation, it can be noted that the virtual strains are obtained from the virtual displacements satisfying the compatibility of displacements, and the internal stresses are in equilibrium with the prescribed forces.

The application of this principle to derive stiffness matrix equations for finite elements is discussed in Chapter 3. However, here it is used to derive the minimum potential energy theorem.

2.7 MINIMUM POTENTIAL ENERGY THEOREM

In the principle of virtual work discussed above, it is assumed that the virtual work performed by the external forces is stored as virtual strain energy,

$$\delta \mathbf{U} = \delta \mathbf{W}$$

A new potential function **V**, also called 'work potential' of prescribed forces, is defined where

$$\mathbf{V} = -\{\tilde{F}\}^t\{U\} \tag{2.21}$$

and a variation of this **V** function due to virtual displacements $\{\delta U\}$ is

$$\delta \mathbf{V} = -\{\tilde{F}\}^t\{\delta U\}, \tag{2.22}$$

therefore $\delta \mathbf{U} = -\delta \mathbf{V}$, or

$$\delta(\mathbf{U}+\mathbf{V}) = 0 \tag{2.23}$$

If a function $\Phi = \mathbf{U} + \mathbf{V}$ is introduced, then at the equilibrium position of the body, the first variation of the function Φ denoted by $\delta\Phi = 0$, i.e., Φ must be stationary. For a stable equilibrium condition it can be shown that the potential function Φ is a minimum[1,2,3,4]. The principle can be enunciated as:

Of all admissible sets of displacements which satisfy the boundary conditions where the displacements are prescribed, the one which also satisfies the equilibrium equations, has a stationary value for the total potential energy function Φ, *and for stable equilibrium,* Φ *is a minimum.*

This theorem is the basis of the Ritz method used in Chapter 1 to solve some simple beam problems. The **V** function can be more elaborately defined than what is shown above so that all possible prescribed forces can be included in the analysis. If in addition to the discrete forces $\{\tilde{F}\}$, there are prescribed surface tractions $\{\tilde{T}\}$ undergoing displacements $\{u\}$ on the surface and prescribed body forces $\{\tilde{b}\}$ undergoing internal displacements $\{u\}$, then

$$\mathbf{V} = -\{\tilde{F}\}^t\{U\} - \int_S \{\tilde{T}\}^t\{u\}\,dS - \int_{vol}\{\tilde{b}\}^t\{u\}\,dvol \tag{2.24}$$

where S represents the surface where the traction $\{\tilde{T}\}$ is prescribed. Note that the vector $\{u\}$ is different from the nodal displacement vector. Here it contains three displacement functions (u, v, w) representing the displacement at a point within the body.

An example of the use of this principle is already shown with a sample problem of a simply supported beam in Chapter 1, section 1.3.

2.8 TWO DIMENSIONAL ELASTICITY PROBLEMS

There are three special two dimensional problems that are commonly popular in structural mechanics. These problems have many common features; by studying one of them well, one can learn others rather easily. Also, these problems are simplifications of the three-dimensional solid mechanics problems that are indeed more complicated to solve. These problems are:

1. Plane stress problem
2. Plane strain problem
3. Axisymmetric problem

2.8.1 Plane Stress Problem

The plane stress problem deals with stress analysis of a uniform thin plate, in the (x, y) plane which is subjected to loads applied within the plane of the plate, such that

$$\sigma_z = \tau_{xz} = \tau_{yz} = 0 \tag{2.25}$$

where the z-axis is perpendicular to the plane. It is important to note that ε_z is not necessarily equal to zero. The above condition allows us to assume that the problem depends on parameters that are functions of x and y coordinates only, which is true if the plate is relatively thin. Thus, the problem has two displacement functions u and v, three strain parameters ε_x, ε_y γ_{xy} and the corresponding stress components σ_x, σ_y and τ_{xy}. Due to the absence of σ_z, the strain

$$\varepsilon_z = -\mu\,(\varepsilon_x + \varepsilon_y).$$

A plate subjected to in-plane forces having a circular hole is considered to be a classical example for a plane stress problem as shown in Fig. 2.7.

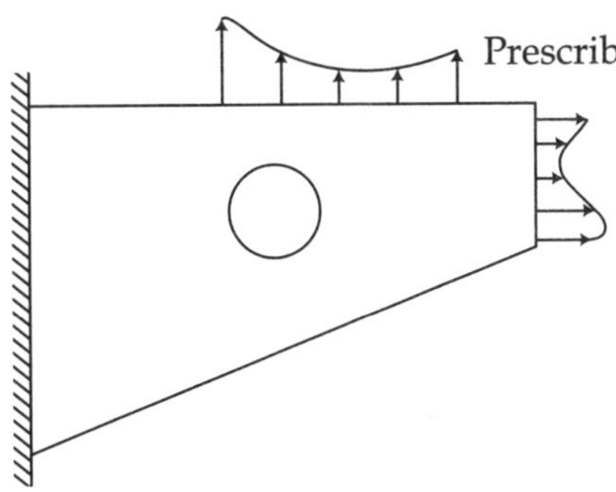

Fig. 2.7 A Thin Plate Subjected to In-plane Forces

From the above three-dimensional stress-strain equations given in Eqs. (2.11) and (2.13) it is possible to derive the following stress-strain relationship for the plane stress problem.

$$\begin{Bmatrix} \varepsilon_x \\ \varepsilon_y \\ \gamma_{xy} \end{Bmatrix} = \frac{1}{E}\begin{bmatrix} 1 & -\mu & 0 \\ -\mu & 1 & 0 \\ 0 & 0 & 2(1+\mu) \end{bmatrix}\begin{Bmatrix} \sigma_x \\ \sigma_y \\ \tau_{xy} \end{Bmatrix} \text{ or}$$

$$\begin{Bmatrix} \sigma_x \\ \sigma_y \\ \tau_{xy} \end{Bmatrix} = \frac{E}{(1-\mu^2)}\begin{bmatrix} 1 & \mu & 0 \\ \mu & 1 & 0 \\ 0 & 0 & \dfrac{(1-\mu)}{2} \end{bmatrix}\begin{Bmatrix} \varepsilon_x \\ \varepsilon_y \\ \gamma_{xy} \end{Bmatrix} \tag{2.26}$$

usually written as $\{\sigma\} = [D]\{\varepsilon\}$, where the matrix $[D]$ is as given above.

2.8.2 Plane Strain Problem

This problem is similar to the plane stress problem; instead of assuming a very thin elastic body, here an infinitely long body in the z-axis subjected to external forces such that the strains

$$\varepsilon_z = \gamma_{xz} = \gamma_{yz} = 0 \tag{2.27}$$

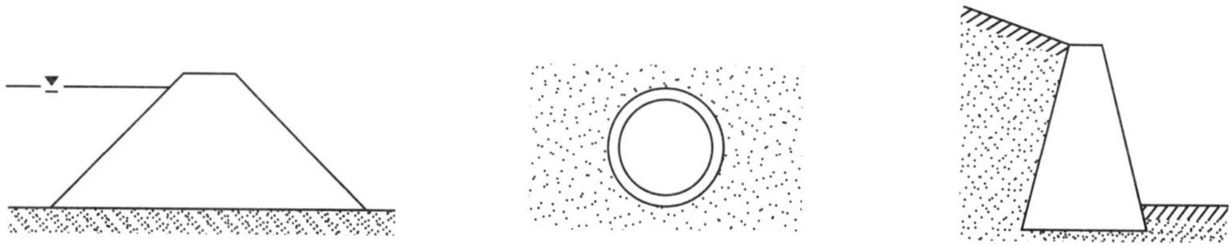

A long earth dam　　A long burried pipe-line　　A long retaining wall

Fig. 2.8 Plane Strain Problems

For this condition, the two displacements u and v, the strains ε_x, ε_y and γ_{xy} and the corresponding stresses σ_x, σ_y and τ_{xy} are all only functions of x and y coordinates. Also σ_z is not necessarily equal to zero and is equal to $-\mu(\sigma_x + \sigma_y)$.

Many problems in geotechnical engineering, such as a long dam, or long buried pipeline or a long continuous footing, etc., can be considered as plane strain problems. Examples are shown in Fig. 2.8. The stress and strain relationships can be derived as:

$$\begin{Bmatrix} \varepsilon_x \\ \varepsilon_y \\ \gamma_{xy} \end{Bmatrix} = \frac{(1-\mu^2)}{E} \begin{bmatrix} 1 & -\dfrac{\mu}{1-\mu} & 0 \\ -\dfrac{\mu}{1-\mu} & 1 & 0 \\ 0 & 0 & \dfrac{2}{1-\mu} \end{bmatrix} \begin{Bmatrix} \sigma_x \\ \sigma_y \\ \tau_{xy} \end{Bmatrix}$$

and

$$\begin{Bmatrix} \sigma_x \\ \sigma_y \\ \tau_{xy} \end{Bmatrix} = \frac{E(1-\mu)}{(1+\mu)(1-2\mu)} \begin{bmatrix} 1 & \dfrac{\mu}{1-\mu} & 0 \\ \dfrac{\mu}{1-\mu} & 1 & 0 \\ 0 & 0 & \dfrac{(1-2\mu)}{2(1-\mu)} \end{bmatrix} \begin{Bmatrix} \varepsilon_x \\ \varepsilon_y \\ \gamma_{xy} \end{Bmatrix} \tag{2.28}$$

which is denoted as $\{\sigma\} = [D]\{\varepsilon\}$, where the matrix $[D]$ is given above for the plane strain case. It is to be noted that both plane stress and plane strain problems are only two dimensional approximations of a real three dimensional problem.

2.8.3 Axisymmetric Problem

This is the third kind of two dimensional problem represented in the cylindrical coordinates (r, θ, z) where the entire body, the prescribed forces and the boundary conditions are axially symmetric with respect to the z-axis such that the displacements u and w and the strain and stress components are only functions of r and z coordinates. Here

$$v = 0, \ \tau_{r\theta} = \tau_{z\theta} = 0$$

The strain displacement equations transformed to the cylindrical coordinates system are:

$$\varepsilon_r = \frac{\partial u}{\partial r}, \ \varepsilon_\theta = \frac{u}{r}, \ \varepsilon_z = \frac{\partial w}{\partial z} \text{ and } \gamma_{rz} = \frac{\partial u}{\partial z} + \frac{\partial w}{\partial r} \tag{2.29}$$

There are many solid mechanics problems of practical interest that come under this category of axisymmetric problems. A circular tank resting on a semi-infinite elastic soil carrying axisymmetric loading is a classical example of an axisymmetric problem. Few simple example problems are illustrated in Fig. 2.9.

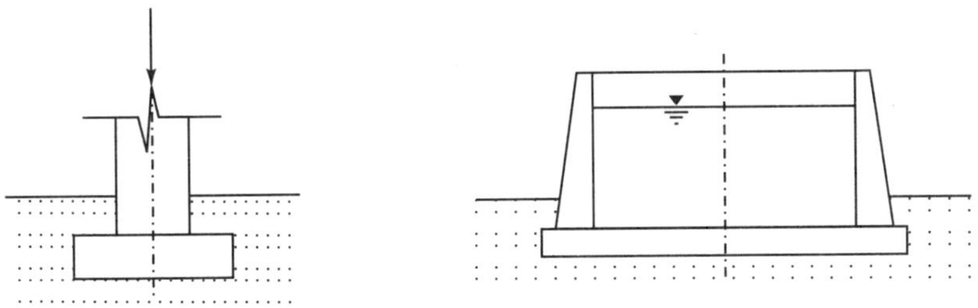

A circular footing on a semi-infinite soil medium

A circular tank on a semi-infinite soil medium

Fig. 2.9 Axisymmetric Problems

The stress-strain relationship in the axisymmetric problem can be derived as

$$\begin{Bmatrix} \sigma_r \\ \sigma_z \\ \sigma_\theta \\ \tau_{rz} \end{Bmatrix} = \frac{E(1-\mu)}{(1+\mu)(1-2\mu)} \begin{bmatrix} 1 & \dfrac{\mu}{1-\mu} & \dfrac{\mu}{1-\mu} & 0 \\ \dfrac{\mu}{1-\mu} & 1 & \dfrac{\mu}{1-\mu} & 0 \\ \dfrac{\mu}{1-\mu} & \dfrac{\mu}{1-\mu} & 1 & 0 \\ 0 & 0 & 0 & \dfrac{1-2\mu}{2(1-\mu)} \end{bmatrix} \begin{Bmatrix} \varepsilon_r \\ \varepsilon_z \\ \varepsilon_\theta \\ \gamma_{rz} \end{Bmatrix} \tag{2.30}$$

which is also given as $\{\sigma\} = [D]\{\varepsilon\}$.

2.9 INITIAL STRESS PROBLEMS

There are many problems in solid mechanics, where the body has some residual stresses before it is subjected to additional forces. Such problems are common in Geomechanics, where initial stresses are caused by existing body forces, movement of rock etc. Problems involving prestressed concrete beams and other solids subjected to stresses during their manufacture by machining, thermal hardening etc. are also considered as initial stress problems. The initial stresses have to be added to the stresses computed from the new strains caused by the new external forces. If $\{\sigma_0\}$ is initial stress vector, then the actual total stress in the body is given as:

$$\{\sigma\} = [D]\,\{\varepsilon\} + \{\sigma_0\} \tag{2.31}$$

2.10 INITIAL STRAIN PROBLEMS

Similar to initial stress problems, there are many instances, where the body can be subjected to initial strains before external forces are applied. The common problem in this category is when some part of the body is subjected to a change in temperature, resulting internal stresses that are known as 'thermal stresses'. If $\{\varepsilon_0\}$ is the initial strain vector, then the corresponding stress is

$$\{\sigma\} = [D]\{\varepsilon - \varepsilon_0\} \tag{2.32}$$

For an isotropic material, shear strain will not be produced by an increase in temperature. If α and Δt represent coefficient of linear expansion and increase in temperature respectively, then for the three dimensional case, the six components of the initial strain vector are:

$$\{\varepsilon_0\}^t = \{\alpha\,\Delta t \;\; \alpha\,\Delta t \;\; \alpha\,\Delta t \;\; 0 \;\; 0 \;\; 0\}^t \tag{2.33}$$

For the plane stress case, the corresponding initial strain vector is

$$\{\varepsilon_0\}^t = \begin{Bmatrix} \alpha\,\Delta t \\ \alpha\,\Delta t \\ 0 \end{Bmatrix} \tag{2.34}$$

and for the plane strain case, the initial strain vector is

$$\{\varepsilon_0\} = (1+\mu)\begin{Bmatrix} \alpha\,\Delta t \\ \alpha\,\Delta t \\ 0 \end{Bmatrix} \tag{2.35}$$

Similarly, for the axisymmetric case, the initial strain vector is

$$\{\varepsilon_0\} = (1+\mu)\begin{Bmatrix} \alpha\,\Delta t \\ \alpha\,\Delta t \\ 0 \end{Bmatrix} \tag{2.36}$$

Equations (2.33), (2.34), (2.35) and (2.36), show the respective components of the initial strain vectors. For the stress strain relations at a point, the initial stress and initial strain problems can be combined together to form a equation of the type,

$$\{\sigma\} = [D]\{\varepsilon - \varepsilon_0\} + \{\sigma_0\}. \tag{2.37}$$

2.11 CONTINUOUS FUNCTIONS AND CONTINUITY BETWEEN ELEMENTS

Ordinarily, the solution in the form of a continuous function or functions in the domain of the body is sought for a linear elastic problem in solid mechanics. In the finite element method, the whole continuum is discretized into discrete elements, and continuous functions are used to represent displacements within the domain of each element. But, the continuity of these functions need not be similar to what exists in the classical closed-form solutions. This situation prompts to define the type of continuity of functions that are used and therefore helps to discuss the continuity between the finite elements.

At this point, it may be interesting to recall the definition of a continuous function in mathematics. A function is continuous in a domain, if there are no jumps, overlaps or gaps in the graph of the function plotted in the domain. In other words, for a given value c for the dependant variable x, there is a unique value of the function $f(x)$. Mathematically, the continuity of a function is stated as having a unique limiting value for the function on both sides of a particular point x within the domain.

An example of a displacement function of a continuous beam is shown in Fig. 2.10 without any hinges in between. The displacement function $v(x)$ is continuous in the domain of the beam $0 \leq x \leq l$, because there is a unique value for the displacement of the beam at any point x within the domain of the beam.

The first derivative of this displacement function yields the slope of the beam, and by inspection, it is observed that the first derivative of the v-function is also a continuous function. But, when the first derivative is differentiated again, it is found that the second derivative of the function is not continuous in this example. At the middle of the beam, there is a jump in the value of the second derivative or the bending moment. This is because of the presence of an applied moment in the middle of the beam as shown. If there were no externally applied moment in the middle of the beam, then the second derivative would also have been a continuous function. Continuing the differentiation, it is observed that the third derivative which represents the distribution of shear force becomes a discontinuous function at many points as shown in the Fig. 2.10. So in this example, the displacement function has continuous derivatives only up to the first derivative and such functions are called "functions of C_1- continuity". If the second derivatives were also continuous, then v-function is identified as one with C_2- continuity. If a function is continuous in a domain, but its first derivative is not continuous everywhere, then that function is identified as one with C_0- continuity. If a continuous beam has a hinge in between its supports, then there is no continuity in slope at the hinge, in other words, the displacement function of that continuous beam is having only C_0-continuity.

In finite element methodology, primarily there are two types of elements used: one with C_0-continuity and the other with C_1-continuity. Examples of C_0-continuity include all problems in solid mechanics excluding beams, thin plates and shell problems where functions with C_1-continuity are desired. Most fluid mechanics and problems other than conventional solid mechanics where the differential equations are of second order, C_0-continuity is sufficient. These ideas would become more apparent as one proceeds further into the theory of the finite element method.

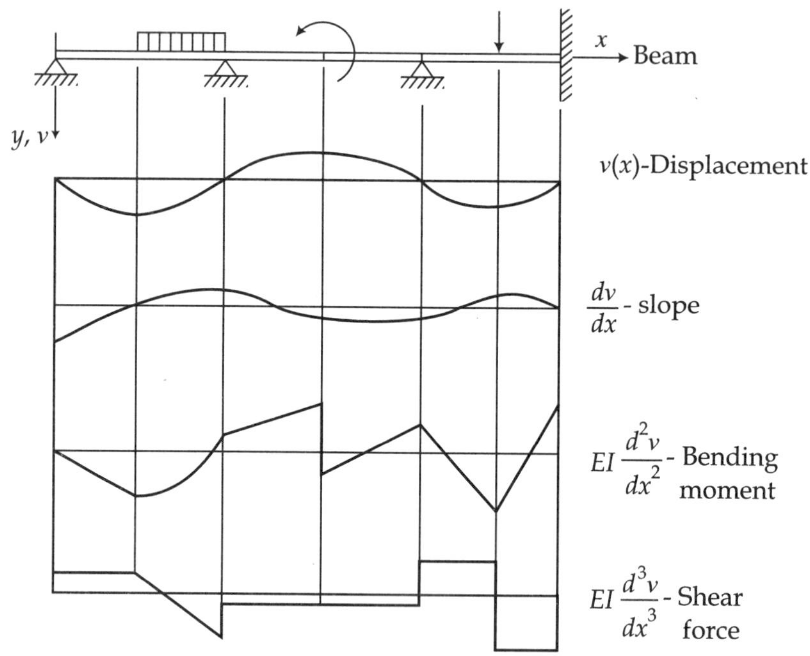

Fig. 2.10 Continuity of Functions

2.12 FINITE ELEMENT METHODOLOGY

A small introduction is given here for the finite element methodology in the form of a sequence of tasks to be performed so as to create the mathematical model and the necessary solution. These tasks also serve as a clue for the development of the flow chart and the necessary subroutines required for the computer code. Six major tasks are required in the formulation and the solution of the finite element model; they are:

Task 1: Discretization of the body into finite elements.

Task 2: Development of the element stiffness matrix equations.

Task 3: Assemblage of all element stiffness matrix equations to form global equations.

Task 4: Incorporation of the prescribed boundary forces and displacements.

Task 5: Solution of the global equations for displacements.

Task 6: Determination of internal strains and stresses.

These tasks are the same for all finite element analysis used in solid mechanics. When the method is extended to solving fluid mechanics or other boundary value problems in general, in the task 1, the domain is discretized into small domains called finite elements. As the term stiffness is used heavily by stress analysts for defining the force-displacement relationships within a solid, it may not have any particular meaning while solving a fluid mechanics or a general boundary value problem. The stiffness matrix equations can then be called system equations. All these tasks are individually explained in the later chapters for different types of finite elements.

2.13 SUMMARY

Equations in solid mechanics are presented using displacements and forces with proper sign conventions. Strain-displacement equations, stress at a point, principal stresses, and strain energy function are introduced. Using principle of virtual work, minimum potential energy theorem is derived. As approximations of three-dimensional solid mechanics problems, two-dimensional problems such as plane stress, plane strain and axisymmetric are introduced with examples. Stress-strain relations are presented including for those associated with initial stress and initial strain problems. Definition of continuity of functions is examined with respect to the characteristics of a continuous beam. The basic steps involved in the finite element methodology are presented.

PROBLEMS

2.1. Determine the principal stresses and principal axes for the stress at a point represented by the conventional stress matrix

$$[\sigma] = \begin{bmatrix} 5 & 3 & -1 \\ 3 & 6 & 2 \\ -1 & 2 & 4 \end{bmatrix}$$

Determine the corresponding von Mises stress.

2.2. Determine the principal stresses and the principal axes for the two dimensional plane stress case given as

$$[\sigma] = \begin{bmatrix} 6 & -5 \\ -5 & 4 \end{bmatrix}$$

Determine the corresponding von Mises stress.

REFERENCES

1. Crandall, S., "Engineering Analysis", McGraw Hill Book Company, 1956.
2. Dym, C.L. and Shames, I.H., "Solid Mechanics: A Variational Approach", McGraw Hill, New York, 1973.
3. Fung, Y.C., "Foundations of Soild Mechanics", Prentice Hall, Englewood Cliffs, N.J, 1965.
4. Langhaar, H.L., "Energy Methods in Applied Mechanics", John Wiley & Sons, New York, 1962.
5. Malvern, L.E., "Introduction to the Mechanics of A Continuous Medium", Prentice Hall, Englewood Cliffs, N.J, 1969.
6. Reddy, J.N., "Energy and Variational Methods in Applied Mechanics", John Wiley & Sons, 1984.
7. Timoshenko, S.P. and Goodier, J.N., "Theory of Elasticity", McGraw Hill, New York, 1970.
8. Vallabhan, C.V. Girija, "A New Formula for Principal Axes", Mechanics Research Communications, Vol. 24, No. 4, pp 443-445, Elsevier Science Ltd., 1997.
9. Washizu, K.,"Variational Principles in Elasticity and Plasticity", Pergamon Press, Oxford, England, 1982.

3

CHAPTER

One-Dimensional Elements: C_0-Continuity

3.1 INTRODUCTION

The simplest of the finite elements is the one-dimensional bar element. A complete study of this simple element from beginning of the formulation to the end of the analysis is extremely useful to understand the basic theory, formulation and methodology of all finite elements. Concepts on shape functions, techniques for formulating the basic stiffness matrix equations, assembling the finite element stiffness matrix equations using the element number-nodal number approach, incorporating the prescribed boundary conditions, the solution of global equations, etc. are presented in detail here. Even though there are more complex techniques used for forming global stiffness matrix equations, only a very simple technique is presented here. With a sound knowledge of the techniques presented for a simple element such as a bar element, further study of finite element analysis using all types of complex finite elements becomes rather easy and possible with very minor modifications.

This basic one-dimensional finite element consists of a bar or a rod of finite length and these elements are interconnected end-to-end to represent a rod with variable area of cross section. See Figs. 3.1 (a) and (b). This simple bar element has been used by structural engineers in the past for analyses of two- and three- dimensional trusses. The components of the trusses are assumed to be prismatic bar elements, hinged together at the joints of the truss, called nodes or nodal points as shown in Fig. 3.2. After a formal presentation of the finite element formulation for a longitudinal bar, the analysis of a truss is presented later in this chapter.

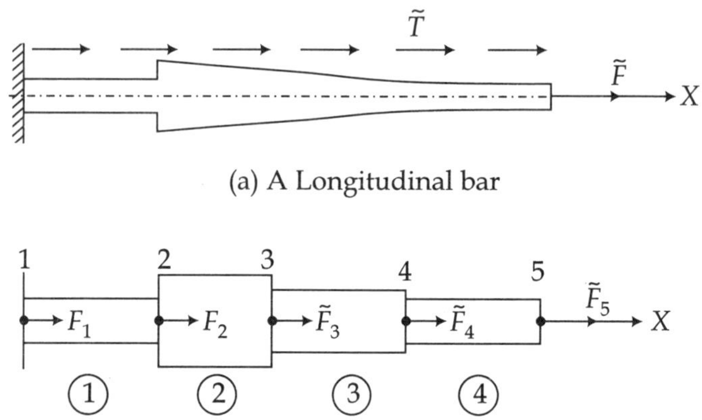

(a) A Longitudinal bar

(b) Finite element discretization

Fig. 3.1

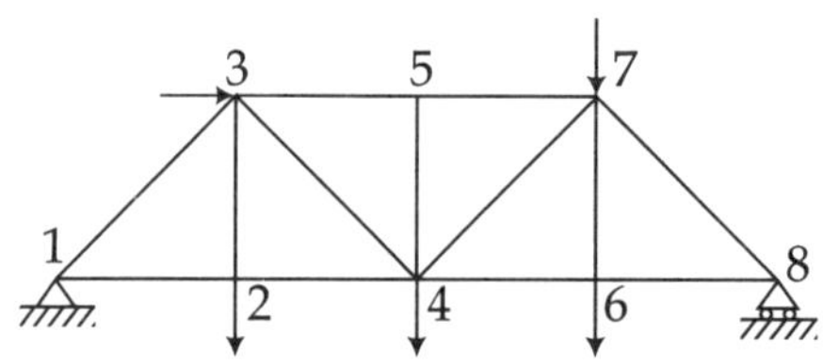

Fig. 3.2 A Two Dimensional Truss

3.2 FINITE ELEMENT ANALYSIS OF A LONG BAR

As mentioned in Chapter 2, the formulation and solution of a finite element problem is completed in six tasks which are discussed in detail here for a one-dimensional bar problem. An example problem of the longitudinal bar is shown in Fig. 3.1(a) with the prescribed forces, surface traction and body forces. This problem is introduced here only to help the beginner to learn the finite element technique in a simple way with least number of equations and with step by step procedures.

The aim in this problem is to obtain the displacements and stresses in the bar which has a uniform or variable cross section. The center of gravity of the cross section at any x (global coordinate) coincides with the longitudinal axis of the bar such that the applied axial force does not cause any bending moment in the bar.

Task 1: Finite Element Discretization

This task consists of dividing the full bar into small segments called 'finite elements'. For simplicity of formulation, it is assumed that each of these small finite elements has a uniform cross section. This assumption is not a major limitation to the overall model with variable cross section; as the number of elements increases, the results obtained from the finite element model of a bar with variable cross section will converge to its actual behavior. Also, an assumption of having uniform cross section within the element is not necessary as can be seen in more sophisticated bar elements later in this chapter. For simplicity, the longitudinal bar is discretized into five bar elements as shown in Fig. 3.1(b). The nodal numbers are shown at the ends of the bar elements, and element numbers are shown in circles. The information connecting the element numbers and the corresponding nodal numbers is cleverly stored in a matrix in the computer program; a discussion on these techniques is given later in section 3.9. It is to be noted that in the discretization shown in Fig. 3.1(b) each element has a constant cross sectional area which is an average area of cross section of the element, if the cross sectional area varies from one node to the other.

Task 2: Stiffness Matrix Equations of a Bar Element

Consider a bar element of length l with a uniform area of cross section A as shown in Fig. 3.3. The material property of the bar is denoted by Young's modulus of elasticity E assuming that the bar behaves as a linear elastic material. For the derivation of the stiffness matrix equations, a local coordinate system is selected for convenience. For any type of element, the four steps required to compute the stiffness matrix are:

1. Assume displacement function(s)
2. Determine internal strain(s)
3. Determine internal stress(es)
4. Develop the element stiffness matrix equations.

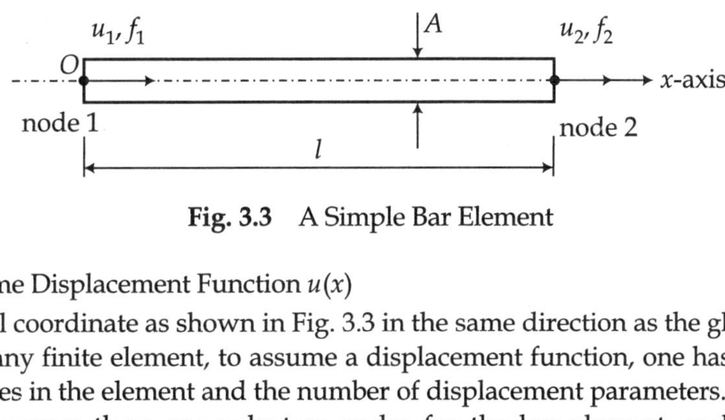

Fig. 3.3 A Simple Bar Element

Step 1. Assume Displacement Function $u(x)$

Here x is a local coordinate as shown in Fig. 3.3 in the same direction as the global coordinate. In general for any finite element, to assume a displacement function, one has to establish the number of nodes in the element and the number of displacement parameters. In this case, it is very simple, because there are only two nodes for the bar element, each node has one displacement parameter. Correspondingly, there will be only one force parameter at each node. The two nodal displacements are u_1 and u_2 at nodes 1 and 2 in the local coordinate system and they can be represented by a vector

$$\{u\} = \begin{Bmatrix} u_1 \\ u_2 \end{Bmatrix} \tag{3.1}$$

The corresponding nodal forces are f_1 and f_2, and they are represented by a vector

$$\{f\} = \begin{Bmatrix} f_1 \\ f_2 \end{Bmatrix} \tag{3.2}$$

It is important to distinguish the difference between the nodal displacement vector $\{u\}$ and the $u(x)$ which is a displacement function.

Since the bar element has two displacement parameters, it is said to have two degrees of freedom (d.o.f.). For the local coordinate, the origin of the local axis is set at node 1. The internal displacement $u(x)$ within the element can at best be approximated by a polynomial in x having two arbitrary parameters α_1 and α_2 as shown in Eq. (3.3). Thus the assumed displacement function $u(x)$ is

$$u(x) = \alpha_1 + \alpha_2 x \tag{3.3}$$

The considerations for the selection of the polynomial as shown above are of utmost importance in the finite element methodology. If $\alpha_2 = 0$ in Eq. (3.3), then $u(x) = \alpha_1$, meaning that the bar element will undergo a rigid body motion (a translation in the x-direction in this case). If $\alpha_1 = 0$, then the strain $\varepsilon_x = \dfrac{du}{dx} = \alpha_2$ allows a constant state of strain within the element. These two characteristics are necessary for the assumed displacement function. Here the parameters α_1 and α_2 are called 'generalized coordinates' which can be expressed in terms of the two nodal displacements u_1 and u_2, which are also identified as another set of generalized coordinates. In the finite element method, the stiffness equations are based on the nodal displacements; therefore it is necessary to express $u(x)$ in terms of u_1 and u_2, instead of α_1 and α_2. Thus,

$$\begin{aligned} &@\, x = 0,\ u_1 = \alpha_1, \text{ and} \\ &@\, x = l,\ u_2 = \alpha_1 + \alpha_2 l \end{aligned} \tag{3.4}$$

Therefore,

$$\alpha_1 = u_1 \text{ and } \alpha_2 = \frac{(u_2 - u_1)}{l} \tag{3.5}$$

and

$$u(x) = u_1 + \frac{(u_2 - u_1)x}{l} \tag{3.6}$$

$$= \left(1 - \frac{x}{l}\right) u_1 + \frac{x}{l} u_2$$

Substituting $N_1 = \left(1 - \dfrac{x}{l}\right)$ and $N_2 = \dfrac{x}{l}$, then

$$u(x) = N_1 u_1 + N_2 u_2 \tag{3.7}$$

In matrix form,

$$u(x) = [N_1\ N_2]\{u\} = [N]\{u\}$$

Here N_1 and N_2 are called 'shape functions', the matrix $[N]$ contains the shape functions[1,2,5]. Plots of these functions are shown in Figs. 3.4(a), (b) and (c), indicate the shape of the assumed displacement function. The value of $u(x)$ at any x can be interpolated from its nodal values, and therefore, these functions are also called 'interpolation functions'. It is interesting to note that the values of these shape functions N_i have values equal to 1 at node i and equal to zero at the other, e.g.,

$$N_1(\text{node } 1) = 1 \text{ and } N_1(\text{node } 2) = 0$$

$$N_2(\text{node } 1) = 0 \text{ and } N_2(\text{node } 2) = 1$$

At this point, another set of functions L_1 and L_2 are introduced where

$$L_1 = N_1 = \left(1 - \frac{x}{l}\right) \text{ and } L_2 = N_2 = \frac{x}{l} \tag{3.8}$$

which are called 'natural coordinates' and have the same properties of N_1 and N_2 which are the shape functions. The natural coordinates are used in integration formulas as well as development of higher order shape functions.

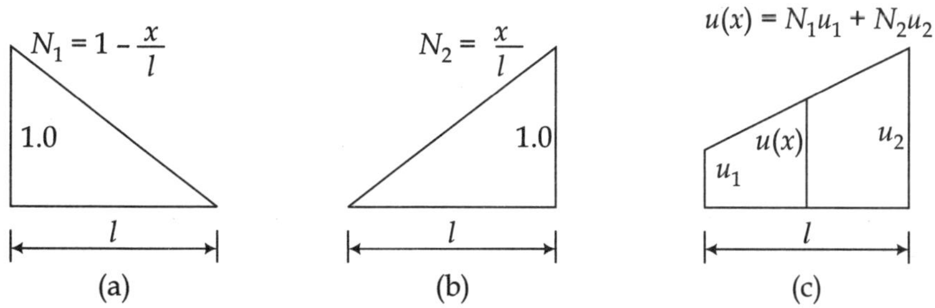

Fig. 3.4 Plot of the Shape Functions and Displacement

Step 2. Internal strain ε_x

Using the assumed displacement function, the internal strain within the element is calculated using the strain- displacement equation, i.e.,

$$\varepsilon_x = \frac{du}{dx} = \frac{1}{l}(u_2 - u_1) \tag{3.9}$$

In matrix notation,

$$\varepsilon_x = \begin{bmatrix} -\frac{1}{l} & \frac{1}{l} \end{bmatrix} \begin{Bmatrix} u_1 \\ u_2 \end{Bmatrix} = [B]\{u\} \tag{3.10}$$

It is to be noted that the longitudinal strain in each element is a constant, hence the element is called a constant strain element.

Step 3. Internal Stress σ_x

The internal stress σ_x is calculated by multiplying the internal strain by Young's modulus of elasticity E, i.e.,

$$\sigma_x = E\,\varepsilon_x = E[B]\{u\} \tag{3.11}$$

Step 4. Equilibrium of forces in the element

The equations of the equilibrium of the forces are the stiffness matrix equations for the finite element. These equations can be developed in three different ways. The first way is to use simple equilibrium at the nodes, and this is applicable to this element only. Element equations are written using internal forces as shown in Fig. 3.3.

Taking the equilibrium of forces at nodes separately,

at node 1, $-\sigma_x A = f_1$

and at node 2, $\sigma_x A = f_2$ (3.12)

and substituting σ_x in terms of nodal displacements using Eq. (3.11) and rewriting them in matrix form,

$$\frac{EA}{l}\begin{bmatrix} 1 & -1 \\ -1 & 1 \end{bmatrix}\begin{Bmatrix} u_1 \\ u_2 \end{Bmatrix} = \begin{Bmatrix} f_1 \\ f_2 \end{Bmatrix},$$

in other words, $\{k\}\{u\} = \{f\}$ (3.13)

The matrix $[k] = \begin{bmatrix} k_{11} & k_{12} \\ k_{21} & k_{22} \end{bmatrix} = \frac{EA}{l}\begin{bmatrix} 1 & -1 \\ -1 & 1 \end{bmatrix}$

is called the element stiffness matrix. From Eq. (3.13) it is seen that the stiffness matrix is a symmetric one.

For complex finite elements considering equilibrium at the nodes intuitively may not be feasible; in such instances, one has to employ the second method where the potential energy function $\Phi = \mathbf{U} + \mathbf{V}$ for the finite element is minimized. The use of this technique is more general and powerful. The Φ function for one element subjected to elemental nodal forces as shown in Fig. 3.3 is given as

$$\Phi = \frac{1}{2}\int_{Vol} \sigma_x \varepsilon_x \, dvol - \{f\}^t \{u\}$$

$$= \frac{E}{2}\left(\frac{u_2 - u_1}{l}\right)^2 Al - f_1 u_1 - f_2 u_2 \qquad (3.14)$$

Minimizing with respect to u_1 and u_2 and setting each equal to zero and representing them in a matrix form,

$$\frac{EA}{l}\begin{bmatrix} 1 & -1 \\ -1 & 1 \end{bmatrix}\begin{Bmatrix} u_1 \\ u_2 \end{Bmatrix} - \begin{Bmatrix} f_1 \\ f_2 \end{Bmatrix} = \begin{Bmatrix} 0 \\ 0 \end{Bmatrix}$$

and these equations are the same as given in Eq. (3.13). As discussed in Eq. (1.15), the stiffness coefficients can be expressed as

$$k_{ij} = \frac{\partial^2 \mathbf{U}}{\partial u_i \, \partial u_j} \qquad \text{for } i \text{ and } j = 1,2$$

and element nodal forces $f_i = -\dfrac{\partial \mathbf{V}}{\partial u_i}$ for $i = 1, 2$ (3.15)

The third and widely used method of determining the stiffness matrix equation is by employing the principle of virtual work. Equating the virtual strain energy to the virtual work done by the prescribed forces,

$\delta\mathbf{U} = \delta\mathbf{W}$, where

$$\delta \mathbf{U} = \int_{vol} \sigma_x \delta \varepsilon_x \, dvol = \int_0^l A \{\delta u\}^t [B]^t E [B] \{u\} \, dx$$

and $$\delta \mathbf{W} = \{\delta u\}^t \{f\} \tag{3.16}$$

the virtual work done by the nodal forces.

Now letting, $\delta\mathbf{U} - \delta\mathbf{W} = 0$ and taking $\{\delta u\}^t$ as a common factor,

$$\{\delta u\}^t \left(\int_0^l EA \, [B]^t [B] \{u\} \, dx - \{f\} \right) = 0 \tag{3.17}$$

Since, $\{\delta u\}^t$ is a virtual displacement vector, its components are arbitrary, the left-hand side of Eq. (3.17) within the brackets must be equal to 0. Thus,

$$EA \int_0^l [B]^t [B] \{u\} \, dx - \{f\} = 0 \tag{3.18}$$

Upon integrating the above equation becomes the same as in Eq. (3.13). It is to be noted that the element stiffness matrix is symmetric and singular. The singularity of the matrix results from the fact that the bar element has the ability to undergo a rigid body motion. Mathematically speaking, the element stiffness matrix is a positive semi-definite matrix.

Task 3: Assembling the Stiffness Matrix Equations

The principles involved in assembling all the element equations together are:

1. Uniqueness of displacements between the elements,
2. Equilibrium of forces at the nodes.

The first condition is easily satisfied by assuming a single displacement with respect to the global axis at the common node. Global forces are actually prescribed nodal forces and they are denoted with global numbers. But the nodal forces in each element are designated by 1 and 2 as subscripts and by superscripts having the element number in a circle. This is illustrated in Figs. 3.1(a), (b) and Fig. 3.5. Let us for convenience substitute for i-th element,

$$\left(\frac{EA}{l} \right)_i = k_i \tag{3.19}$$

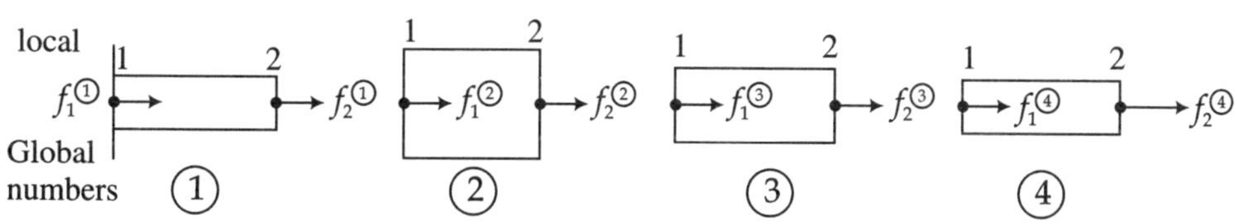

Fig. 3.5 Nodal Forces in each Element

Equilibrium of forces at each node is accomplished by adding the elemental nodal forces at each node and equating with the prescribed global force at that node. For example, let us take node 2. The element forces at node 2 from elements 1 and 2 are added together, and thus the corresponding stiffness equations are added and equated with the prescribed global force $\tilde{F}_2$ at node 2. For node 1, since the displacement is prescribed, there is an unknown reaction R there in addition to any prescribed global force $\tilde{F}_1$. The equilibrium equations at the nodes therefore are:

$$\begin{aligned}
&\text{@node 1,} \quad \tilde{F}_1 + R = f_1^{(1)} &&= k_1(u_1 - u_2)\\
&\text{@node 2,} \quad \tilde{F}_2 = f_2^{(1)} + f_1^{(2)} &&= k_1(-u_1 + u_2) + k_2(u_2 - u_3)\\
&\text{@node 3,} \quad \tilde{F}_3 = f_2^{(2)} + f_1^{(3)} &&= k_2(-u_2 + u_3) + k_3(u_3 - u_4)\\
&\text{@node 4,} \quad \tilde{F}_4 = f_2^{(3)} + f_1^{(4)} &&= k_3(-u_3 + u_4) + k_4(u_4 - u_5)\\
&\text{@node 5,} \quad \tilde{F}_5 = f_2^{(4)} &&= k_4(-u_4 - u_5)
\end{aligned} \tag{3.20}$$

These equations are written in a matrix form as shown on the next page.

$$\begin{bmatrix} k_1 & -k_1 & & & \\ -k_1 & k_1+k_2 & -k_2 & & \\ & -k_2 & k_2+k_3 & -k_3 & \\ & & -k_3 & k_3+k_4 & -k_4 \\ & & & -k_4 & k_4 \end{bmatrix} \begin{Bmatrix} u_1 \\ u_2 \\ u_3 \\ u_4 \\ u_5 \end{Bmatrix} = \begin{Bmatrix} \tilde{F}_1 + R \\ \tilde{F}_2 \\ \tilde{F}_3 \\ \tilde{F}_4 \\ \tilde{F}_5 \end{Bmatrix} \tag{3.21}$$

These are the global equations and they can be symbolically represented as

$$[K]\{U\} = \{\tilde{F}\} \tag{3.22}$$

where K, U and $\tilde{F}$ are the global stiffness matrix, displacement vector and prescribed force vector respectively.

In this finite element assemblage, there are prescribed discrete forces as nodal forces. But in general there can be prescribed distributed forces that have to be lumped on to the nodes. For example, let the element be subjected to a surface traction per unit area of the surface, $\tilde{T}$, and a body force per unit volume of the body, $\tilde{b}$. If the perimeter of the bar is p, by assuming intuitively that the total applied distributed force is equally divided between the nodes as equivalent prescribed nodal forces pertaining to this element, then

$$\tilde{f}_1 = \tilde{f}_2 = \tilde{T}\,p\frac{l}{2} + \tilde{b}\,A\frac{l}{2} \tag{3.23}$$

These prescribed force components have to be added together with other prescribed nodal forces to form the global force vector $\{F\}$ in the global stiffness matrix equation.

It is very important to distinguish between the components of element nodal force vector f_1 and f_2 and the components of the equivalent prescribed nodal force vector $\tilde{f}_1$ and $\tilde{f}_2$ representing the prescribed distributed forces in an element. f_1 and f_2 are internal forces that keep the corresponding element in equilibrium. The sum of all equivalent prescribed nodal forces along with the reactions keep the entire assemblage of elements in equilibrium.

3.2.1 Assembling using Global Potential Energy Function

Equations of global stiffness matrix can be derived by minimizing the 'Total Potential Energy Function' for the entire collection of elements and prescribed forces. This technique has the added advantage of perceiving the entire assemblage of elements, prescribed forces including nodal forces and other distributed forces within the elements. For complex finite elements with varying distributed forces, calculating the equivalent nodal forces intuitively may not be accurate and taking equilibrium at the nodes may not be feasible; in such instances, the total potential energy for the entire assemblage has to be minimized to get the stiffness matrix equation. The Φ function for the entire assemblage of elements is given as

$$\Phi = \sum_{i=1}^{N}\left\{\frac{1}{2}\int_{Vol}\sigma_x\varepsilon_x\,dvol - \int_0^1 \tilde{T}\,pudx - \int_0^1 \tilde{b}\,Audx\right\}_i - \{\tilde{F}\}^t\{U\} \tag{3.24}$$

Here l, A and p are the individual length, area of cross section and perimeter of the each bar element i, respectively and N is the total number of elements. Since all elements are of same length, area etc., just for this example, the total strain energy function is obtained as,

$$\mathbf{U} = \frac{EA}{2l}\,\{(u_2 - u_1)^2 + (u_3 - u_2)^2 + (u_4 - u_3)^2 + (u_5 - u_4)^2\}$$

Differentiating with respect to u_i, $i = 1, ..., 5$, the same stiffness matrix as given in Eq. (3.14) is obtained. The equivalent nodal forces for the prescribed distributed forces on the i-th element are obtained from the work potential due to distributed forces as:

$$\mathbf{V} = -\int_0^l \tilde{T}\,pu\,dx - \int_0^l \tilde{b}\,Au\,dx$$

$$= -\int_0^l \tilde{T}\,p(N_1u_1 + N_2u_2)dx - \int_0^l \tilde{b}\,A(N_1u_1 + N_2u_2)\,dx$$

And,

$$\tilde{f}_1 = -\frac{\partial \mathbf{V}}{\partial u_1} = p\,\frac{l}{2} + \tilde{b}\,A\,\frac{l}{2}, \tag{3.25}$$

$$= \tilde{f}_2$$

Note that the subscripts 1 and 2 in Eq. (3.25) on the force and displacement vectors refer to the local numbers of the i-th element. It is interesting to note that here the equivalent element nodal forces according to the energy principle are identically equal to the lumped forces as shown in Eq. (3.23). If there are additional prescribed discrete nodal forces, the work potential due to discrete prescribed nodal forces for the entire assemblage is,

$$\mathbf{V} = -\{\tilde{F}\}^t\{U\}$$

Differentiating the above work potential with respect to u_i, $i = 1,...5$, the nodal prescribed force vector is obtained. These force vector components have to be added to the components of the distributed force vector to get the final global force vector.

3.2.2 Assemblage using Principle of Virtual Work

The third and widely used method of determining the stiffness matrix equation is by employing the principle of virtual work. Writing the virtual strain energy for the total assemblage and equating to the virtual work done by the prescribed forces, the same equations are obtained. For all elements,

$$\delta \mathrm{U} = \sum_{i=1}^{N} \int_{vol} (\delta\varepsilon_x \, \sigma_x)_i \, dvol = \sum_{i=1}^{N} \int_0^l (A\{\delta u\}^t \, [B]^t \, E \, [B] \, \{u\})_i \, dx \tag{3.26}$$

and therefore, the element stiffness matrix for the i-th element is

$$[k]_i = \int_0^l (A \, [B]^t \, E \, [B])_i \, dx \tag{3.27}$$

and virtual work done by the prescribed distributed forces on all elements is

$$\delta \mathrm{W} = \sum_{i=1}^{N} \int_0^l \{\delta u\}^t ([N]^t \, \tilde{T} \, p + [N]^t \, \tilde{b} A)_i \, dx \tag{3.28}$$

Therefore, the equivalent nodal force vector due to the prescribed distributed forces for i-th element is

$$\{\tilde{f}\}_i = \int_0^l ([N]^t \, \tilde{T} \, p + [N]^t \, \tilde{b} \, A)_i \, dx \tag{3.29}$$

which is the same as in Eq. (3.23). All these methods result in the same set of equations.

Task 4: Incorporation of Prescribed Displacements

To start with, a simple method is shown here for achieving this. For the given problem, the displacement $\tilde{u}_1$ at node 1 is prescribed as equal to 0 and therefore there has to be a reaction at node 1, say = R in addition to the prescribed force $\tilde{F}_1$, as shown in Eqs. (3.20) and (3.21). But, the reaction in a static problem is always not known or cannot be easily calculated as in this case. Here the value of R can be easily calculated from the equilibrium of all the forces in the X-direction, and substituting the value of R into the global equation given by Eq. (3.21), it may appear that there are five unknowns and five equations, and therefore the five displacements can be calculated. However, the stiffness matrix in the global equation is still a singular matrix. Since R is an unknown and $\tilde{u}_1 = 0$, the first equation in the global system can be ignored. Substituting the value of $\tilde{u}_1$ in the second equation, the global equation is revised as a 4×4 matrix with 4 unknown displacements and then the matrix is no longer singular. This is shown below.

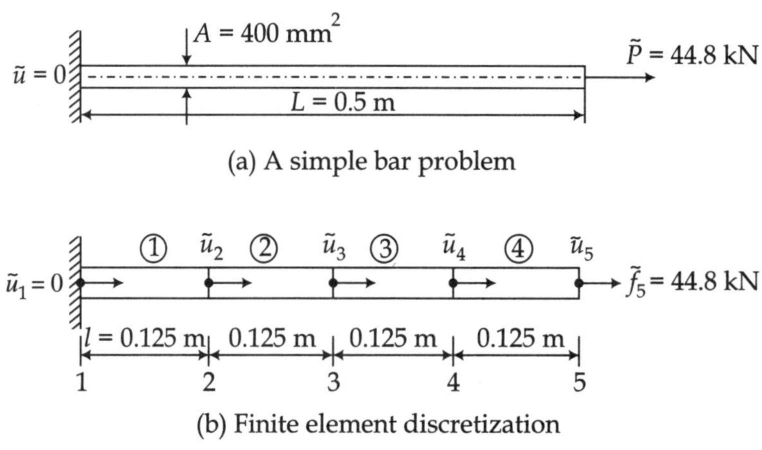

Fig. 3.6

$$\begin{bmatrix} k_1+k_2 & -k_2 & & \\ -k_2 & k_2+k_3 & -k_3 & \\ & -k_3 & k_3+k_4 & -k_4 \\ & & -k_4 & k_4 \end{bmatrix} \begin{Bmatrix} u_2 \\ u_3 \\ u_4 \\ u_5 \end{Bmatrix} = \begin{Bmatrix} \tilde{F}_2 \\ \tilde{F}_3 \\ \tilde{F}_4 \\ \tilde{F}_5 \end{Bmatrix} \tag{3.30}$$

Task 5: Solution of the Global Equations

After prescribing the displacement boundary conditions, the matrix is no longer singular, and the revised equations can be solved by any linear simultaneous equation solver. There are several methods for solving these equations in numerical analysis; but for solving finite element equations, the LDU decomposition method which is explained in the Appendix I is very popular. Here, the global matrix can be formed as a banded matrix and since the matrix is symmetric, only the half-banded matrix needs to be stored on the computer. For small matrices, it is immaterial what method of solution is used. However, for a large system of equations to make the solution computationally efficient, either a half banded solver or a skyline solver can be used. These two techniques are briefly described in the Appendix I. Using one of these techniques, the global displacements are obtained as

$$\{U\} = [K]^{-1}\{\tilde{F}\} \tag{3.31}$$

Task 6: Calculation of Internal Strains and Stresses

The internal strains within each element are constants for this type of element and for the i-th element,

$$(\varepsilon_x)_i = \left(\frac{u_2 - u_1}{l}\right)_i = [B]_i\{u\}_i \tag{3.32}$$

and the corresponding stresses are:

$$(\sigma_x)_i = E(\varepsilon_x)_i \text{ for } i = 1,2,\ldots, n. \tag{3.33}$$

Sometimes, the analyst might be interested to compute the reactions where the displacements are prescribed. The reaction R, in this case can be calculated from the first equation of Eq. (3.20).

Example Problem 3.1

An example of a simple longitudinal bar is shown in Fig. 3.7, with all the dimensions, material properties and loading. These problems are presented only for getting some practice for beginners before they get involved with more complex and larger finite element equations. For convenience, the bar is assumed to be uniform and all the elements are made equal. The properties of the elements are:

$$l = 0.125 \text{ m.}, A = 400 \text{ mm}^2 \text{ and } E = 70 \text{ GPa}.$$

The element stiffness matrix for all elements in the local coordinate system is therefore,

$$k\begin{bmatrix} 1 & -1 \\ -1 & 1 \end{bmatrix}\begin{Bmatrix} u_1 \\ u_2 \end{Bmatrix} = \begin{Bmatrix} f_1 \\ f_2 \end{Bmatrix} \tag{3.34}$$

where $k = \dfrac{EA}{l} = 224 \times 10^3$ kN/m. The applied force at node 5 is 44.8 kN and the global matrix before prescribing the displacements is

$$224 \times 10^3 \begin{bmatrix} 1 & -1 & & & \\ -1 & 2 & -1 & & \\ & -1 & 2 & -1 & \\ & & -1 & 2 & -1 \\ & & & -1 & 1 \end{bmatrix} \begin{Bmatrix} u_1 \\ u_2 \\ u_3 \\ u_4 \\ u_5 \end{Bmatrix} = \begin{Bmatrix} R \\ 0 \\ 0 \\ 0 \\ 44.8 \end{Bmatrix} \tag{3.35}$$

and after prescribing displacements,

$$224 \times 10^3 \begin{bmatrix} 2 & -1 & & \\ -1 & 2 & -1 & \\ & -1 & 2 & -1 \\ & & -1 & 1 \end{bmatrix} \begin{Bmatrix} u_2 \\ u_3 \\ u_4 \\ u_5 \end{Bmatrix} = \begin{Bmatrix} 0 \\ 0 \\ 0 \\ 44.8 \end{Bmatrix}$$

Solving the above equations, the displacements are:

$$u_1 = 0,\ u_2 = 0.2,\ u_3 = 0.4,\ u_4 = 0.6 \text{ and } u_5 = 0.8 \text{ mm.}$$

The corresponding strains and stresses within the elements are of course all equal as shown in Fig. 3.7. Naturally, the method yields exact solution, because the problem is of constant strain and the finite element model used constant strain elements. Even one element or unequal elements would yield the exact behavior of the bar. The reaction at node 5 is given by Eq. (3.20) as

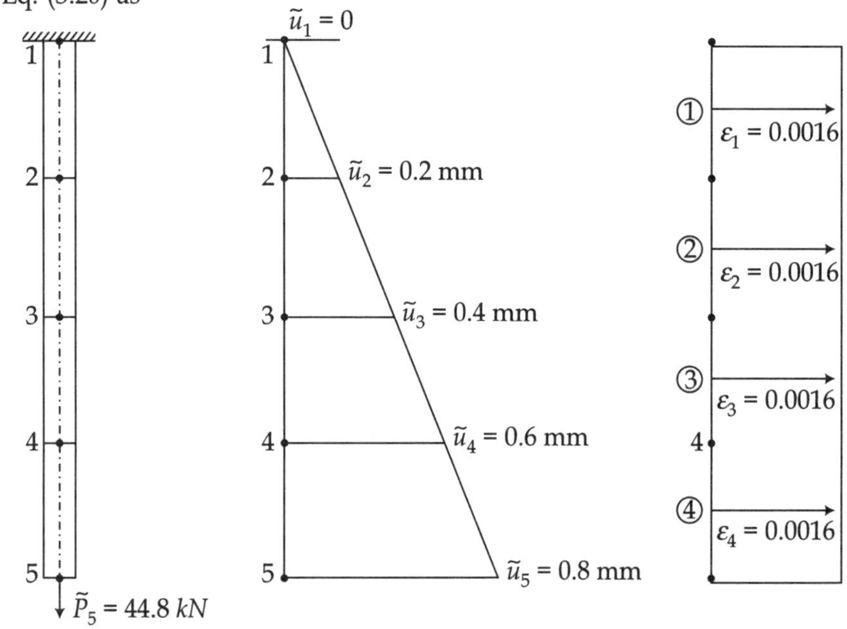

(a) Example problem of a long bar (b) Nodal Displacements (c) Element Strains

Fig. 3.7

$$R = k(u_1 - u_2) = -44.8 \text{ kN},$$

which is the exact value of the reaction.

Example Problem 3.2 This example contains a traction force, $\tilde{T} = 448.0$ kPa, on the surface as shown in Fig. 3.8(a). Here the dimensions of the elements are the same as given in Problem No. 1, and the perimeter of the bar element is 80.0 mm. It is required to determine the nodal displacements, element strains, element stresses and reaction at the support.

Again using equal elements for illustration purposes, the element stiffness matrix is obtained as the same, but the nodal forces have to be calculated using Eq. (3.23). For the element in the local coordinate system, the equivalent nodal forces are:

$$\tilde{f}_1 = 2.24 \text{ and } \tilde{f}_2 = 2.24 \text{ kN, respectively.}$$

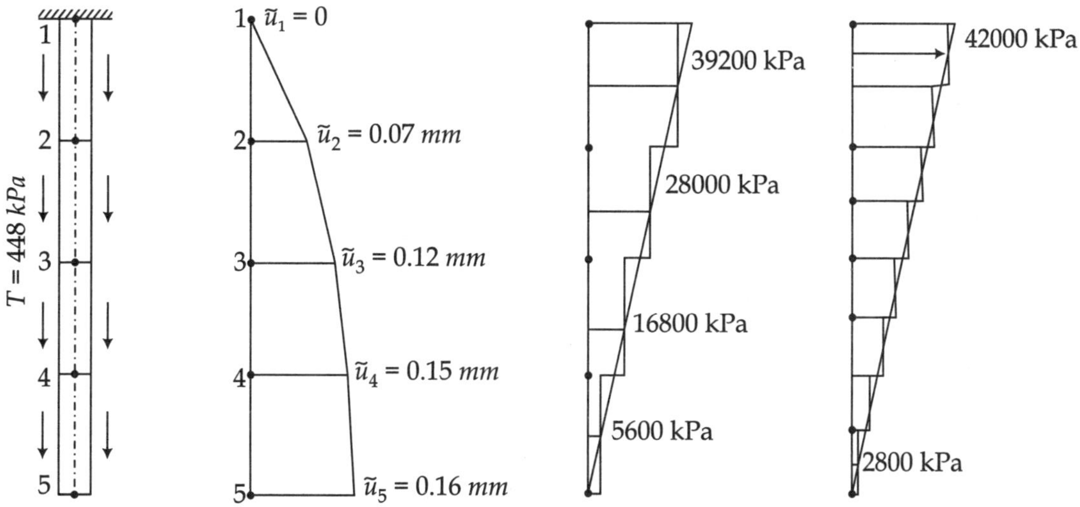

(a) Problem (b) Displacements (c) Stresses (4 elements) (d) Stresses (8 elements)

Fig. 3.8

Substituting these into the global equations and prescribing the displacement boundary conditions, the final equation is as shown below:

$$224 \times 10^3 \begin{bmatrix} 2 & -1 & & \\ -1 & 2 & -1 & \\ & -1 & 2 & -1 \\ & & -1 & 1 \end{bmatrix} \begin{Bmatrix} u_2 \\ u_3 \\ u_4 \\ u_5 \end{Bmatrix} = \begin{Bmatrix} 4.48 \\ 4.48 \\ 4.48 \\ 2.24 \end{Bmatrix}$$

Solving the above equations, the unknown displacements are obtained as

$$u_1 = 0,\ u_2 = 0.07,\ u_3 = 0.12,\ u_4 = 0.15 \text{ and } u_5 = 0.16 \text{ mm.}$$

The nodal displacements and the element stresses are calculated as shown in Figs. 3.8(b), (c) and (d). The reaction R at node 1 can be calculated from the first equation of the global equations, and the exact value of the reaction is obtained as shown below.

$$R + 2.24 = k(u_1 - u_2)$$
$$= -15.68 \text{ kN}$$
$$R = -17.92 \text{ kN},$$

which is the exact value of the reaction. Mathematically speaking, the displacement in the bar varies quadratically, while in the finite element model, the displacement varies linearly within each element. Further, the strains and stresses are supposed to vary linearly, and the finite element analysis yields constant strains and stresses within the elements. Now, if the number of elements for this problem are increased, the discrepancy between the nodal stresses can be reduced as can be seen in the Fig. 3.8(d).

3.3 CONTINUITY BETWEEN ELEMENTS

The title of this chapter contains a word 'C_0-continuity', which was discussed in Chapter 2. In the bar element only nodal displacements are used as unknown parameters, and therefore only compatibility of the displacements is guarenteed. The strains and stresses at the common nodes are not compatible. This is seen in the results of Example 3.2. Such a finite element is identified as an element with C_0- continuity. In order to make the first derivatives at the interelement nodes to be compatible, the first derivatives at nodes in addition to nodal displacements are required as nodal parameters; such an element will be identified as one with C_1-continuity. For the analysis of the bending of beams C_1-continuity elements are necessary. This is discussed in Chapter 8. Now if displacement and its first and second derivatives are used as nodal parameters in the formulation of a finite element, continuity up to second derivatives between the elements can be achieved, and such an element will be identified as one with C_2-continuity. However, such an element is not used often in structural mechanics.

3.4 RELATIONSHIP WITH RITZ METHOD IN THE GLOBAL COORDINATES

Since the displacement functions are assumed in the local coordinate system and the total energy function Φ is minimized locally, the finite element model can be said to be a Ritz model in the local coordinate system. Now, for the first example problem, if a set of five continuous functions ψ_i, $i = 1,\ldots, 5$ with C_0-continuity in the global coordinates as shown in Fig. 3.9 are used to define the displacement function $u(X)$ in the global coordinates, then $u(X)$ can be written as

$$u(X) = \sum_{i=1}^{5} u_i \, \psi_i \tag{3.36}$$

with five unknown displacement parameters, u_i, $i = 1,\ldots,5$.

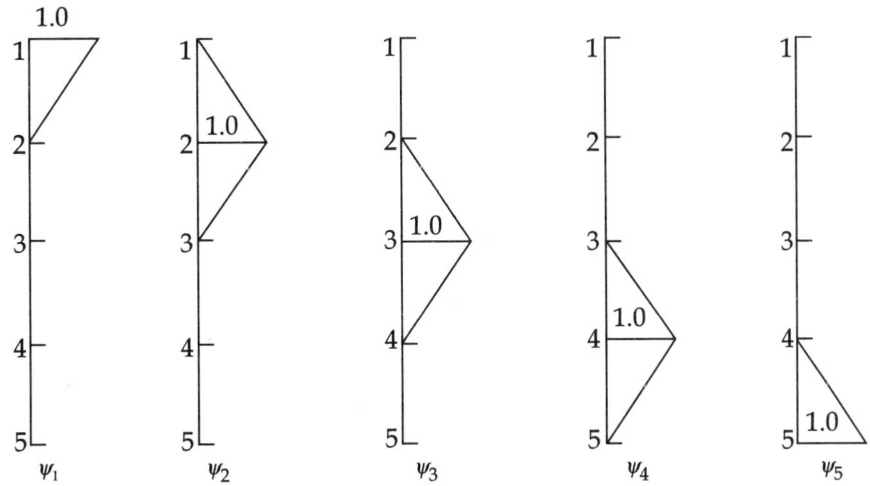

Fig. 3.9 A Set of C_0- Continuous Functions as Basis

Expressing the energy function Φ in terms of the global displacement parameters and minimizing the Φ function with respect to these parameters, exactly the same set of equations are obtained for the Example problem 3.1 in Eq. (3.21). This fact clearly demonstrates that the finite element method is nothing but a Ritz method in the local as well as in the global coordinates. The set of functions ψ_i, $i = 1,\dots, 5$ is said to belong to the Sobolov space in functional analysis.[3]

3.5 EQUIVALENT FORCE VECTOR

The equivalent nodal forces are computed by differentiating the work potential function **V** with respect to the nodal displacements. i.e.,

$$\tilde{f}_i = -\frac{\partial \mathbf{V}}{\partial u_i}, \quad i = 1, 2 \tag{3.37}$$

If $\tilde{T}$ is the distributed surface forces per unit area of the surface, and $\tilde{b}$ is the body force per unit volume of the element, then

$$\mathbf{V} = -\int_0^l \tilde{T}pu\,dx - \int_0^l \tilde{b}Au\,dx = -\int_0^l \tilde{T}p\sum_i^2 N_i\,u_i\,dx - \int_0^l \tilde{b}A\sum_i^2 N_i\,u_i\,dx$$

$$\tilde{f}_i = \frac{\partial \mathbf{V}}{\partial u_i} = \int_0^l \tilde{T}pN_i\,dx + \int_0^l \tilde{b}A\,N_i\,dx\,,\; i = 1, 2 \tag{3.38}$$

Here p and A are the perimeter and area of cross section of the bar element.

If $\tilde{T}$ and $\tilde{b}$ are constants, then $\tilde{f}_1 = \tilde{f}_2 = \frac{1}{2}\tilde{T}\,pl + \frac{1}{2}\tilde{b}Al$ as indicated in Eq. (3.23).

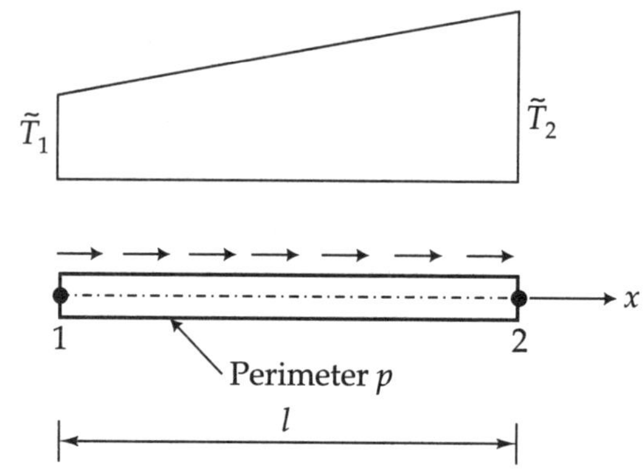

Fig. 3.10 Linear Variation of Traction on Bar Element

But, if $\tilde{T}$ is linearly varying from $\tilde{T}_1$ to $\tilde{T}_2$ along the element as shown in Fig. 3.10, then $\tilde{T}$ at any x can be interpolated using the shape functions N_1 and N_2 respectively, i.e. **V** can be expressed as

$$\mathbf{V} = \int_0^l (N_1\,\tilde{T}_1 + N_2\,\tilde{T}_2)\,p\,(N_1\,u_1 + N_2\,u_2)\,dx \tag{3.39}$$

$$\tilde{f}_1 = -\frac{\partial \mathbf{V}}{\partial u_1} = \frac{(2\tilde{T}_1 + \tilde{T}_2)\,pl}{6}$$

and

$$\tilde{f}_2 = -\frac{\partial \mathbf{V}}{\partial u_2} = \frac{(\tilde{T}_1 + 2\tilde{T}_2)\,pl}{6} \tag{3.40}$$

Similar expressions can be derived for the equivalent body force vector also.

3.6 LINEAR STRAIN BAR ELEMENT

So far constant strain bar element is studied and when used to solve Example Problem 3.2, where the strain varies linearly, there is substantial incompatibility between strains and stresses at the common nodes. This discrepancy can be minimized by employing a better finite element where the internal strain varies linearly within the element.

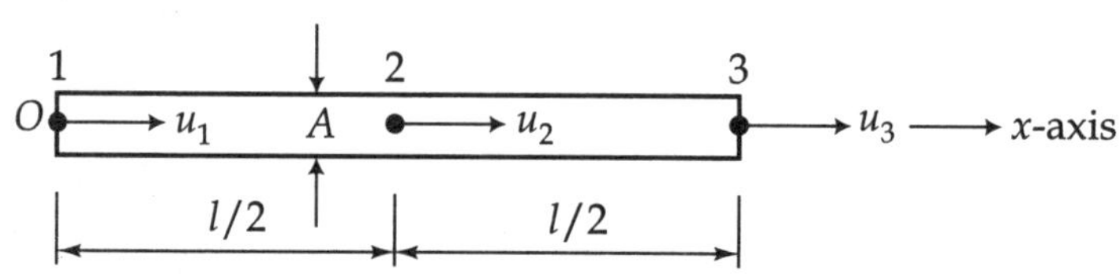

Fig. 3.11 A Linear Strain Element

The derivation of the stiffness matrix for this element is as follows.

Step 1. Assume Displacement Function $u(x)$

A linear variation of strain within the element can be accomplished by using a quadratic function in x, such that the function accommodates the required rigid body motion and constant strain condition within the element, i.e.,

$$u(x) = \alpha_1 + \alpha_2 x + \alpha_3 x^2 \tag{3.41}$$

There are three adjustable parameters, hence three displacement parameters are necessary and this can be accomplished by providing three nodes in the element. These nodes are conveniently placed at the ends and at the center of the element in the local coordinate system as shown in Fig. 3.11. Substituting the displacement values u_1, u_2 and u_3 in Eq. (3.41) and solving for α_1, α_2 and α_3.

$$\alpha_1 = u_1$$

$$\alpha_2 = \frac{1}{l} = (-3u_1 + 4u_2 - u_3)$$

$$\alpha_3 = \frac{2}{l^2}(u_1 - 2u_2 + u_3)$$

Substituting in Eq. (3. 41),

$$u(x) = N_1 u_1 + N_2 u_2 + N_3 u_3 \tag{3.42}$$

$$= [N_1 \; N_2 \; N_3]\{u\} \quad \text{where}$$

$$N_1 = \left(1 - \frac{3x}{l} + \frac{2x^2}{l^2}\right), \tag{3.43}$$

$$N_2 = 4\left(\frac{x}{l} - \frac{x^2}{l^2}\right)$$

and

$$N_3 = \left(-\frac{x}{l} + \frac{2x^2}{l^2}\right)$$

The shape functions for the linear strain bar element is illustrated in Fig. 3.12

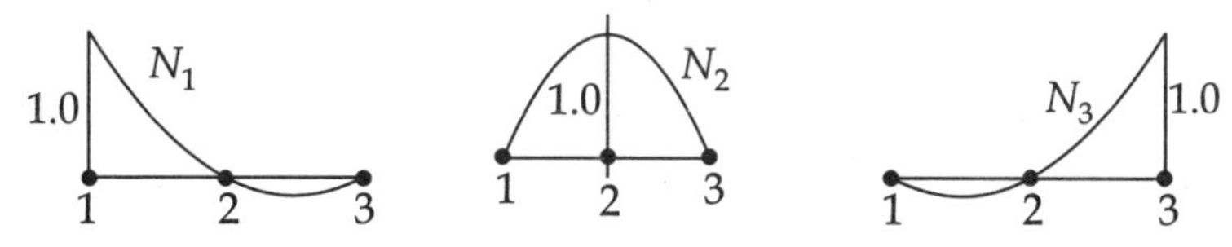

Fig. 3.12 Shape Functions for the Linear Strain Element

Step 2. Internal Strain ε_x

Using the strain displacement equation for the axial strain,

$$\varepsilon_x = \frac{du}{dx}$$

$$= \left[\frac{dN_1}{dx} \; \frac{dN_2}{dx} \; \frac{dN_3}{dx}\right]\{u\}$$

$$= \left[\left(-\frac{3}{l} + \frac{4x}{l^2} \right) \left(\frac{4}{l} - \frac{8x}{l^2} \right) \left(-\frac{1}{l} + \frac{4x}{l^2} \right) \right] \{u\}$$

$$= [B]\{u\} \tag{3.44}$$

Step 3. Internal Stress σ_x

The internal stress

$$\sigma_x = E\,\varepsilon_x$$

$$= E\,[B]\{u\} \tag{3.45}$$

Step 4. Apply Minimum Potential Energy Theorem

The potential energy function, $\mathbf{\Phi} = \mathbf{U} + \mathbf{V}$, and differentiating **U** and **V** with respect to u_i, the stiffness matrix and the equivalent nodal force vector for prescribed distributed forces respectively are obtained.

$$\mathbf{U} = \frac{1}{2}\int_0^l \sigma_x\,\varepsilon_x\,A\,dx = \frac{1}{2}\int_0^l \{u\}^t\,\{B\}^t\,E\,\{B\}\,\{u\}\,A\,dx$$

and the stiffness matrix

$$[k] = \int_0^l E\,A\,[B]^t\,[B]\,dx$$

$$= EA\int_0^l \begin{bmatrix} \left(\frac{dN_1}{dx}\right)^2 & \frac{dN_1}{dx}\frac{dN_2}{dx} & \frac{dN_1}{dx}\frac{dN_3}{dx} \\ & \left(\frac{dN_2}{dx}\right)^2 & \frac{dN_2}{dx}\frac{dN_3}{dx} \\ sym. & & \left(\frac{dN_3}{dx}\right)^2 \end{bmatrix} dx \tag{3.46}$$

where

$$k_{ij} = \int_0^l E\,A\,\frac{dN_i}{dx}\,\frac{dN_j}{dx}\,dx$$

Integrating,

$$[k] = \frac{EA}{3l}\begin{bmatrix} 7 & -8 & 1 \\ -8 & 16 & -8 \\ 1 & -8 & 7 \end{bmatrix}$$

If a uniform traction $\tilde{T}$ is prescribed on the surface of the element, the corresponding force vector is

$$\tilde{f}_1 = -\frac{\partial \mathbf{V}}{\partial u_1} = +\frac{\partial}{\partial u_1}\int_0^l \tilde{T}p\,(N_1\,u_1 + N_2\,u_2 + N_3\,u_3)dx$$

$$= \tilde{T}p\int_0^l N_1\,dx = \frac{\tilde{T}pl}{6} \tag{3.47}$$

Similarly $$\tilde{f}_2 = -\frac{\partial \mathbf{V}}{\partial u_2} = \tilde{T}p\int_0^l N_2\,dx = \frac{4\tilde{T}pl}{6}, \quad \text{and}$$

$$\tilde{f}_3 = \tilde{f}_1$$

3.6.1 Strains and Stresses at the Nodes of the Element

Using Eq. (3.44), i.e., $\varepsilon_x = [B]\{u\}$ and substituting the nodal displacements, the equations for the nodal strains for each element are as follows:

$$@x = 0,\ (\varepsilon_x)_1 = \frac{1}{l}(-3u_1 + 4u_2 - u_3)$$

$$@x = \frac{l}{2},\ (\varepsilon_x)_2 = \frac{1}{l}(-u_1 + u_3),\ \text{and} \tag{3.48}$$

$$@x = l,\ (\varepsilon_x)_3 = \frac{1}{l}(u_1 - 4u_2 + 3u_3)$$

The nodal stresses can be computed by multiplying the respective strain components by E.

Example Problem 3.3 The same Example problem 3.2 with prescribed surface traction as shown in Fig. 3.13 (a) is solved here with four linear strain elements. The stiffness coefficient $k = (EA/3l)$ of each element = 74.6667×10^3. The element force vector is

$$(0.74667\ 2.98667\ 0.74667)^t\ \text{kN}.$$

The global stiffness matrix and the force vector are assembled as given below:

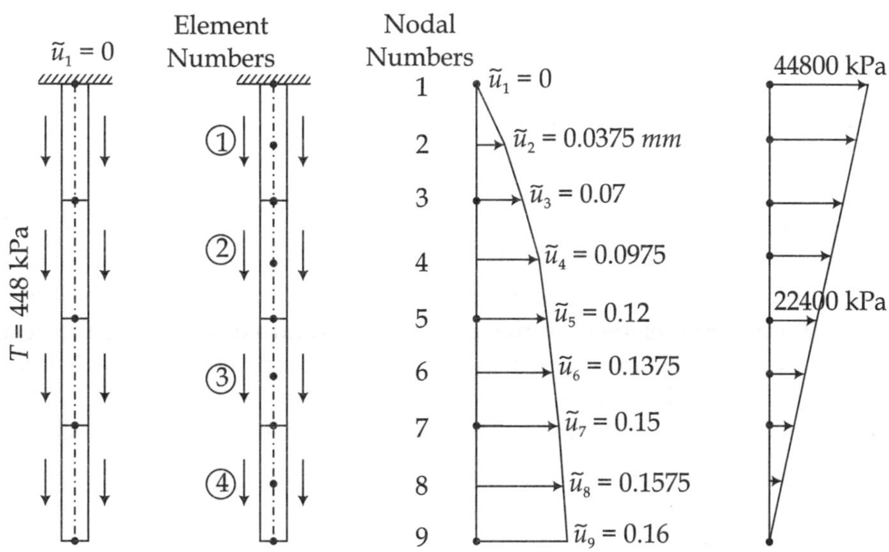

(a) Problem (b) Dicretization (c) Displacements (d) Stresses

Fig. 3.13

$$74.6667 \times 10^3 \begin{bmatrix} 7 & -8 & 1 & & & & & & \\ -8 & 16 & -8 & & & & & & \\ 1 & -8 & 14 & -8 & 1 & & & & \\ & & -8 & 16 & -8 & & & & \\ & & 1 & -8 & 14 & -8 & 1 & & \\ & & & & -8 & 16 & -8 & & \\ & & & & 1 & -8 & 14 & -8 & 1 \\ & & & & & & -8 & 16 & -8 \\ & & & & & & 1 & -8 & 7 \end{bmatrix} \begin{Bmatrix} u_1 \\ u_2 \\ u_3 \\ u_4 \\ u_5 \\ u_6 \\ u_7 \\ u_8 \\ u_9 \end{Bmatrix} = \begin{Bmatrix} 0.74667 + R \\ 2.98667 \\ 1.49333 \\ 2.98667 \\ 1.49333 \\ 2.98667 \\ 1.49333 \\ 2.98667 \\ 0.74667 \end{Bmatrix}$$

After prescribing the boundary displacement $\tilde{u}_1 = 0$, on to the global equations as discussed earlier, the equations are solved for the nodal displacements. The nodal displacements and the calculated element nodal stresses are plotted in Figs. 3.13 (c) and (d). The reaction R at node 1 can be calculated from the first equation of the original global equation, that is before prescribing the boundary displacements. In other words,

$$\frac{EA}{3l}(7u_1 - 8u_2 + u_3) = 0.74667 + R,$$

and substituting the values of u_1, u_2 and u_3, R is obtained as –17.92 kN, which is the exact value of the reaction. The equivalent nodal force from the distributed traction force has to be included in the calculation as shown above in order to obtain the correct value of the reaction R.

Since a linear strain element was used to represent a bar problem, where the internal strain was varying linearly, the solution resulted with perfect compatibility of strains at the common nodes for this problem. This problem can also be solved by using only one linear bar element or combining with unequal linear bar elements, and in both cases, the exact solution is obtained. If the loading or the cross section of the bar are varying, such that there are discontinuities of strains, the finite element solution with linear bar elements would yield more accurate results than the solution with constant strain elements. However, there is no requirement to have compatability of strains at the interelement nodes for the linear strain element. For the same number of elements, the linear bar element solution requires more degrees of freedom and bandwidth, but linear element solution will normally use fewer number of elements. The use of higher order elements for the solution of longitudinal bar problems is not a crucial issue here, as the finite element techniques are primarily used to solve two- or three-dimensional problems of elasticity, and very seldom for one-dimensional problems. The linear strain element in this chapter is introduced to educate the beginner on elements other than the constant strain element. This brings up another issue. Is it possible to develop an element which has a quadratic variation of strain within? Of course, it is possible with a four noded element, and the shape functions can be developed using the same principle indicated here with four adjustable parameters. The Lagrange interpolation function discussed in section 3.8 of this chapter would also help to write the required shape functions directly.

3.7 ELEMENT STIFFNESS MATRICES FOR SPECIAL BAR ELEMENTS

3.7.1 Bar Element with Linearly Varying Area of Cross Section

A simple two degrees of freedom (d.o.f.) bar element has its cross-sectional area varying linearly from A_1 at node 1 to A_2 at node 2 as shown in Fig. 3.14. The four steps discussed for the development of element stiffness matrix earlier are used sequentially to clearly illustrate the methodology.

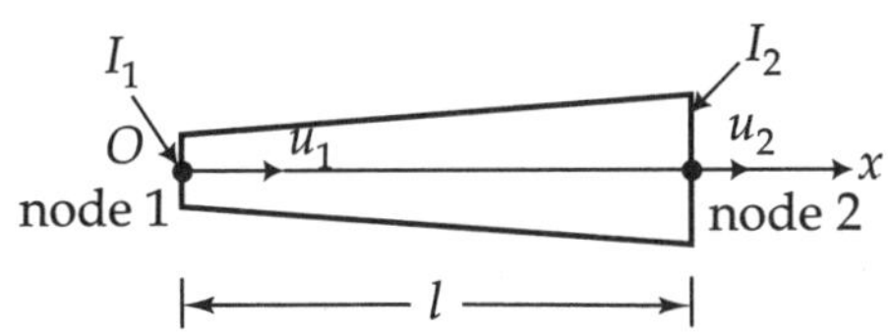

Fig. 3.14 A Bar with Linearly Varying Area of Cross Section

Step 1. Assume Displacement Function $u(x)$

Since there are only two degrees of freedom, $u(x)$ can be assumed to be the same as given in Eqs. (3.4) through (3.7), i.e.,

$$u(x) = [N_1 \; N_2]\{u\}$$

Here, the shape functions are linear functions.

Step 2. Internal Strain ε_x

The equation for the strain is the same as Eq. (3.10), i.e.,

$$\varepsilon_x = [B]\{u\}$$

yielding a constant strain within the element.

Step 3. Internal Stress σ_x

As given in Eq. (3.11),

$$\sigma_x = E\,\varepsilon_x = E[B]\{u\}$$

Step 4. Minimum Potential Energy Theorem

So far all the equations are the same as for the bar element having uniform area of cross section. The total energy function $\mathbf{\Phi} = \mathbf{U} + \mathbf{V}$, where

$$\mathbf{U} = \frac{1}{2}\int_0^l \sigma_x \, \varepsilon_x \, A \, dx$$

$$= \frac{E}{2}\{u\}^t \int_0^l [B]^t [B] (N_1 A_1 + N_2 A_2) dx \{u\}$$

and therefore the element stiffness matrix becomes

$$[k] = E \int_0^l [B]^t [B] (N_1 A_1 + N_2 A_2) dx \tag{3.49}$$

$$= \frac{E(A_1 + A_2)}{2l} \begin{bmatrix} 1 & -1 \\ -1 & 1 \end{bmatrix}$$

The interesting characteristic of the element stiffness matrix is that in the final equation the area of cross section is the average of the nodal areas. If the bar is subjected to a tensile force P, the stress σ_x varies from

$$\sigma_x = \frac{P}{A_1} \text{ at node 1}$$

to

$$\sigma_x = \frac{P}{A_2} \text{ at node 2,}$$

and in between at any x, interpolating the area linearly i.e., using natural coordinates L_1 and L_2,

$$\sigma_x = \frac{P}{\left(1 - \frac{x}{l}\right) A_1 + \frac{x}{l} A_2} = \frac{P}{L_1 A_1 + L_2 A_2}$$

thus the internal stress does not vary linearly, but, in the finite element model, the strain and stress within the element are constants, and would give the best possible average values based on the energy principle. It can be verified that as the number of elements increase, the solution converges to the exact.

3.7.2 A 3 d.o.f. Bar Element with Linearly Varying Area of Cross Section

Here the bar has three d.o.f. and hence has three nodes as given for the linear strain element in Fig. 3.15.

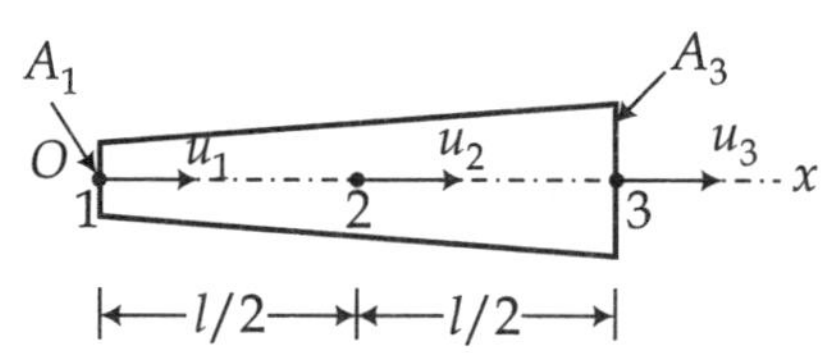

Fig. 3.15 A 3 d.o.f. Bar Element with Linearly Varying Area

The required steps 1, 2 and 3 for the formulation of the element stiffness matrix are the same as given in Eqs (3.41) through (3.46). For step 4, where the strain energy function in $\Phi = \mathbf{U} + \mathbf{V}$

$$\mathbf{U} = \frac{1}{2}\int_0^l \sigma_x \varepsilon_x A_x dx$$

$$= \frac{1}{2}\int_0^l \{u\}^t\,[B]^t\,[B]\,(L_1 A_1 + L_2 A_3)\{u\}\,dx$$

In other words, the displacements are interpolated using quadratic functions given in Eq. (3.43), and the area of cross section is interpolated linearly using the natural coordinates L_1 and L_2 given in Eq. (3.8). Minimizing the **U** function with respect to its nodal displacements, the element stiffness matrix is obtained as

$$[k] = \int_0^l E\,[B]^t\,[B]\,(L_1\,A_1 + L_2\,A_3)\,dx$$

$$= \frac{E}{l}\begin{bmatrix} \frac{11A_1}{6} + \frac{A_3}{2} & -2A_1 - \frac{2A_3}{3} & \frac{A_1 + A_3}{6} \\ & \frac{8}{3}(A_1 + A_3) & -\frac{2A_1}{3} - 2A_3 \\ Symmetric & & \frac{A_1}{2} + \frac{11A_3}{6} \end{bmatrix} \tag{3.50}$$

If the element is subjected to a body force $\tilde{b}$ per unit volume, the work potential

$$\mathbf{V} = -\int_0^l \tilde{b}\,A\,u\,dx$$

$$= -\int_0^l \tilde{b}\,(L_1\,A_1 + L_2\,A_3)\,(N_1\,u_1 + N_2\,u_2 + N_3\,u_3)\,dx \tag{3.51}$$

Differentiating with respect to u_1, u_2 and u_3, the equivalent nodal force vector is obtained as

$$f_1 = -\frac{\partial \mathbf{V}}{\partial u_1} = \int_0^l \tilde{b}\,(L_1 A_1 + L_2\,A_3)\;N_1 dx$$

$$= \frac{\tilde{b}\,A_1\,l}{6}$$

$$f_2 = -\frac{\partial \mathbf{V}}{\partial u_2} = \frac{\tilde{b}l}{3}\,(A_1 + A_3) \tag{3.52}$$

$$f_3 = -\frac{\partial \mathbf{V}}{\partial u_3} = \frac{\tilde{b}l}{6}\,A_3$$

3.8 USE OF LAGRANGE INTERPOLATION FUNCTION

Upon careful examination of the interpolation functions for constant strain and linear strain elements, it can be seen that they are nothing but special cases of the general Lagrange interpolation function, which for the *i*- th node in an element with *n* nodes, is given as,[5]

$$N_i = \frac{(x - x_1)(x - x_2)\ldots(x - x_{i-1})(x - x_{i+1})\ldots(x - x_n)}{(x_i - x_1)(x_i - x_2)\ldots(x_i - x_{i-1})(x_i - x_{i+1})\ldots(x_i - x_n)} \tag{3.53}$$

An easy way to remember this function is that it will have (n-1) terms in the numerator as well as in the denominator, and will not contain the term (x-x_i) in the numerator. If $x = x_i$, i.e., at node i, $N_i = 1.0$, and for all other nodes the values of N_i are identically equal to zero. This property has been noticed earlier among the interpolation functions. This interpolation function is good only for elements with C_0-Continuity. Using this formula, the interpolation functions for a two noded (constant strain) element can be written as,

at node 1, $$N_1 = \frac{(x - l)}{0 - l} = \left(1 - \frac{x}{l}\right) = L_1$$

at node 2, $$N_2 = \frac{(x - 0)}{l - 0} = \frac{x}{l} = L_2$$

The functions L_1 and L_2 are defined earlier as 'natural coordinates', and have the property that $L_1 + L_2 = 1.0$ within the element $0 \le x \le l$. Some of powerful properties of the natural coordinates in creating new higher order shape functions and subsequent numerical integrations are used in the forthcoming chapters.

Now for a three noded (linear strain) element, at nodes 1, 2 and 3,

$$N_1 = \frac{\left(x - \frac{l}{2}\right)(x - l)}{\left(0 - \frac{l}{2}\right)(0 - l)} = \frac{2}{l^2}\left(x - \frac{l}{2}\right)(x - l) = L_1(2L_1 - 1)$$

$$N_2 = \frac{(x - 0)(x - l)}{\left(\frac{l}{2} - 0\right)\left(\frac{l}{2} - l\right)} = -\frac{4x}{l^2}(x - l) = 4L_1L_2$$

$$N_3 = \frac{(x - 0)\left(x - \frac{l}{2}\right)}{(l - 0)\left(l - \frac{l}{2}\right)} = \frac{2x}{l^2}\left(x - \frac{l}{2}\right) = L_2(2L_2 - 1) \tag{3.54}$$

and these equations are the same as given in Eq. (3.43). Now, using this formula, the shape functions for a four-noded element with interior nodes, say at arbitrary points as shown in Fig. 3.16, can be written as

$$N_1 = \frac{(x-a_2)(x-a_3)(x-a_4)}{(a_1-a_2)(a_1-a_3)(a_1-a_4)}$$

$$N_2 = \frac{(x-a_1)(x-a_3)(x-a_4)}{(a_2-a_1)(a_2-a_3)(a_2-a_4)}$$

$$N_3 = \frac{(x-a_1)(x-a_2)(x-a_4)}{(a_3-a_1)(a_3-a_2)(a_3-a_4)}$$

$$N_4 = \frac{(x-a_1)(x-a_2)(x-a_3)}{(a_4-a_1)(a_4-a_2)(a_4-a_3)} \tag{3.55}$$

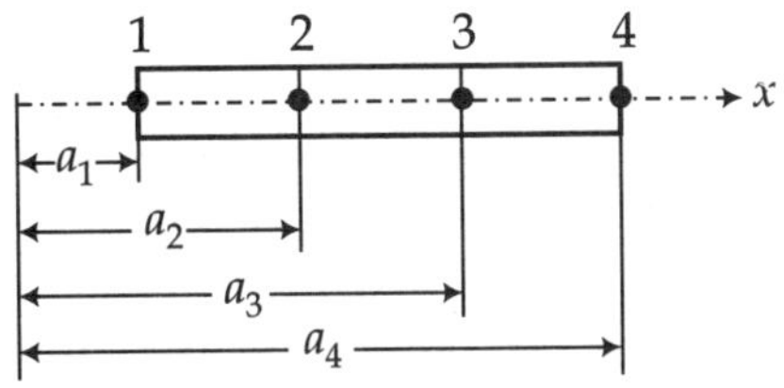

Fig. 3.16 A Bar Element with 4 Nodes

3.8.1 Shape Functions in terms of other Natural Coordinates

In finite element analysis, often a new coordinate system is employed where the parameter ξ defining the coordinate system varies between –1 and + 1. Such a new system is also called natural coordinate system[1] and is very useful for numerical integrations using Gauss Quadrature formulae. And this is shown in Chapter 5. In Fig. 3.17, the origin of the x-coordinate system is O, while the origin of the new coordinate system is A. For one-dimensional elements with two degrees of freedom, the interpolation functions in terms of ξ, are

$$N_1(\xi) = \frac{\xi-1}{-1-1} = \frac{1-\xi}{2} \text{ and } N_2(\xi) = \frac{\xi+1}{1+1} = \frac{1+\xi}{2} \tag{3.56}$$

where
$$\xi = \frac{2(x-x_1)}{x_2-x_1} - 1 \text{ and } x = \frac{\xi(x_2-x_1)+(x_2+x_1)}{2}$$

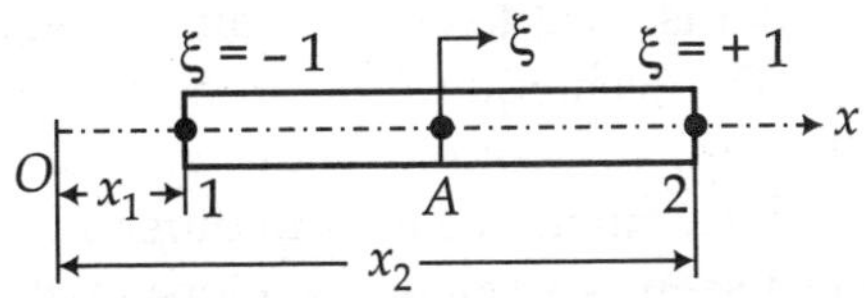

Fig. 3.17 Natural Coordinates for the Bar Element

For the three degrees of freedom elements, having quadratic variations, the new shape functions can be formulated using the Lagrange interpolation formulae as

$$N_1(\xi) = \frac{\xi}{2}(\xi - 1) \tag{3.57}$$
$$N_2(\xi) = (1 + \xi)(1 - \xi)$$
$$N_3(\xi) = \frac{\xi}{2}(\xi + 1)$$

Note that from nodes 1 to 2, ξ varies linearly between –1 and + 1.

3.9 TWO-DIMENSIONAL TRUSSES

Structural engineers are familiar with analysis of a two-dimensional truss; here, a finite element approach is presented. A truss is composed of several one-dimensional bar elements,[4] hinged together at nodes as shown in Fig. 3.2. The prescribed loads are applied at the nodes. It is customary to perform a static analysis by the method of joints if the truss is a statically determinate, where the internal forces are computed without calculating the nodal displacements. If the nodes 5 and 6 in Fig. 3.2 are connected together, then the truss becomes a statically indeterminate, and one has to use energy principles to analyze the forces in the members. A total displacement approach is used in the finite element analysis. The technique is applicable for analysis of statically determinate or indeterminate trusses. The method of developing global equations and the solution are given with the sequence of tasks as discussed earlier.

Task 1. Finite Element Discretization

In this task, the entire truss is identified by the total number of elements, number of nodes and the element number- nodal number data. This data is stored in the computer as a matrix usually called connectivity matrix. At this point, it is advantageous to start studying the computer coding for a finite element problem, the task discussed here is closely associated with the preparation of the input data for the problem. Let

NELM - the number of finite elements,

NNODE - the number of nodes,

NNCON - the number of nodes where loads are prescribed,

NNDIS - the number of nodes where displacements are prescribed,

X(i),Y(i) - nodal coordinates, and

NOD(i,j) - the connectivity matrix for the truss element,

F(i) - component of nodal force vector.

The most important data that is used to connect all the finite elements is stored in the connectivity matrix, NOD(i,j). The technique used in assembling the elements using this matrix is easily understood through an example. Let us consider a simple truss as shown in Fig. 3.18, the element numbers are written in circles to differentiate from the nodal numbers. The connectivity matrix and its use in assembling the global matrix will be evident in Task 3. One can use any computer language for the finite element code, the actual flow chart and structure are the same for all codes. A Fortran computer code to read and store the required data is given below:

```
    READ ( , ) NELM,NNODE,NNCON,NNDIS
    DO 40 I = 1, NNODE
40  READ ( , ) X(I),Y(I)
    DO 50 I = 1,NELM
50  READ ( , ) J, NOD(I,1),NOD(I,2)
    DO 60 I = 1,NNCON
60  READ ( , ) K, F(2*K-1),F(2*K)
```

Routine to Read Input Data

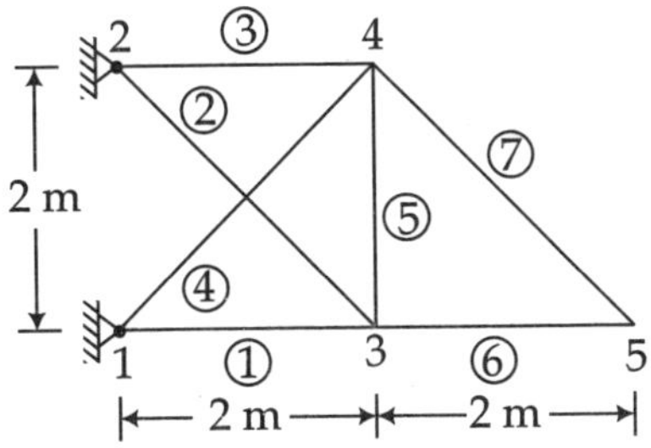

Fig. 3.18 A Simple Two-D Truss

From the nodal coordinates (x_i, y_i) the length and the orientation of each bar element have to be computed. Let us consider an arbitrary bar element as shown in Fig. 3.19. with $\bar{x}$, $\bar{y}$ as its local coordinate axes. If θ is the angle that $\bar{x}$ makes with the global X-axis, then

$$c = \cos\theta = \frac{(x_2 - x_1)}{l}$$

$$s = \sin\theta = \frac{(y_2 - y_1)}{l},$$

where

$$l = \sqrt{(x_2 - x_1)^2 + (y_2 - y_1)^2} \tag{3.58}$$

Task 2: Element Stiffness Matrix

The element stiffness matrix $[\bar{k}]$ of this element in the $(\bar{x}, \bar{y})$ axes is obtained as in Eq. (3.14).

i.e.
$$[\bar{k}] = \frac{EA}{l}\begin{bmatrix} 1 & -1 \\ -1 & 1 \end{bmatrix}$$

But, these equations have to be transformed into the global axes (X,Y) shown in Fig. 3.19, so that the transformed matrix equations can be easily assembled to form the global stiffness matrix equations. Suppose the nodal displacement in the local coordinate system are $\bar{u}_1$, $\bar{u}_2$ respectively along the $\bar{x}$-axis as shown in Fig. 3.19. Their corresponding components in the global coordinates have to have 4 parameters, (u_1, v_1) at node 1 and (u_2, v_2) at node 2, as shown in the figure.

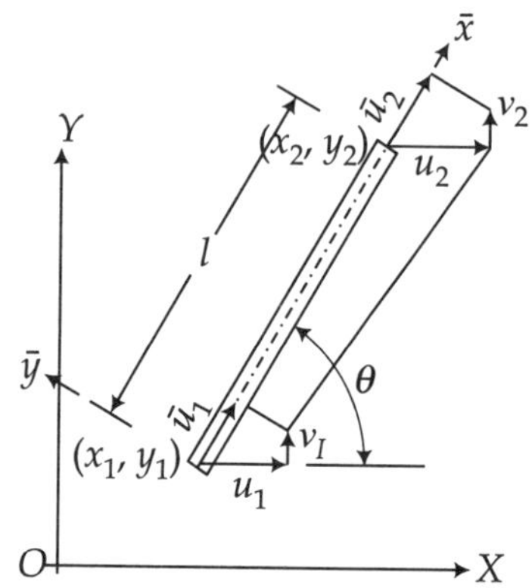

Fig. 3.19 Transformation of Coordinates

From simple geometry,

$$\bar{u}_1 = u_1 \cos\theta + v_1 \sin\theta$$

$$\bar{u}_2 = u_2 \cos\theta + v_2 \sin\theta \tag{3.59}$$

and representing in matrix form,

$$\begin{Bmatrix} \bar{u}_1 \\ \bar{u}_2 \end{Bmatrix} = \begin{bmatrix} c & s & 0 & 0 \\ 0 & 0 & c & s \end{bmatrix} \begin{Bmatrix} u_1 \\ v_1 \\ u_2 \\ v_2 \end{Bmatrix} = [T]\{u\} \tag{3.60}$$

where $\{u\} = \{u_1\ v_1\ u_2\ v_2\}$ and $[T]$ is a transformation matrix. An easier method to transform the element stiffness matrix from local to global coordinates is to equate the strain energy expressed in both coordinate systems. Strain energy with respect to $(\bar{x}, \bar{y})$ axes is

$$\mathbf{U} = \frac{1}{2} \begin{Bmatrix} \bar{u}_1 \\ \bar{u}_2 \end{Bmatrix}^t [\bar{k}] \begin{Bmatrix} \bar{u}_1 \\ \bar{u}_2 \end{Bmatrix}$$

$$= \frac{1}{2} \{u\}^t [T]^t [\bar{k}][T]\{u\}.$$

Since the strain energy of the global system is

$$\mathbf{U} = \frac{1}{2} \{u\}^t [k] \{u\},$$

equating both energies, the global stiffness matrix $[k]$ is obtained as,

$$[k] = [T]^t [\bar{k}] [T] \tag{3.61}$$

$$= \left(\frac{EA}{l}\right)_i \begin{bmatrix} c^2 & cs & -c^2 & -cs \\ cs & s^2 & -cs & -s^2 \\ -c^2 & -cs & c^2 & cs \\ -cs & -s^2 & cs & s^2 \end{bmatrix}.$$

Task 3: Assembling Element Stiffness Equations

Unlike the one-dimensional bar problem shown in Fig. 3.1, here the element matrices are of the order of 4, hence, to assemble the global stiffness matrix equations some special attention is necessary. Compatibility of displacements is achieved by assuming a unique value of displacements at common nodes, and the equilibrium of forces at the nodes is represented in the equations by adding the contributions of the stiffness coefficients from each element at a particular node. The subroutine for assembling element matrices can be understood by placing the stiffness coefficients in the global matrix as shown in Fig. 3.20, for the example problem in Fig. 3.18. During the assembling, a connectivity matrix NOD (n x 2) is used to determine the global nodal numbers of each element, i.e., i = NOD(n,1) and j = NOD(n,2) as stored in the connectivity matrix (see Table 3.1).

Table 3.1 Elements of Connectivity Matrix

Element No.	*Local No.*	*Global No.*	*Connectivity Matrix*
n	1	i	NOD (n, 1)
	2	j	NOD (n, 2)

Then, the element stiffness matrix in the local coordinate system can be placed in the global coordinate system as shown below.

$$\begin{bmatrix} k_{11} & k_{12} & k_{13} & k_{14} \\ k_{21} & k_{22} & k_{23} & k_{24} \\ k_{31} & k_{32} & k_{33} & k_{34} \\ k_{41} & k_{42} & k_{43} & k_{44} \end{bmatrix} = \begin{bmatrix} K_{2i-1,2i-1} & K_{2i-1,2i} & K_{2i-1,2j-1} & K_{2i-1,2j} \\ K_{2i,2i-1} & K_{2i,2i} & K_{2i,2j-1} & K_{2i,2j} \\ K_{2j-1,2i-1} & K_{2j-1,2i} & K_{2j-1,2j-1} & K_{2j-1,2j} \\ K_{2j,2i-1} & K_{2j,2i} & K_{2j,2j-1} & K_{2j,2j} \end{bmatrix}$$

Stiffness matrix in local coordinates *Stiffness matrix in the Global coordinates*

Fig. 3.20 Local Element Stiffness Matrix Transformed to Global Matrix

Table 3.2 Nodal Connectivity Matrix for the Truss Problem Fig. 3.18

Element No(n)	*NOD (n, 1)*	*NOD (n, 2)*
1	1	3
2	3	2
3	2	4
4	1	4
5	3	4
6	3	5
7	4	5

To assemble all the element stiffness matrices of the members of the truss shown in Fig. 3.18, the connectivity matrix coefficients are shown in the Table. 3.2. In the assembly procedure the subroutine BAR2D for each element is called which will supply the local stiffness matrix (SE) and the local force vector (FE) if there are traction or body forces in the element. Using the nodal connectivity matrix NOD, the elements of the local stiffness matrix and force vector are placed in the proper cells of the global stiffness matrix GS and global force vector F as shown in the Figs. 3.20 and 3.21. A careful examination and comprehension of the figure enable to understand the subroutines used for the assembly of the global matrix. On the left side the subroutine corresponds to the assembly of the full global matrix while on the right side is made to create the half-banded global matrix.

```
      DO 200 N = 1, NELM
      CALL BAR2D
      II = 2 * NOD(N,1)
      JJ = 2 * NOD(N,2)
      IG(1) = II - 1
      IG(2) = II
      IG(3) = JJ - 1
      IG(4) = JJ
      DO 150 I = 1,4
      K = IG(I)
      F(K ) = F(K) + FE(I)
      DO 150 J = 1,4
      L = IG(J)
150   GS(K,L) = GS(K,L) + SE(I,J)
200   CONTINUE
```

(a) For full matrix

```
      DO 200 N = 1, NELM
      CALL BAR2D
      II = 2 * NOD(N,1)
      JJ = 2 * NOD(N,2)
      IG(1) = II - 1
      IG(2) = II
      IG(3) = JJ - 1
      IG(4) = JJ
      DO 150 I = 1,4
      K = IG(I)
      F(K) = F(K) +FE(I)
      DO 150 J = 1,4
      L = IG(J)
      IF (L. LT. K) GO TO 150
      JJ = L - K + 1
150   GS(K,JJ)=GS(K,JJ)+SE(I,J)
200   CONTINUE
```

(b) For half-banded matrix

Subroutines for Assembling the Global Stiffness Matrix

Task 4: Incorporation of Prescribed Displacements

There are different methods by which the prescribed displacements can be incorporated into the global equations. Suppose the *i*-th component of the displacement u_i has to be prescribed with a specific value $\tilde{u}_i$ at a particular nodal point. This type of constraint is called 'single point constraint'. The other type of boundary constraint that occurs often in finite element stress analysis is called 'multi-point constraint' which is discussed at the end of this section.

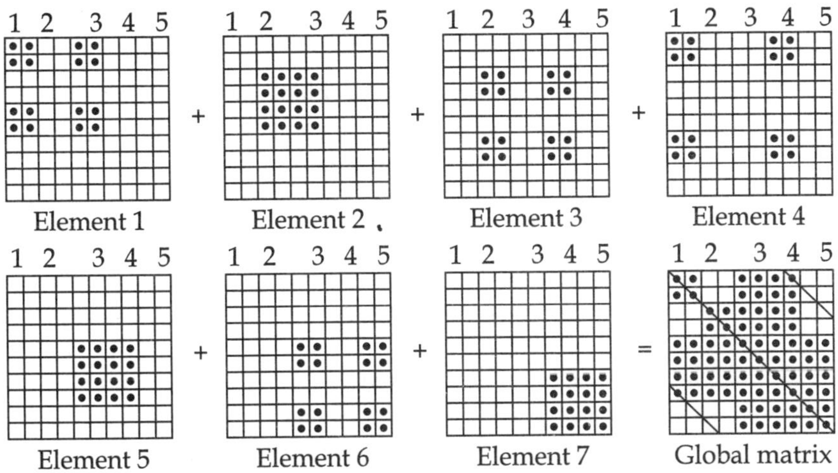

Fig. 3.21 Assembling Element Stiffness Matrices to form Global Stiffness Matrix using Connectivity Matrix

Single point constraint can be made relatively easily by using the concept described in section 3.2, Task 4. In other words, prescribing a particular displacement $\tilde{u}_i$ at nodal point i is accomplished by modifying all the coefficients of the global stiffness matrix along the i-th row as zero except the diagonal coefficient k_{ii}, which is set equal to 1.0, and simultaneously the corresponding force vector f_i is replaced by the actual prescribed value $\tilde{u}_i$. During the solution of the global equations, this procedure guarantees the value of u_i to be equal to the prescribed value $\tilde{u}_i$. If this technique is to be used on a half-banded matrix equation, special additional modifications have to be made. This procedure is illustrated in Figs. 3.22 (a,b,c and d).

Suppose the displacement component u_5 has to be prescribed as $\tilde{u}_5$. If all the coefficients of the matrix in the 5-th row are replaced by zeros except the diagonal element K_{55} which is set equal to 1.0 and the value of $f_5 = \tilde{u}_5$, then the solution of this equation will automatically guarantee the value of u_5 as specified. For a full matrix, this operation is sufficient, and the standard procedure of the solution of simultaneous equations can be used to obtain the final displacement vector.

But in finite element methodology a half-banded matrix technique (or other schemes as given in the Appendix-II) is popular because these techniques require considerably less storage space to deal with large number of equations. In Fig. 3.22b, as the terms in row 5 are altered, symmetry of the global matrix is destroyed and for the half band matrix solver, the global matrix has to be symmetric. Now, as the terms on the right side of the matrix in 5-th row are replaced as zeros, the corresponding terms on the 5-th column which are associated with the prescribed displacement $\tilde{u}_5$ are also made equal to zero by subtracting $k_{i5}\tilde{u}_5$ from the corresponding component of the force vector f_i as illustrated in Fig. 3.22c, which makes the global stiffness matrix a symmetric one. Now the global matrix is ready to be stored in the half banded form as shown in Fig. 3.22(d).

A simple subroutine is given below. NUM is the total number of global stiffness equations and MBAND is the half-band width of the matrix. The control parameters are:

NBOU : Number of nodes where the displacements are prescribed

NBO (m, i) : A control equal to 0 or nonzero for i = 1, 2 to define if displacements are prescribed or not in x-and y-directions

BU (m, i) : i = 1, 2 define the values of the prescribed displacements in the x-and y- directions.

NBO (m, i) : is used to denote the direction or directions of the prescribed displacements. If NBO (m, 1) is equal to zero, then the displacement in x-direction in not prescribed, if it is nonzero, then value of the prescribed displacement is stored in BU(m,1). Similarly for NBO (m, 2).

```
      DO 21 I = 1, NBOU
      N = 2 * NF(I) - 1
      DO 20 L = 1, 2
      IF (NBO(I,L)) 7, 16, 7
   7  DO 15 M = 1, NUM
      GS(N,M) = 0.0
  15  CONTINUE
      GS(N,M) = 1.0
      U(N) = BU(I,L)
  16  N = N + 1
  20  CONTINUE
  21  CONTINUE
```

(a) For a full global matrix

```
      DO 21 I = 1, NBOU
      N = 2* NF(I) - 1
      IF (NBO(I,L)) 7,16,7
   7  FF = BU(I,L)
      DO 15 M = 2, MBAND
      K = N-M+1
      IF (K) 10,10,9
   9  F(K) = F(K)-GS(K,M) * FF
      GS(K,M) = 0.0
  10  K = N+M-1
      IF (NUM-K) 15,12,12
  12  F(K) = F(K)-GS(N,M) * FF
      GS(N,M) = 0.0
  15  CONTINUE
      GS(N,1) = 1.0
      F(N) = FF
  17  N = N+ 1
  20  CONTINUE
  21  CONTINUE
```

(b) For a half-banded matrix

Subroutines for Prescribing Displacements

Another technique to achieve the same result is to add a large number C (of the order of 10^{15} or 10^{20} times k_{55}) to the diagonal coefficient k_{55} making it equal to $k_{55} + C$, and replace f_5 by $f_5 + C\,\tilde{u}_5$. All coefficients in the 5-th row and 5-th column become small in comparison with the diagonal term and therefore during the Gauss elimination solution process, those numbers become insignificant and the final value of u_5 is maintained at its prescribed value. The advantage of this technique is that the matrix remains symmetric, and no alteration of the coefficients of the matrix and the force vector is necessary to maintain symmetry. Also, if the equation corresponding to node 5 is divided by C the resulting coefficients are the same as those in the first method, i.e., with 1.0 on the main diagonal. This technique can be explained easily using a variational approach called 'Penalty Function Method'.

$$
\left[\begin{array}{cccccccccccccc}
\times&\times&\times&\times&\times&\times&&&&&&&& \\
\times&\times&\times&\times&\times&\times&\times&&&&&&& \\
\times&\times&\times&\times&\times&\times&\times&\times&&&&&& \\
\times&\times&\times&\times&\times&\times&\times&\times&\times&&&&& \\
\times&\times&\times&\times&\times&\times&\times&\times&\times&\times&&&& \\
\times&\times&\times&\times&\times&\times&\times&\times&\times&\times&\times&&& \\
&\times&\times&\times&\times&\times&\times&\times&\times&\times&\times&\times&& \\
&&\times&\times&\times&\times&\times&\times&\times&\times&\times&\times&\times& \\
&&&\times&\times&\times&\times&\times&\times&\times&\times&\times&\times&\times \\
&&&&\times&\times&\times&\times&\times&\times&\times&\times&\times&\times \\
&&&&&\times&\times&\times&\times&\times&\times&\times&\times&\times \\
&&&&&&\times&\times&\times&\times&\times&\times&\times&\times \\
&&&&&&&\times&\times&\times&\times&\times&\times&\times \\
&&&&&&&&\times&\times&\times&\times&\times&\times
\end{array}\right]
\begin{Bmatrix} u_1\\ u_2\\ u_3\\ u_4\\ u_5\\ u_6\\ u_7\\ u_8\\ u_9\\ u_{10}\\ u_{11}\\ u_{12}\\ u_{13}\\ u_{14} \end{Bmatrix}
=
\begin{Bmatrix} f_1\\ f_2\\ f_3\\ f_4\\ f_5\\ f_6\\ f_7\\ f_8\\ f_9\\ f_{10}\\ f_{11}\\ f_{12}\\ f_{13}\\ f_{14} \end{Bmatrix}
$$

(a) Original Matrix Equation in Full Form

$$
\left[\begin{array}{cccccccccccccc}
\times&\times&\times&\times&\times&\times&&&&&&&& \\
\times&\times&\times&\times&\times&\times&\times&&&&&&& \\
\times&\times&\times&\times&\times&\times&\times&\times&&&&&& \\
\times&\times&\times&\times&\times&\times&\times&\times&\times&&&&& \\
0&0&0&0&1&0&0&0&0&0&&&& \\
\times&\times&\times&\times&\times&\times&\times&\times&\times&\times&\times&&& \\
&\times&\times&\times&\times&\times&\times&\times&\times&\times&\times&\times&& \\
&&\times&\times&\times&\times&\times&\times&\times&\times&\times&\times&\times& \\
&&&\times&\times&\times&\times&\times&\times&\times&\times&\times&\times&\times \\
&&&&\times&\times&\times&\times&\times&\times&\times&\times&\times&\times \\
&&&&&\times&\times&\times&\times&\times&\times&\times&\times&\times \\
&&&&&&\times&\times&\times&\times&\times&\times&\times&\times \\
&&&&&&&\times&\times&\times&\times&\times&\times&\times \\
&&&&&&&&\times&\times&\times&\times&\times&\times
\end{array}\right]
\begin{Bmatrix} u_1\\ u_2\\ u_3\\ u_4\\ u_5\\ u_6\\ u_7\\ u_8\\ u_9\\ u_{10}\\ u_{11}\\ u_{12}\\ u_{13}\\ u_{14} \end{Bmatrix}
=
\begin{Bmatrix} f_1\\ f_2\\ f_3\\ f_4\\ \tilde{u}_5\\ f_6\\ f_7\\ f_8\\ f_9\\ f_{10}\\ f_{11}\\ f_{12}\\ f_{13}\\ f_{14} \end{Bmatrix}
$$

(b) Prescribing $\tilde{u}_5$ in the Equation (Note Symmetry is Lost)

$$
\left[\begin{array}{cccccccccccccc}
\times & \times & \times & \times & 0 & \times & & & & & & & & \\
\times & \times & \times & \times & 0 & \times & \times & & & & & & & \\
\times & \times & \times & \times & 0 & \times & \times & \times & & & & & & \\
\times & \times & \times & \times & 0 & \times & \times & \times & \times & & & & & \\
0 & 0 & 0 & 0 & 1 & 0 & 0 & 0 & 0 & 0 & & & & \\
\times & \times & \times & \times & 0 & \times & \times & \times & \times & \times & \times & & & \\
 & \times & \times & \times & 0 & \times & \times & \times & \times & \times & \times & \times & & \\
 & & \times & \times & 0 & \times & \times & \times & \times & \times & \times & \times & \times & \\
 & & & \times & 0 & \times & \times & \times & \times & \times & \times & \times & \times & \times \\
 & & & & 0 & \times & \times & \times & \times & \times & \times & \times & \times & \times \\
 & & & & & \times & \times & \times & \times & \times & \times & \times & \times & \times \\
 & & & & & & \times & \times & \times & \times & \times & \times & \times & \times \\
 & & & & & & & \times & \times & \times & \times & \times & \times & \times \\
 & & & & & & & & \times & \times & \times & \times & \times & \times
\end{array}\right]
\left\{\begin{array}{c} u_1 \\ u_2 \\ u_3 \\ u_4 \\ u_5 \\ u_6 \\ u_7 \\ u_8 \\ u_9 \\ u_{10} \\ u_{11} \\ u_{12} \\ u_{13} \\ u_{14} \end{array}\right\}
=
\left\{\begin{array}{c}
f_1 - k_{15}\tilde{u}_5 \\
f_2 - k_{25}\tilde{u}_5 \\
f_3 - k_{35}\tilde{u}_5 \\
f_4 - k_{45}\tilde{u}_5 \\
\tilde{u}_5 \\
f_6 - k_{56}\tilde{u}_5 \\
f_7 - k_{57}\tilde{u}_5 \\
f_8 - k_{58}\tilde{u}_5 \\
f_9 - k_{59}\tilde{u}_5 \\
f_{10} - k_{5,10}\tilde{u}_5 \\
f_{11} \\
f_{12} \\
f_{13} \\
f_{14}
\end{array}\right\}
$$

(c) Procedure to make the Coefficient Matrix Symmetric

$$
\left[\begin{array}{cccccc}
\times & \times & \times & \times & 0 & \times \\
\times & \times & \times & 0 & \times & \times \\
\times & \times & 0 & \times & \times & \times \\
\times & 0 & \times & \times & \times & \times \\
1 & 0 & 0 & 0 & 0 & 0 \\
\times & \times & \times & \times & \times & \times \\
\times & \times & \times & \times & \times & \times \\
\times & \times & \times & \times & \times & \times \\
\times & \times & \times & \times & \times & \times \\
\times & \times & \times & \times & \times & \\
\times & \times & \times & \times & & \\
\times & \times & \times & & & \\
\times & \times & & & & \\
\times & & & & &
\end{array}\right]
\quad
\left\{\begin{array}{c}
f_1 - k_{15}\tilde{u}_5 \\
f_2 - k_{25}\tilde{u}_5 \\
f_3 - k_{35}\tilde{u}_5 \\
f_4 - k_{45}\tilde{u}_5 \\
\tilde{u}_5 \\
f_6 - k_{56}\tilde{u}_5 \\
f_7 - k_{57}\tilde{u}_5 \\
f_8 - k_{58}\tilde{u}_5 \\
f_9 - k_{59}\tilde{u}_5 \\
f_{10} - k_{5,10}\tilde{u}_5 \\
f_{11} \\
f_{12} \\
f_{13} \\
f_{14}
\end{array}\right\}
$$

(d) Modification of the Load Vector and the Half-banded Matrix

Fig. 3.22 Prescribing Displacements in Half-banded Matrix Equation

3.10 MULTI-POINT CONSTRAINT BY PENALTY FUNCTION METHOD

An example of a multi point constraint[3] is given in Fig. 3.23, when the support conditions allow movement along a plane, say in the direction of a new axis $\bar{x}$, but does not allow any movement normal to the plane in the direction of $\bar{y}$. The condition for this case at a node i can be represented by an equation connecting the two displacements U_{2i-1} and U_{2i} in terms of the angle θ, i.e., for the boundary condition given in Fig. 3.23,

$$-U_{2i-1} \sin \theta + U_{2i} \cos \theta = 0 \tag{3.62}$$

i.e., the displacement in the direction of $\bar{y}$ axis is zero. This condition can be generalized as

$$\alpha U_{2i-1} + \beta U_{2i} = \tilde{U}_p \tag{3.63}$$

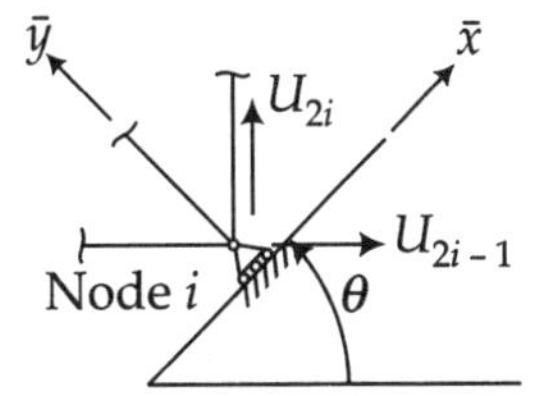

Fig. 3.23. Prescribed Displacement on an Inclined Plane

where $\tilde{U}_p$ is a prescribed value of displacement in the $\bar{y}$ axis. In the penalty function method, the total potential energy function is augmented by the strain energy of a fictitious spring with a very large spring constant C in the direction of the prescribed displacement. Let

$$d = \alpha U_{2i-1} + \beta U_{2i} - \tilde{U}_p \tag{3.64}$$

as the deformation in the fictitious spring, the augmented total potential energy function becomes

$$\Phi_n = \frac{1}{2}\{U\}^t [K]\{U\} + \frac{1}{2}Cd^2 + \mathbf{V} \tag{3.65}$$

$$= \frac{1}{2}\{U\}^t [K]\{U\} + \frac{1}{2}C(\alpha U_{2i-1} + \beta U_{2i} - \tilde{U}_p)^2 + \mathbf{V}$$

Differentiating the new potential energy function with respect to the U_{2i-1} and U_{2i}, the necessary coefficients of the global stiffness equation at node i are obtained. Only the original four global stiffness matrix coefficients corresponding to node i (as shown below), without the fictitious spring will be modified, i.e.,

$$\begin{bmatrix} K_{2i-1,2i-1} & K_{2i-1,2i} \\ K_{2i,2i-1} & K_{2i,2i} \end{bmatrix} \text{ becomes } \begin{bmatrix} K_{2i-1,2i-1} + C\alpha^2 & K_{2i-1,2i} + C\alpha\beta \\ K_{2i,2i-1} + C\alpha\beta & K_{2i,2i} + C\beta^2 \end{bmatrix}$$

and the corresponding force vector will be modified:

$$\begin{Bmatrix} F_{2i-1} \\ F_{2i} \end{Bmatrix} \text{ becomes } \begin{Bmatrix} F_{2i-1} + C\alpha\tilde{U}_p \\ F_{2i} + C\beta\tilde{U}_p \end{Bmatrix} \tag{3.66}$$

A very large value for the fictitious spring constant C, physically means that a very stiff spring support in the direction of the $\bar{y}$ axis is provided. Thus, the value of the d in Eq. (3.64) can be maintained as a small quantity. In other words, Eq. (3.63) is satisfied.

If one divides the corresponding equations by C, then all the off diagonal elements in rows $2i$-1 and $2i$ will be almost equal to zero, except the four elements shown in Eq. (3.65). The corresponding four elements of the new equations for the global matrix will be:

$$\begin{bmatrix} \alpha^2 & \alpha\beta \\ \alpha\beta & \beta^2 \end{bmatrix} \text{ and the force vector is } \begin{Bmatrix} \alpha\tilde{U}_p \\ \beta\tilde{U}_p \end{Bmatrix} \tag{3.67}$$

If $\alpha = \cos\theta$ and $\beta = -\sin\theta$ and $\tilde{U}_p = 0$ then for small values of θ, the Eq. (3.66) can make the global stiffness matrix unstable. In such circumstances, it is better to keep the equation as sgiven in Eq. (3.65).

Example Problem 3.4

This example shown in Fig. 3.18 is provided here for practice of assembling and prescribing displacements etc. for a four-degrees of freedom element considering X and Y axes. The values of EA of all horizontal elements are 2 k, while that of all inclined bars are $2\sqrt{2}$ k, so that the values of EA/l becomes equal to k. The following are the stiffness matrices of all elements. For the horizontal elements,

$$[k_1] = [k_3] = [k_6] = k\begin{bmatrix} 1.0 & 0 & -1.0 & 0 \\ 0 & 0 & 0 & 0 \\ -1.0 & 0 & 1.0 & 0 \\ 0 & 0 & 0 & 0 \end{bmatrix},$$

for the vertical element,

$$[k_5] = k\begin{bmatrix} 0 & 0 & 0 & 0 \\ 1.0 & 0 & -1.0 & 0 \\ 0 & 0 & 0 & 0 \\ -1.0 & 0 & 1.0 & 0 \end{bmatrix},$$

for the inclined elements,

$$[k_2] = [k_7] = k\begin{bmatrix} 0.5 & -0.5 & -0.5 & 0.5 \\ -0.5 & 0.5 & 0.5 & -0.5 \\ -0.5 & 0.5 & 0.5 & -0.5 \\ 0.5 & -0.5 & -0.5 & 0.5 \end{bmatrix} \quad \text{and}$$

$$[k_4] = k \begin{bmatrix} 0.5 & 0.5 & -0.5 & -0.5 \\ 0.5 & 0.5 & -0.5 & -0.5 \\ -0.5 & -0.5 & 0.5 & 0.5 \\ -0.5 & -0.5 & 0.5 & 0.5 \end{bmatrix}$$

The global stiffness matrix is obtained as:

$$K \begin{bmatrix} 1.5 & 0.5 & 0 & 0 & -1.0 & 0 & -0.5 & -0.5 & 0 & 0 \\ 0.5 & 0.5 & 0 & 0 & 0 & 0 & -0.5 & -0.5 & 0 & 0 \\ 0 & 0 & 1.5 & -0.5 & -0.5 & 0.5 & -1.0 & 0 & 0 & 0 \\ 0 & 0 & -0.5 & 0.5 & 0.5 & -0.5 & 0 & 0 & 0 & 0 \\ -1.0 & 0 & -0.5 & 0.5 & 2.5 & -0.5 & 0 & 0 & -1.0 & 0 \\ 0 & 0 & 0.5 & -0.5 & -0.5 & 1.5 & 0 & -1.0 & 0 & 0 \\ -0.5 & -0.5 & -1.0 & 0 & 0 & 0 & 2.0 & 0 & -0.5 & 0.5 \\ -0.5 & -0.5 & 0 & 0 & 0 & -1.0 & 0 & 2.0 & 0.5 & -0.5 \\ 0 & 0 & 0 & 0 & -1.0 & 0 & -0.5 & 0.5 & 1.5 & -0.5 \\ 0 & 0 & 0 & 0 & 0 & 0 & 0.5 & -0.5 & -0.5 & 0.5 \end{bmatrix} \begin{Bmatrix} u_1 \\ u_2 \\ u_3 \\ u_4 \\ u_5 \\ u_6 \\ u_7 \\ u_8 \\ u_9 \\ u_{10} \end{Bmatrix} = \begin{Bmatrix} R_1 \\ R_2 \\ R_3 \\ R_4 \\ 0 \\ 0 \\ 0 \\ 0 \\ 0 \\ -50 \end{Bmatrix}$$

In these equations global numbers are used as that is what is to be followed in the finite element analysis computer program. In the force vector the first 4 quantities are reactions at nodes 1 and 2 respectively. Applying the prescribed displacement boundary conditions and solving for displacements, the results are shown in the following Tables 3.3 and 3.4.

Table 3.3 Nodal Displacements

Node	*Displacements/k*	
	along x-axis	*along y-axis*
1	0.0	0.0
2	0.0	0.0
3	-71.43	-114.29
4	78.57	-135.72
5	-121.43	-435.74

The following are the member forces in Table 3.4.

Table 3.4 Forces in Members

Member No.	1	2	3	4	5	6	7
Forces in kN	-71.43	30.31	78.57	-40.41	-21.43	-50.02	70.71

3.11 ANALYSIS OF THREE-DIMENSIONAL TRUSSES

The use of the one-dimensional bar element can be extended to the analysis of three-dimensional trusses, where the individual elements of the truss are connected together forming a hinge at the common nodes.[4] In the 3-D analysis, each node has three degrees of freedom. But the element being either in tension or compression, has only two degrees of freedom $\bar{u}_1$ and $\bar{u}_2$ in its own local coordinate system. If $\bar{u}_1$ and $\bar{u}_2$ are the components of the displacement vector in the local coordinates of the element, then the nodal displacement vector can be transformed into the corresponding global components just like the transformation for the two dimensional truss. Here the components of the global displacement vector are u_1, v_1, w_1 at node 1 and similar three components at node 2 as shown in Fig. 3.23. Using simple geometry, the global displacement components can be related to the local components in the following manner.

$$\begin{Bmatrix} \bar{u}_1 \\ \bar{u}_2 \end{Bmatrix} = \begin{bmatrix} l & m & n & 0 & 0 & 0 \\ 0 & 0 & 0 & l & m & n \end{bmatrix} \begin{Bmatrix} u_1 \\ v_1 \\ w_1 \\ u_2 \\ v_2 \\ w_2 \end{Bmatrix} = [T]\{u\} \tag{3.68}$$

where $[T]$ is the transformation matrix, l, m, and n are the direction cosines of the element with respect to the global axes. If the nodal coordinates are x_1, y_1 and z_1 for node 1 and x_2, y_2 and z_2 for node 2,from simple geometry the direction cosines can be calculated as follows.

$$l = \frac{x_2 - x_1}{L}, \; m = \frac{y_2 - y_1}{L} \text{ and } n = \frac{z_2 - z_1}{L}$$

and

$$L = \sqrt{(x_2 - x_1)^2 + (y_2 - y_1)^2 + (z_2 - z_1)^2} \tag{3.69}$$

is the length of the element. Following the equations used to transform the element stiffness matrix to the global coordinates as given in Eq. (3.61), the global stiffness matrix $[k]$ is obtained as:

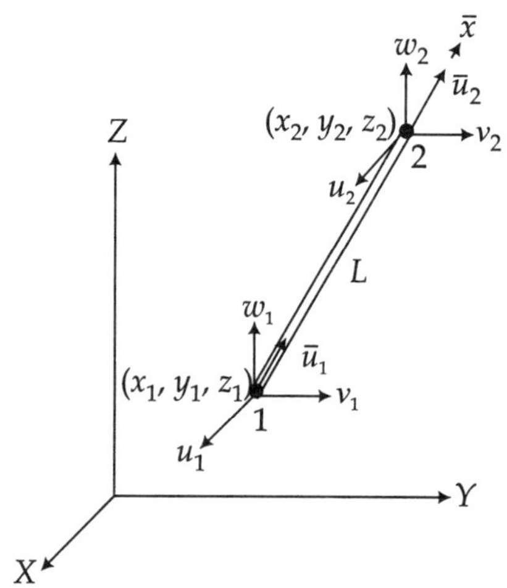

Fig. 3.24 A 3-D Bar Element

$$[k] = [T]^t [\bar{k}] [T] = \frac{EA}{L} \begin{bmatrix} l^2 & lm & ln & -l^2 & -lm & -ln \\ lm & m^2 & mn & -lm & -m^2 & -mn \\ ln & mn & n^2 & -ln & -mn & -n^2 \\ -l^2 & -lm & -ln & l^2 & lm & ln \\ -lm & -m^2 & -mn & lm & m^2 & mn \\ -ln & -mn & -n^2 & ln & mn & n^2 \end{bmatrix} \tag{3.70}$$

Assembly and incorporation of prescribed displacement boundary conditions are very similar to the two-dimensional case, hence the development is left as an exercise to the reader.

3.12 THERMAL STRESS ANALYSIS

The finite element analysis of a one-dimensional element system due to temperature variations is achieved by considering thermal expansion in each element as an initial strain within that element. An equivalent nodal force vector is computed due to this initial strain condition in each element by employing the principle of virtual work. In Eq. (2.32) in Chapter 2, the stress strain relationships in the case of thermal strain or initial strain condition are given for two- and three-dimensional conditions. For one-dimensional state of strain, the thermal strain ε_0 due a temperature differential of ΔT for a material with a coefficient of linear thermal expansion α is given as

$$\varepsilon_0 = \alpha \Delta T \tag{3.71}$$

and the stress-strain relationship can be written as

$$\sigma = E(\varepsilon - \varepsilon_0) \tag{3.72}$$

The virtual strain energy in the element is

$$\begin{aligned} \delta \mathbf{U} &= \int_{vol} \delta\varepsilon\, \sigma\, dvol \\ &= \int_{vol} \delta\varepsilon\, E\, (\varepsilon - \varepsilon_0)\, dvol \\ &= \int_{vol} \{\delta u\}^t [B]^t E(\varepsilon - \varepsilon_0)\, dvol \\ &= \{\delta u\}^t \left(\int_{vol} E\, [B]^t [B] \{u\}\, dvol - \int E\, [B]^t\, \varepsilon_0\, dvol \right) \\ &= \{\delta u\}^t ([k]\{u\} - \{f_T\}) \end{aligned} \tag{3.73}$$

where $$\{f_T\} = E \int_{vol} [B]^t\, \varepsilon_0\, dvol$$

$$= EA\alpha\Delta T \begin{Bmatrix} -1 \\ 1 \end{Bmatrix} \tag{3.74}$$

which is the equivalent nodal force vector for an element. These nodal forces are added to the respective global force vector components and solved for the global displacements of the finite element model. In thermal stress problems, care should be taken to calculate the internal stress within the element, in other words, Eq. (2.32) has to be used, i.e.,

$$\sigma = E(\varepsilon - \alpha\Delta T)$$

3.13 SUMMARY

Finite element method is introduced to analyse a longitudinal bar using simple bar elements. The basic four steps used here to derive the element stiffness matrix is the same for all finite elements. Shape functions, equivalent load vector, higher order bar elements, continuity between elements, Lagrangian interpolation functions etc. are presented. Assembly of elements using equilibrium principles, is accomplished by employing an element number/nodal number concept. Use of the bar element is extended to solve two- and three-dimensional truss problems. Elementary principles for computer coding is shown here in FORTRAN; however, the basic equations remain similar for any other computer language.

PROBLEMS

3.1 Write a computer program to analyze the displacements and forces in a

(a) two dimensional truss,

(b) three dimensional truss.

3.2 An aluminum and a steel bar connected together to form a composite structural member as shown below. The modulus of elasticity of aluminum, E_a is 100 GPa, the modulus of elasticity of steel, E_s is 200 GPa, and their cross-sectional areas are A_a = 10 cm^2 and A_s = 5 cm^2.

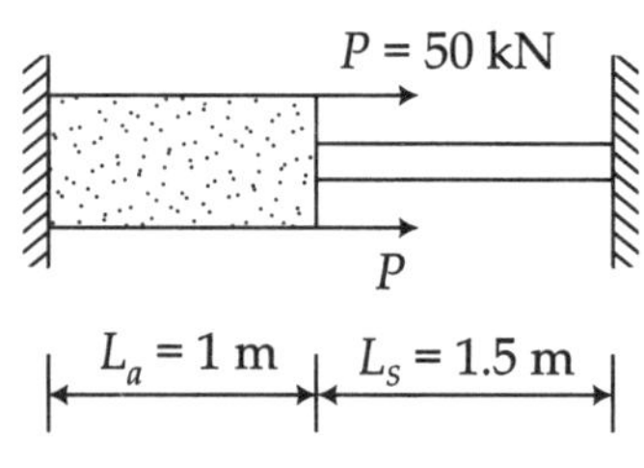

Fig. P3.1

Find the displacements, reactions, strains and stresses by using constant strain bar element.

3.3 Calculate the element stiffness matrices and by assembling them obtain the linear system of equations for the following tapered bar: (a) using constant strain elements and (b) using linear strain elements. Assume that the bar consists of three elements. What is the tip displacement and stress at $x = 2L/3$ in each case. Write your comments by comparing the results. The area of cross section varies from 10cm^2 at the left end to 2 cm^2 at the tip. E = 150GPa.

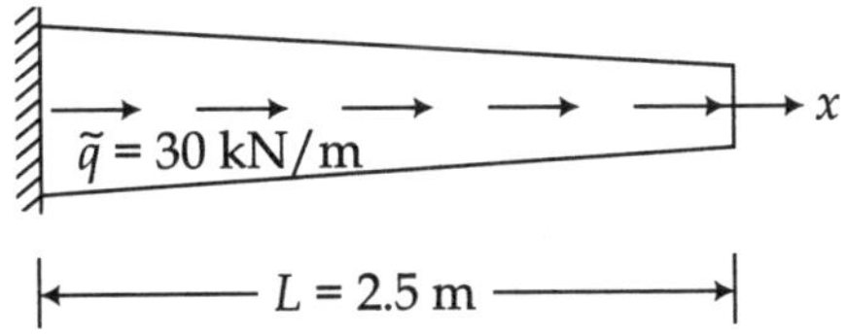

Fig. P3.2

3.4. Write computer program to solve the structure given in Problems 1 and 2.

3.5. Analyze the following trusses using the finite element method if E = 200 GPa and area of cross section A = 2 cm^2 for all members.

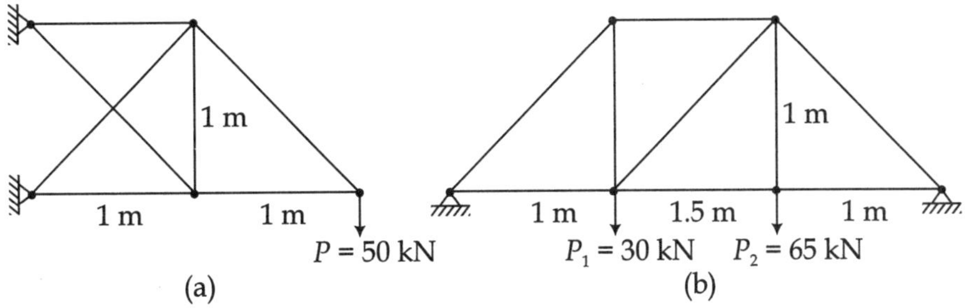

Fig. P3.3

REFERENCES

1. Chandrupatla, T.R. and Belegundu, A.D., "Introduction to Finite Elements in Engineering", Prentice Hall, Inc., New Jersey, 2nd Edition, 1997.
2. Reddy, J.N., "An Introduction to Finite Element Method", McGraw Hill Book Co., New York, 2nd Edition, 1993.
3. Oden, J.T., "Applied Functional Analysis ", Prentice Hall, Englewood Cliffs, New Jersey, 1979.
4. Weaver, W. and Gere J.M., "Matrix Analysis of Framed Structures", Van Nostrand Reinhold, New York, 1990.
5. Zienkiewicz, O.C. and Taylor, R.L., "The Finite Element Method" Vols. I and II, McGrawHill Book Company, 1991.

4

CHAPTER

Two Dimensional Problems-I: C_0-Triangular Elements

4.1 INTRODUCTION

Finite element method as it is known today started with triangular elements to solve two-dimensional problems of elasticity.[7] A brief introduction of two-dimensional problems is already given in Chapter 2. In this and the next two chapters, the finite element method is shown to solve these problems by discretizing the domain into small elements such as triangular, rectangular, quadrilateral or even curved elements[1,2,5,8] as shown in Figs. 4.1(a), (b) and (c). Moreover, readers are advised to study some of the classical solutions of two-dimensional problems of elasticity in order that they get some in-depth feeling for what they are accomplishing with the finite element solutions. Soon they will realize that closed-form solution to many of the two-dimensional elasticity problems exist only for very simple geometry, material property and boundary conditions. It is then, the power of the finite element method as a comprehensive powerful stress analysis technique is understood where complex geometry, material property and boundary conditions can be included easily in the overall solution procedure.

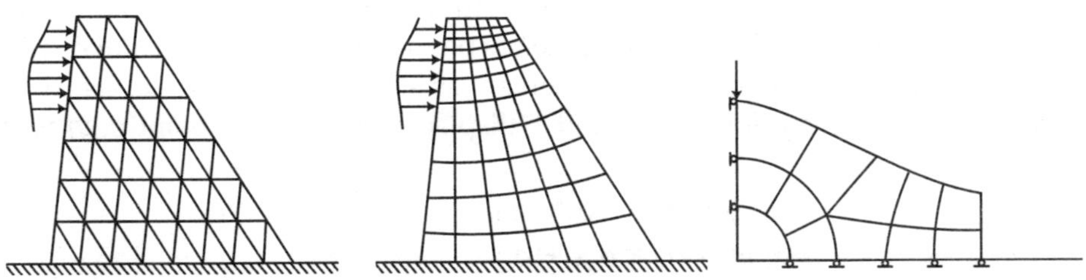

a) Triangular elements b) Quadrilateral elements c) Curved elements

Fig. 4.1 Domain Discretization Using Triangular or Quadrilateral Elements

4.2 CONSTANT STRAIN TRIANGULAR ELEMENT (CST)

The simplest of all two-dimensional finite elements is the straight triangular element. This element was the first element developed which was called a 'finite element'.[3] The development of the global system equations for the discrete assemblage of elements, follows the same set of tasks described before for the bar elements, and they are repeated here also since they are slightly different for the two - dimensional problems.

Task 1: Discretization of the Body and the Input data

First and foremost, one has to discretize the domain into finite elements. The regions where there are high stress or strain gradients, smaller elements are to be employed and vice versa. The input data starts with the material properties such as modulus of elasticity E, Poisson ratio ν and the thickness. For plane stress problems the thickness is the actual thickness t, and for plane strain problems, the thickness is taken as equal to one. After discretization of the domain, the input data should contain, the total number of elements, total number of nodes and the corresponding values of the (x, y) coordinates, the number of nodes where the displacements are prescribed, along with an identification of the nodes and corresponding displacements, and the number and identification of the nodes where the forces are prescribed and corresponding magnitudes of those forces. The details on how the input data is prepared for the computer code is given in task 3, which is used to assemble the element stiffness matrices to form the global stiffness matrix and the global force vector. Some of these input parameters are required for the task 4, which is used to modify the global stiffness matrix to prescribe the displacement boundary conditions.

Task 2: Development of Element Stiffness Matrix Equations

The most important task in the finite element methodology is the development of the element stiffness matrix equations. The rest of the tasks can be repeated in the same way that have been described previously for the bar elements. Information that is necessary here are the values of the nodal coordinates, (x_i, y_i) for $i = 1, 2, 3$ and the thickness t. For simplicity, it is assumed that the material is linear, elastic, homogeneous and isotropic with material properties of the element E and μ; however this is not a limitation of the method. The four steps required for the development of the stiffness matrix are followed here also.

Step 1. Assumed Displacement Functions

Following the two dimensional problems of elasticity, there are two displacement parameters u and ν at every point in the domain of the problem. Let us assume that the corners of the triangle are nodes and these elements are interconnected at their common nodes. With three nodes for the element, and with two displacement parameters per node, there are six nodal displacement parameters, making this element as 6 d.o.f. element as shown in Fig. 4.2. Let us try a simple linear polynomial in x and y for both u and ν displacements at any point (x,y) within the triangle as shown:

$$\begin{aligned} u(x, y) &= \alpha_1 + \alpha_2 x + \alpha_3 y \\ \nu(x, y) &= \alpha_4 + \alpha_5 x + \alpha_6 y \end{aligned} \tag{4.1}$$

i.e.,

$$\begin{Bmatrix} u(x,y) \\ v(x,y) \end{Bmatrix} = \begin{bmatrix} 1 & x & y & 0 & 0 & 0 \\ 0 & 0 & 0 & 1 & x & y \end{bmatrix} \begin{Bmatrix} \alpha_1 \\ \alpha_2 \\ \alpha_3 \\ \alpha_4 \\ \alpha_5 \\ \alpha_6 \end{Bmatrix}$$

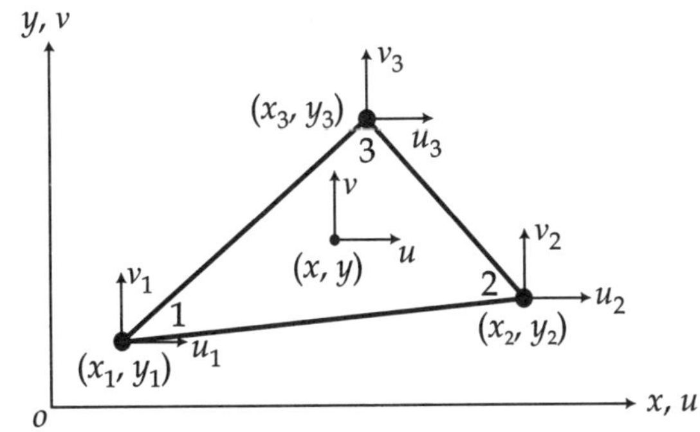

Fig. 4.2 A Triangular Element (Constant Strain)

Here, there are six α- parameters and this number tallies with the 6 d.o.f. of the element. The parameters α_1 and α_4 satisfy the rigid body motions(translations) of the element in the x and y directions while the terms α_3 and α_5 represent the rigid body rotation of the element in the x-y plane. The components of strain at a point within the element, such as ε_x, ε_y and γ_{xy} are α_2, α_6 and $\alpha_3 + \alpha_5$ respectively and therefore a state of constant strain within the element is achieved by the selection of the displacement function. Now, having achieved the compatibility of displacements at the common nodes, the compatibility of displacements along the interelement boundaries of the triangle is satisfied since the displacements vary linearly along the boundaries of all elements. Since there are no displacement derivatives as nodal parameters, this element has only $\mathbf{C_0}$- continuity. As one progresses more on the theory of finite element methods, it will be clear that the question of interelement compatibility is a major issue in the finite element methodology. For this element, it was quite an easy job to assume the displacement functions which satisfy the interelement compatibility. Now, it is necessary to represent them in the form of shape or interpolation functions as was done before for the case of the bar element. From the nodal coordinates, the nodal displacement parameters can be written as:

$$\begin{Bmatrix} u_1 \\ u_2 \\ u_3 \end{Bmatrix} = \begin{bmatrix} 1 & x_1 & y_1 \\ 1 & x_2 & y_2 \\ 1 & x_3 & y_3 \end{bmatrix} \begin{Bmatrix} \alpha_1 \\ \alpha_2 \\ \alpha_3 \end{Bmatrix} \text{ and } \begin{Bmatrix} v_1 \\ v_2 \\ v_3 \end{Bmatrix} = \begin{bmatrix} 1 & x_1 & y_1 \\ 1 & x_2 & y_2 \\ 1 & x_3 & y_3 \end{bmatrix} \begin{Bmatrix} \alpha_4 \\ \alpha_5 \\ \alpha_6 \end{Bmatrix} \quad (4.2)$$

Solving for the α- parameters using the Cramer's rule,

$$\alpha_1 = \frac{1}{2A}\begin{vmatrix} u_1 & x_1 & y_1 \\ u_2 & x_2 & y_2 \\ u_3 & x_3 & y_3 \end{vmatrix}$$

$$= \frac{1}{2A}\left[(x_2\,y_3 - x_3\,y_2)u_1 + (x_3\,y_1 - x_1\,y_3)u_2 + (x_1\,y_2 - x_2\,y_1)u_3\right]$$

$$= \frac{1}{2A}(a_1\,u_1 + a_2\,u_2 + a_3\,u_3)$$

$$\alpha_2 = \frac{1}{2A}\begin{vmatrix} 1 & u_1 & y_1 \\ 1 & u_2 & y_2 \\ 1 & u_3 & y_3 \end{vmatrix} = \frac{1}{2A}\left[(y_2 - y_3)u_1 + (y_3 - y_1)u_2 + (y_1 - y_2)u_3\right]$$

$$= \frac{1}{2A}(b_1\,u_1 + b_2\,u_2 + b_3\,u_3)$$

$$\alpha_3 = \frac{1}{2A}\begin{vmatrix} 1 & x_1 & u_1 \\ 1 & x_2 & u_2 \\ 1 & x_3 & u_3 \end{vmatrix} = \frac{1}{2A}\left[(x_3 - x_2)u_1 + (x_1 - x_3)u_2 + (x_2 - x_1)u_3\right]$$

$$= \frac{1}{2A}(c_1\,u_1 + c_2\,u_2 + c_3\,u_3) \tag{4.3}$$

where $$2A = 2(\text{area of the triangle}) = \begin{vmatrix} 1 & x_1 & y_1 \\ 1 & x_2 & y_2 \\ 1 & x_3 & y_3 \end{vmatrix} \tag{4.4}$$

Here
$$a_i = x_j\,y_k - x_k\,y_j$$
$$b_i = y_j - y_k \quad \text{and} \tag{4.5}$$
$$c_i = x_k - x_j$$

where i, j, k vary in a cyclic order. Now, the displacement functions can be written as,

$$u(x, y) = N_1\,u_1 + N_2\,u_2 + N_3\,u_3$$
$$v(x, y) = N_1\,v_1 + N_2\,v_2 + N_3\,v_3 \tag{4.6}$$

where $$N_i = \frac{1}{2A}(a_i + b_i x + c_i y).$$

Now the internal displacements can be written in terms of nodal displacements as:

$$\begin{Bmatrix} u(x,y) \\ v(x,y) \end{Bmatrix} = \begin{bmatrix} N_1 & 0 & N_2 & 0 & N_3 & 0 \\ 0 & N_1 & 0 & N_2 & 0 & N_3 \end{bmatrix} \{u\} = [N]\{u\} \tag{4.7}$$

where $\{u\}$ is the element nodal displacement vector.

Step 2. Internal Strains

Using the strain-displacement equations[5] discussed in Chapter 2,

$$\begin{Bmatrix} \varepsilon_x \\ \varepsilon_y \\ \gamma_{xy} \end{Bmatrix} = \frac{1}{2A} \begin{bmatrix} b_1 & 0 & b_2 & 0 & b_3 & 0 \\ 0 & c_1 & 0 & c_2 & 0 & c_3 \\ c_1 & b_1 & c_2 & b_2 & c_3 & b_3 \end{bmatrix} \begin{Bmatrix} u_1 \\ v_1 \\ u_2 \\ v_2 \\ u_3 \\ v_3 \end{Bmatrix} \tag{4.8}$$

and in matrix form,

$$\{\varepsilon\} = [B]\{u\}.$$

Step 3. Internal Stresses

Using the stress-strain relationship given for the plane elasticity problems, from Eq. (2.18), for the plane stress case

$$[D] = \frac{E}{(1-\mu^2)} \begin{bmatrix} 1 & \mu & 0 \\ \mu & 1 & 0 \\ 0 & 0 & \dfrac{(1-\mu)}{2} \end{bmatrix}$$

and, for the plane strain case

$$[D] = \frac{E(1-\mu)}{(1+\mu)(1-2\mu)} \begin{bmatrix} 1 & \dfrac{\mu}{1-\mu} & 0 \\ \dfrac{\mu}{1-\mu} & 1 & 0 \\ 0 & 0 & \dfrac{(1-2\mu)}{2(1-\mu)} \end{bmatrix}$$

By substituting $E = E/(1-\mu^2)$ and $\mu = \mu/(1-\mu)$, the plane stress equations shown above can be converted to plane strain equations.

Step 4. Minimum Potential Energy Theorem

The potential energy function $\mathbf{\Phi} = \mathbf{U} + \mathbf{V}$ which is a minimum for static equilibrium condition is

$$\Phi = \frac{1}{2}\int_{vol} \{\varepsilon\}^t \{\sigma\}\, dvol - \{f\}^t \{u\}$$

$$= \frac{1}{2}\int_{vol} \{u\}^t [B]^t [D] [B] \{u\}\, dvol - \{f\}^t \{u\} \tag{4.9}$$

Here $\{f\}$ is the nodal force vector. From Eq. (4.9), the element stiffness matrix is obtained as,

$$[k] = \int_{vol} [B]^t [D] [B] \, dvol$$

$$= At[B]^t [D] [B] \tag{4.10}$$

because all the quantities to be integrated are constants.

The same set of equations can be derived using the principle of virtual work, i.e.,

virtual strain energy $\quad \delta \mathbf{U} = \int_{vol} \{\delta\varepsilon\}^t \{\sigma\} \, dvol$

$$= \int_{vol} \{\delta u\}^t [B]^t [D] [B] \{u\} \, dvol \tag{4.12}$$

virtual work done by the nodal forces,

$$\delta \mathbf{W} = \{\delta u\}^t \{f\} \tag{4.13}$$

setting $\quad \delta \mathbf{U} - \delta \mathbf{W} = 0$

$$= \{\delta u\}^t \left(\int_{vol} [B]^t [D] [B] \, dvol \, \{u\} - \{f\} \right) \tag{4.14}$$

For arbitrary virtual displacements, the quantity within the brackets must vanish, this procedure yields the same equations as before. It may be noted that from now on, step 4 can be omitted, using the matrices $[B]$ and $[D]$, the stiffness matrix can be written as

$$[k] = \int_{vol} [B]^t [D] [B] \, dvol$$

Thus, the development of the element stiffness matrix boils down to the steps 1, 2 and 3.

The four steps used in the development of the element stiffness matrix are shown in a simple flow chart, as this technique is the same for all finite elements discussed from now on.

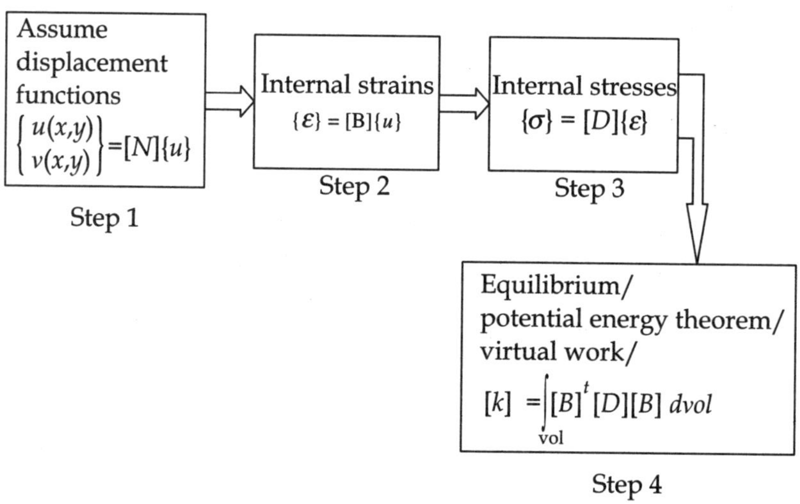

4.2.1 Shape Functions and Area Coordinates

Some properties of the shape functions[6,8] need to be mentioned. Upon examination of all the three shape functions N_i, one can see that N_1 at node 1 is equal to 1 and varies linearly from 1 to zero at both nodes 2 and 3, and, N_2 and N_3 have similar characteristics, and therefore, by inspection, it is easy to verify that

$$N_1 + N_2 + N_3 = 1.0 \tag{4.15}$$

at any point (x,y) within the triangle. Observing this linear variation of the shape functions within the triangle, one can attribute some geometrical properties to N_1, N_2 and N_3, i.e.,

$$N_1 = \frac{A_1}{A} = L_1$$

$$N_2 = \frac{A_2}{A} = L_2 \tag{4.16}$$

$$N_3 = \frac{A_3}{A} = L_3,$$

where A_1, A_2 and A_3 are the areas of the triangles opposite to nodes 1,2 and 3 as illustrated in Fig. 4.3. These area ratios identified as L_1, L_2 and L_3 are called 'area coordinates' or 'natural coordinates', such that

$$L_1 + L_2 + L_3 = 1.0 \tag{4.17}$$

It may be noted that the coordinates x and y can also be interpolated from the nodal coordinates using the area coordinates, where

$$x = L_1\, x_1 + L_2\, x_2 + L_3\, x_3$$

$$y = L_1\, y_1 + L_2\, y_2 + L_3\, y_3$$

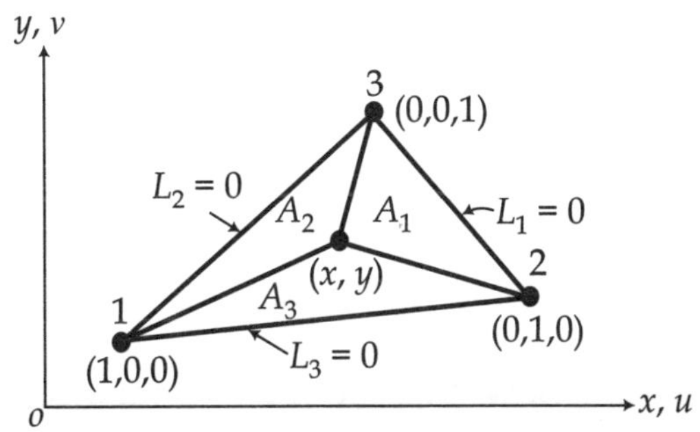

Fig. 4.3 Area or Natural Coordinates

at every point within the triangle. Given any three values of a function at the nodes, one can interpolate the value of the function within the triangle linearly using the area coordinates. In this case of constant strain triangle, it so happens that the area coordinates and the shape functions N_1, N_2, and N_3 are the same, but need not be so for other sophisticated triangular elements, and hence they are designated differently here. One of the great advantage of the area coordinate concept is the use of two explicit integration formulae[8]

$$\int_A L_1^a L_2^b L_3^c \, dA = \frac{a!\ b!\ c!}{(a+b+c+2)!} 2A \tag{4.18}$$

$$\int_{side23} L_2^a L_3^b \, dl = \frac{a!\ b!}{(a+b+1)!}(l_{23}) \tag{4.19}$$

These formulae will be very handy to integrate explicitly some of the element stiffness matrix coefficients and force vector components. Besides these explicit integrations, there are many numerical integration formulae that are very well used in finite element calculations.

Example Problem 4.1 Integration of a Polynomial in a Triangle

Determine the value of the integral $I = \int_\Delta xy^2 \, dA$ in the triangle shown in Fig. 4.4 below using the formula given in Eq. (4.18).

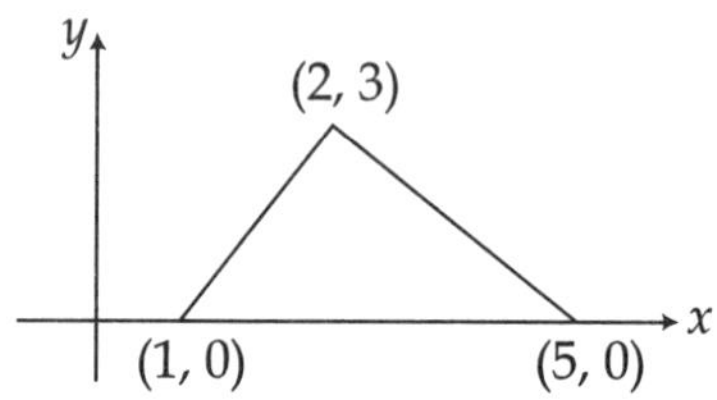

Fig. 4.4

Using area coordinates,

$x = L_1 + 5L_2 + 2L_3$ and $y = 3L_3$ and noting that the area of the triangle = 6,

$$I = \iint_A (L_1 + 5L_2 + 2L_3)(3L_3)^2 \, dA$$

$$= 9\iint_A (L_1L_3^2 + 5L_2L_3^2 + 2L_3^3) \, dA.$$

$$= 21.6, \quad \text{using Eq. (4.18).}$$

4.2.2 Equivalent Nodal Force Vector for Prescribed Forces

There are three types of prescribed external forces.

1. Prescribed Nodal Forces in x and y Directions

These force components are directly read into the global force vector from the input data such as $\tilde{f}_{2i-1}$ and $\tilde{f}_{2i}$ at node i in the x and y directions respectively.

2. Prescribed Body Force Vector $\{\tilde{b}\}$ per Unit Volume of the Body

Here the work potential due to the body force vector $\{\tilde{b}\} = \{\tilde{b}_x \;\; \tilde{b}_y\}^t$ per unit volume of the body is written as

$$\mathbf{V} = -\int_A \{\tilde{b}\}^t \begin{Bmatrix} u \\ v \end{Bmatrix} t\, dx\, dy \tag{4.20}$$

$$= -\int_A \{\tilde{b}\}^t \, [N] \, \{u\} t\, dx\, dy$$

Hence the equivalent body force vector is obtained as

$$\{\tilde{f}\} = \frac{1}{3} At \{\tilde{b}_x \;\; \tilde{b}_y \;\; \tilde{b}_x \;\; \tilde{b}_y \;\; \tilde{b}_x \;\; \tilde{b}_y\}^t \tag{4.21}$$

3. Prescribed Surface Traction $\{\tilde{T}\}$ per Unit Area of the Surface

Prescribed surface force is shown for side 2-3 in Fig. 4.5 (a). The corresponding work potential is written as

$$\mathbf{V} = -\int_{side23} \begin{Bmatrix} \tilde{t}_x \\ \tilde{t}_y \end{Bmatrix}^t \begin{Bmatrix} u \\ v \end{Bmatrix} t\, dl \tag{4.22}$$

$$= -\int_{2-3} \begin{Bmatrix} \tilde{t}_x \\ \tilde{t}_y \end{Bmatrix}^t [N] \, \{u\} t\, dl$$

hence the equivalent nodal force vector, if $\tilde{t}_x$ and $\tilde{t}_y$ are constants, is

$$\{\tilde{f}\} = \frac{lt}{2} \{0 \;\; 0 \;\; \tilde{t}_x \;\; \tilde{t}_y \;\; \tilde{t}_x \;\; \tilde{t}_y\}^t \tag{4.23}$$

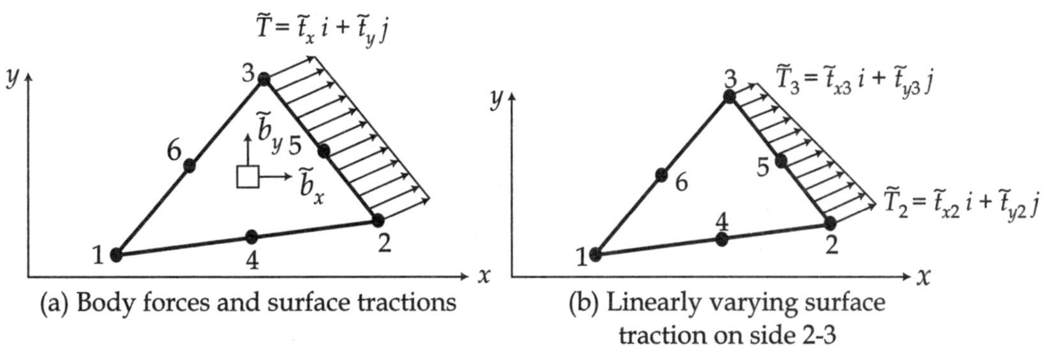

Fig. 4.5 Forces Acting on the Side of an Element

And if $\tilde{t}_x$ and $\tilde{t}_y$ vary linearly along the side of the triangle, then interpolating $\tilde{t}_x$ and $\tilde{t}_y$ using the area coordinates,

$$\tilde{t}_x = L_2\,\tilde{t}_{x2} + L_3\,\tilde{t}_{x3} \text{ and}$$
$$\tilde{t}_y = L_2\,\tilde{t}_{y2} + L_3\,\tilde{t}_{y3} \tag{4.24}$$

$$\{\tilde{f}\} = \frac{lt}{6}\{0 \quad 0 \quad (2\tilde{t}_{x2} + \tilde{t}_{x3}) \quad (2\tilde{t}_{y2} + \tilde{t}_{y3}) \quad (\tilde{t}_{x2} + 2\tilde{t}_{x3}) \quad (\tilde{t}_{y2} + 2\tilde{t}_{y3})\}^t \tag{4.25}$$

In Eqs. (4.23) and (4.25), l is the length of the side 2-3.

Task 3. Assembling the Element Stiffness Equations

Now the global stiffness matrix and the corresponding global force vector have to be formed by assembling the individual element stiffness matrices and prescribed force vectors. The key to the assembly procedure is the nodal connectivity matrix. To understand the equivalence of the element numbers in the local and global coordinates, let the 6 × 6 element stiffness matrix be divided into 2 × 2 submatrices. Let i, j, k be the global nodal numbers of the element n and the corresponding local numbers are 1, 2, 3. The correspondence of the local nodal numbers with the global nodal numbers and the respective displacement identifications using subscripts are shown in Table. 4.1. Odd subscripts correspond to the displacements in the x-axis while the even subscripts correspond to those in the y-axis. The correspondence of the submatrices (both diagonal and offdiagonal) to the global submatrices are shown in Fig. 4.6.

Table 4.1

Local Node		*Local Disp. Comp*		*Global Node*		*Global Disp. Comp*
1	→	$\begin{Bmatrix}1\\2\end{Bmatrix}$	⇒	i	→	$\begin{Bmatrix}2i-1\\2i\end{Bmatrix}$
2	→	$\begin{Bmatrix}3\\4\end{Bmatrix}$	⇒	j	→	$\begin{Bmatrix}2j-1\\2j\end{Bmatrix}$
3	→	$\begin{Bmatrix}5\\6\end{Bmatrix}$	⇒	k	→	$\begin{Bmatrix}2k-1\\2k\end{Bmatrix}$

A simple example of a five element discretization and its connectivity matrix is shown in Fig. 4.7. Using the connectivity matrix and the information given in Table 4.1, Figs. 4.6 and 4.7, the five element stiffness matrices are assembled together as shown pictorially in Fig. 4.8.

$$\begin{bmatrix}k_{11} & k_{12}\\ k_{21} & k_{22}\end{bmatrix} \Rightarrow \begin{bmatrix}k_{2i-1,2i-1} & k_{2i-1,2i}\\ k_{2i,2i-1} & k_{2i,2i}\end{bmatrix}$$

$$\begin{bmatrix}k_{13} & k_{14}\\ k_{23} & k_{24}\end{bmatrix} \Rightarrow \begin{bmatrix}k_{2i-1,2j-1} & k_{2i-1,2j}\\ k_{2i,2j-1} & k_{2i,2j}\end{bmatrix}$$

$$\begin{bmatrix}k_{15} & k_{16}\\ k_{25} & k_{26}\end{bmatrix} \Rightarrow \begin{bmatrix}k_{2i-1,2k-1} & k_{2i-1,2k}\\ k_{2i,2k-1} & k_{2i,2k}\end{bmatrix}$$

$$\begin{bmatrix} k_{33} & k_{34} \\ k_{43} & k_{44} \end{bmatrix} \Rightarrow \begin{bmatrix} k_{2j-1,2j-1} & k_{2j-1,2j} \\ k_{2j,2j-1} & k_{2j,2j} \end{bmatrix}$$

$$\begin{bmatrix} k_{35} & k_{36} \\ k_{45} & k_{46} \end{bmatrix} \Rightarrow \begin{bmatrix} k_{2j-1,2k-1} & k_{2j-1,2k} \\ k_{2j,2k-1} & k_{2j,2k} \end{bmatrix}$$

$$\begin{bmatrix} k_{55} & k_{56} \\ k_{65} & k_{66} \end{bmatrix} \Rightarrow \begin{bmatrix} k_{2k-1,2k-1} & k_{2k-1,2k} \\ k_{2k,2k-1} & k_{2k,2k} \end{bmatrix}$$

Submatrices in Local Coordinates *Submatrices in Global Coordinates*

Fig. 4.6 Equivalence of Local and Global Stiffness Matrices

Even though, this is a repetition of what has been shown for assembling 4×4 matrices for the case of the truss elements, it is felt that the assembly technique is quite important for the beginner, so that he may write subroutines to assemble more complex finite elements. There are of course more sophisticated techniques used to assemble element stiffness matrices as used in multipurpose finite element codes, where the prescribed displacement components are deleted from the global vector, using a procedure called static condensation, thus reducing the number of unknowns in the solution process. This technique is not presented here.

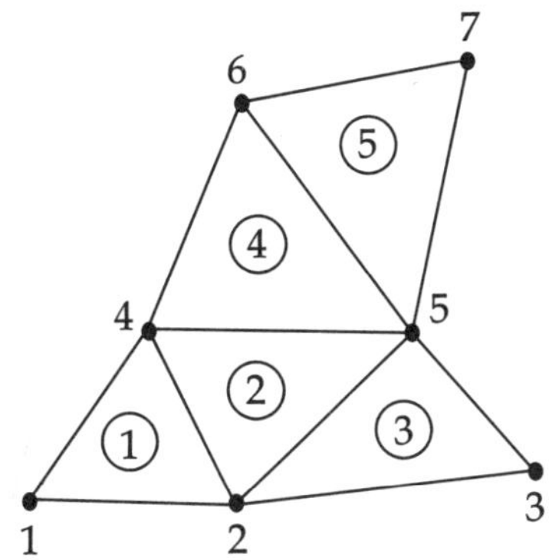

Element No.	Nodal connectivity matrix		
n	NOD(n,1)	NOD(n,1)	NOD(n,3)
1	1	2	4
2	2	5	4
3	2	3	5
4	4	5	6
5	5	7	6

Fig. 4.7 A Simple Discretization and Connectivity Matrix

Using the technique described for the assembly of the truss elements, the triangular elements are also assembled and this is pictorially shown in Fig. 4.8. From this figure it is observed that the global stiffness matrix is banded and being symmetric, only half of the matrix needs to be stored on the computer to save storage as well as computational time. The half band width, MBAND depends on the arrangement of the numbering of the nodes, and it can be seen that MBAND can be calculated from the connectivity matrix as

$$\text{MBAND} = \text{DOF} * (\text{MD} + 1) \tag{4.26}$$

where DOF is the degree of freedom per node, and MD is the maximum difference between nodal numbers for all elements. For the CST elements, DOF = 2 and for the given example MD = 3, and hence the half band width, MBAND = 8. Let

NELM Total number of finite elements

NOD(n, 3) Nodal connectivity matrix.

A subroutine is presented below for the assembly of triangular elements as a half-banded (rectangular) matrix. ELSTIFF is a subroutine which will develop the element stiffness matrix of a triangular element and is used as many times as the number of elements NELM. Of the two numbers (3 and 2) that you see in the subroutine, the first one (=3, for triangular elements) represent the number of nodes in the element and the other number (=2) is the number of degrees of freedom per node. Perhaps, it may be worthwhile to go through the do-loops of the subroutines in order to understand how the matrix is assembled in a half-banded form. Once this subroutine is understood, one can write any subroutine to assemble any type of finite element described in the following chapters.

```
      DO 270 NN = 1, NELM
      CALL ELSTIFF
      DO 220 I = 1, 3
      LM(I) = 2*(NOD(NN,I)-1)
220   CONTINUE
      DO 260 I = 1, 3
      DO 260 K = 1, 2
      II = LM(I) + K
      KK = 2*(I-1) + K
      DO 230 J = 1, 3
      DO 230 L = 1, 2
      JJ = LM(J) + L - II + 1
      LL = 2*(J-1) + L
      IF (JJ. LE. 0) GO TO 230
      GS(II,JJ) = GS(II,JJ) + SE(KK,LL)
230   CONTINUE
260   CONTINUE
270   CONTINUE
```

Subroutine for Assembling Half-banded Global Stiffness Matrix For Constant Strain Triangular Elements

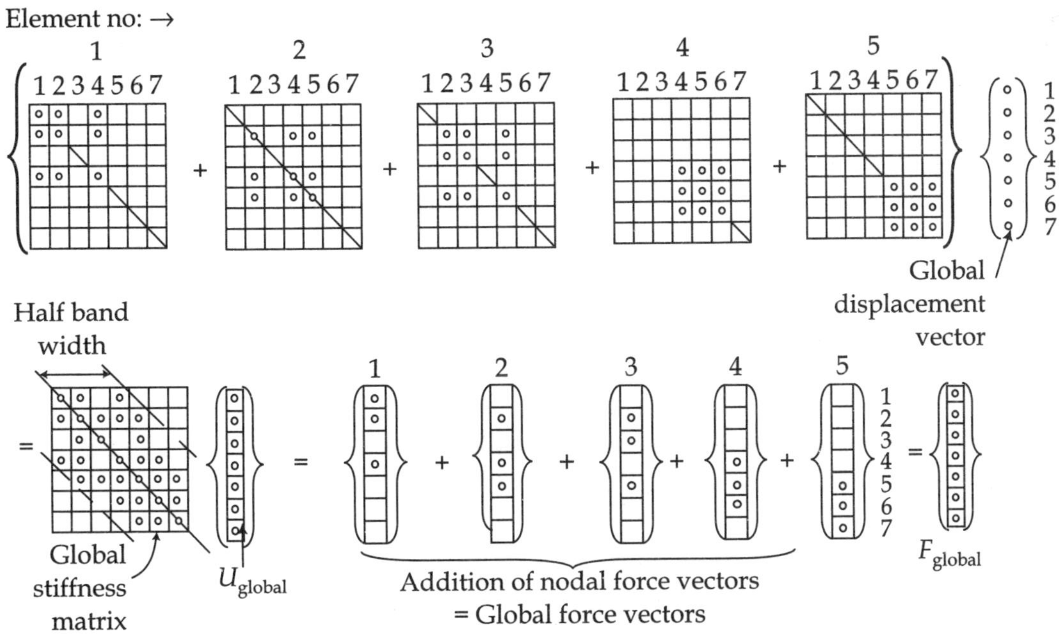

Fig. 4.8 Assembly of element stiffness Matrix Equations. (Consider each circle in the stiffness matrix as 2 x 2)

Task 4: Incorporation of Prescribed Displacements on the Boundary

This procedure is same as discussed previously and hence not repeated here. The subroutine shown in Chapter 3 for prescribing displacement boundary conditions can be used here also and hence not shown here. For a two-dimensional plane elasticity problem, at least three boundary displacements are to be prescribed, out of which at least one should be in each direction. This way all the three rigid body motions can be restrained. If by mistake, only displacements in one direction are specified, the global stiffness matrix will be still singular and one cannot obtain a solution.

Task 5: Solution of the Modified Global Equations

For this, either the half band solver or the skyline solver can be used and these are described in the Appendix I. The preference of the method depends on the numbering of the nodes. If the numbering of the nodes are such that there are too many zeros inside the half banded matrix (perhaps due to poor numbering of nodes), the skyline solver is to be preferred. The outcome of the solution routine is the global displacement vector.

Task 6: Calculation of the Element Strains and Stresses

Again, using the connectivity matrix the local numbers of the elements are determined and the internal strain components are computed using the local $[B]$ matrix as shown in Eq. (4.8). Using Eq. (2.18) all the components of the stress are also computed.

These six tasks complete the finite element solution of a plane elasticity problem. It is interesting to note that a single computer program can be used for both plane stress and plane strain problems, by properly choosing the correct $[D]$ matrix for the stress-strain relationships as given in Eq. (2.18). Incidently, it has to be noted that since the material properties are controlled through the [*D*] matrix. Anisotropic properties can also be incorporated and have to be introduced for every element independently while calculating the element stiffness matrix before assembling the global stiffness matrix.

Example Problem 4.2 Two examples are shown here. One is a very simple assemblage of two elements as shown in Figs. 4.9 (a) and (b). These examples are shown only for the benefit of the beginner to develop the stiffness matrix equations and learn to assemble them, put boundary conditions and solve them. Once beginners dwell deep into the finite element computer programs, they can lose track of the simple fundamentals used in the methodology.

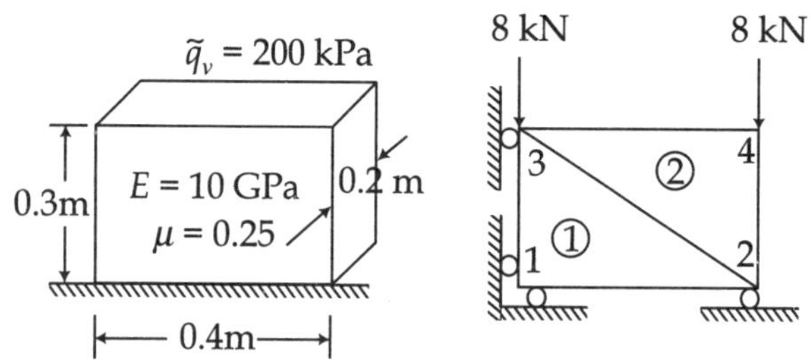

(a) A rectangular block resting on a smooth plane

(b) Finite element idealization

Fig. 4.9 A Simple Plane Stress Problem

The coordinates, material properties and the loading are shown in the figure. For element one, the [*D*] and [*B*] matrices are:

$$[D] = 1.0667 \times 10^4 \begin{bmatrix} 1 & .25 & 0 \\ .25 & 1 & 0 \\ 0 & 0 & .375 \end{bmatrix}, \text{ and}$$

$$[B] = \frac{1}{0.12} \begin{bmatrix} -0.3 & 0 & 0.3 & 0 & 0 & 0 \\ 0 & -0.4 & 0 & 0 & 0 & 0.4 \\ -0.4 & -0.3 & 0 & 0.3 & 0.4 & 0 \end{bmatrix}$$

The element stiffness matrix becomes:

$$[k] = At\,[B]^t\,[D]\,[B]$$

$$= 8.889 \times 10^3 \begin{bmatrix} .15 & .075 & -.09 & -.045 & -.06 & -.03 \\ 0.75 & .19375 & -.03 & -.03375 & -.045 & -.16 \\ -.09 & -.03 & .09 & 0 & 0 & .03 \\ -.045 & -.03375 & 0 & .03375 & .045 & 0 \\ -.06 & -.045 & 0 & .045 & .06 & 0 \\ -.03 & -.16 & .03 & 0 & 0 & .16 \end{bmatrix}$$

Similarly, the element stiffness matrix for the second element is obtained as:

$$[k] = 8.889 \times 10^3 \begin{bmatrix} .06 & 0 & -.06 & -.045 & 0 & .045 \\ 0 & .16 & -.03 & -.16 & -.03 & 0 \\ -.06 & -.03 & .15 & .075 & -.09 & -.045 \\ -.045 & -.16 & .075 & .19375 & -.03 & -.03375 \\ 0 & .03 & -.09 & -.03 & .09 & 0 \\ -.045 & 0 & -.045 & -.03375 & 0 & .03375 \end{bmatrix}$$

Now the two matrices are assembled together to form an 8×8 global stiffness matrix as show below:

$$[k] = 8.889 \times 10^3 \begin{bmatrix} .15 & .075 & -.09 & -.045 & -.06 & -.03 & & \\ .075 & .19375 & -.03 & -.03375 & -.045 & -.16 & & \\ -.09 & -.03 & .15 & 0 & 0 & .075 & -.06 & -.045 \\ -.045 & -.03375 & 0 & 1.9375 & .075 & 0 & -.03 & -.16 \\ -.06 & -.045 & 0 & .075 & .15 & 0 & -.09 & -.03 \\ -.03 & -.16 & 0.75 & 0 & 0 & .19375 & -.045 & -.03375 \\ & & -.06 & -.03 & -.09 & -.045 & .15 & .075 \\ & & -.045 & -.16 & -.03 & -.03375 & .075 & .19375 \end{bmatrix}$$

After prescribing the displacement boundary conditions, stiffness matrix can be written as a 4×4 matrix in a reduced form and is as shown below:

$$8.889 \times 10^3 \begin{bmatrix} .15 & .075 & -.06 & -.045 \\ .075 & .19375 & -.045 & -.03375 \\ -.06 & -.045 & .15 & .075 \\ -0.045 & -.03375 & .075 & .19375 \end{bmatrix} \begin{Bmatrix} u_3 \\ u_6 \\ u_7 \\ u_8 \end{Bmatrix} = \begin{Bmatrix} 0 \\ -8 \\ 0 \\ -8 \end{Bmatrix}$$

The solution of these equations yields:

$u_3 = 0.002$ m.

$u_6 = -\ 0.006$ m.

$u_7 = 0.002$ m., and

$u_8 = -\ 0.006$ m.

The element strains and stresses for both elements are:

$$\begin{Bmatrix} \varepsilon_x \\ \varepsilon_y \\ \gamma_{xy} \end{Bmatrix} = \begin{Bmatrix} 0.005 \\ -0.02 \\ 0 \end{Bmatrix}, \text{ and } \begin{Bmatrix} \sigma_x \\ \sigma_y \\ \tau_{xy} \end{Bmatrix} = \begin{Bmatrix} 0 \\ -200 \\ 0 \end{Bmatrix} \text{kPa}$$

The results are exact because the problem is of constant strain and we have employed constant strain elements in the finite element model. Next, example is a problem where the strains vary within the domain.

Example Problem 4.3 A plane strain idealization of a concrete dam as shown in Fig. 4.10a, is used in this example. The cross section of the dam is divided into 4 elements with 6 nodes, such that there are 12 d.o.f. for the entire assemblage as shown in Fig. 4.10b. So far this discretization is intended to do some hand computations. Modifying the global matrix for the prescribed displacement boundary conditions and solving the resulting equations, the displacement vector is obtained. From the element displacements, the element stresses and strains which are constants within each element. are computed. The element stresses for the discretization shown in Fig. 4.10b is shown in Table 4.2 and the displacements are shown in Fig. 4.11. This solution is only for an illustration of what kind of solution one gets from a very coarse discretization. Because of the large discontinuities of the strains and stresses at the interelement boundaries, it is appropriate to consider values of the stresses and strains as values at the respective centroids of the triangular elements. Another option is to average the stresses at the nodal points. Naturally one may ask a question, "is this finite element solution accurate enough", in other words, "are more elements necessary"?

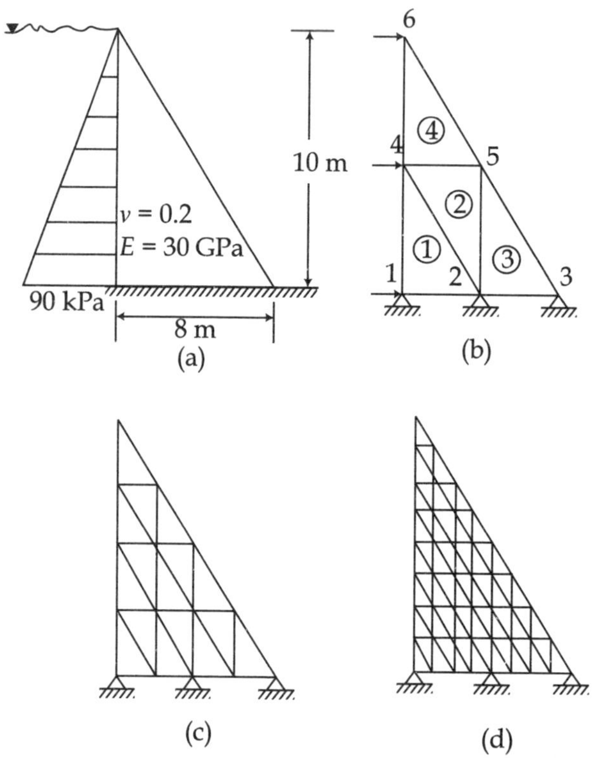

Fig. 4.10 A Simple Example Used for Hand and Computer Solutions

The answer, of course, is 'not accurate enough'. For the example problem, one has to discretize the domain into much smaller elements to obtain a satisfactory solution. A convergence study of the problem for different mesh sizes is shown here. First the number of elements are increased (4 times) and again increased (16 times) as illustrated in the Figs. 4.10c and *d*. The convergence of the horizontal displacement of the water side of the dam for various discretizations are shown in Fig. 4.11. To determine the number of elements required for a problem, it is necessary to perform a convergence study, until one gets some experience with similar problems, and thus it is possible to have an intuitive feeling on the number of elements to be used. It can be observed that the constant strain elements (CST) converge relatively slow to the exact solution, especially when there is considerable variation of stress in the domain of the body.

Table 4.2 Stresses in the Dam for Discretization in Fig. 4.10b

Element no.	$\sigma_x(kPa)$	$\sigma_y(kPa)$	$\tau_{xy}(kPa)$
1	11.993	47.973	59.835
2	-31.984	-27.339	19.628
3	-5.1583	-20.663	51.787
4	-25.149	0	18.750

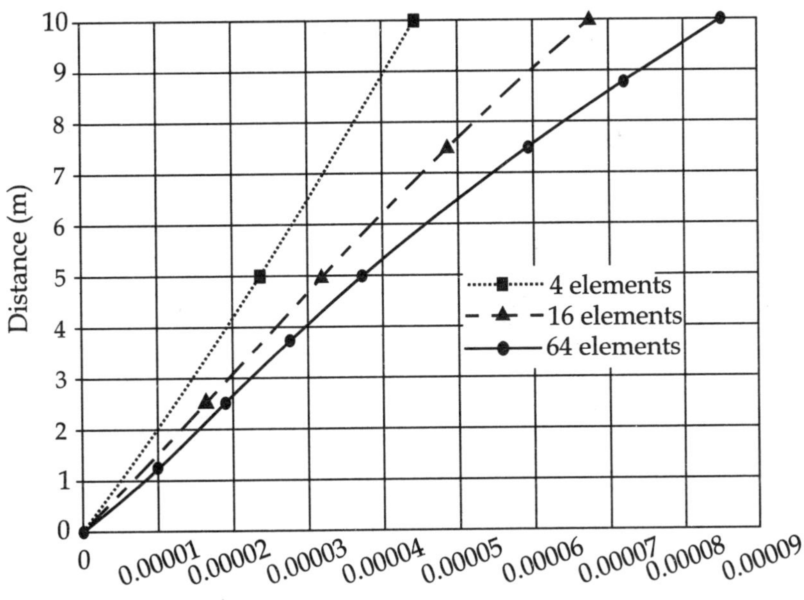

Fig. 4.11 Results for the Problem in Fig. 4.10

Example Problem 4.4 Another classical example of a plane stress problem is the evaluation of stress concentrations in a thin plate.[7] If the plate is infinite with a circular hole in the middle and is subjected to a uniform stress in one direction away from the hole, the stress concentration at the hole is exactly three times the stress at far away. This problem is solved for a finite square plate 16.0 cm. X 16.0 cm with a uniform thickness equal to 0.5 cm and subjected to a uniform stress σ_x = 1.0 kPa on the boundary as shown in Fig. 4.12. The modulus of elasticity of the plate E = 30 GPa and Poisson's ratio ν = 0.25. Only the stress concentration along the vertical axis through the center of the plate is shown here. The stress concentration is almost equal to 3.0. However, the stress distribution in a finite plate and infinite plates are theoretically not the same.

4.3 SPECIAL TRIANGULAR ELEMENTS (CST CONTINUED)

Suppose one has to solve a thin plate problem loaded in its own plane, but the plate has a variable thickness which can be assumed to be uniformly varying within the element. Mathematically, variation of the thickness of the plate is in violation of the plane stress concept. But, using the steps suggested for the development of the element stiffness equations one can very easily obtain the element stiffness matrix and obtain a practical solution for the problem. Let us assume that the triangular plate has thicknesses t_1, t_2 and t_3 at nodal points 1, 2 and 3 respectively as shown in Fig. 4.13.

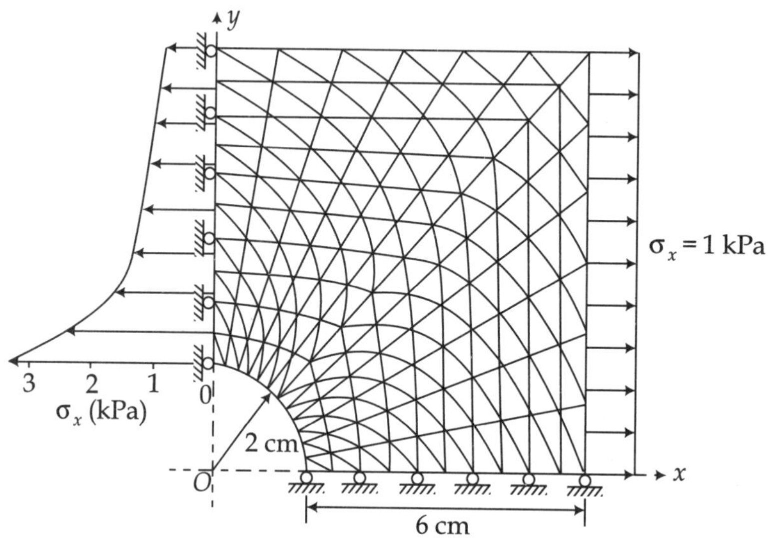

Fig. 4.12 Stress Concentration in a Plate with a Hole in the Middle

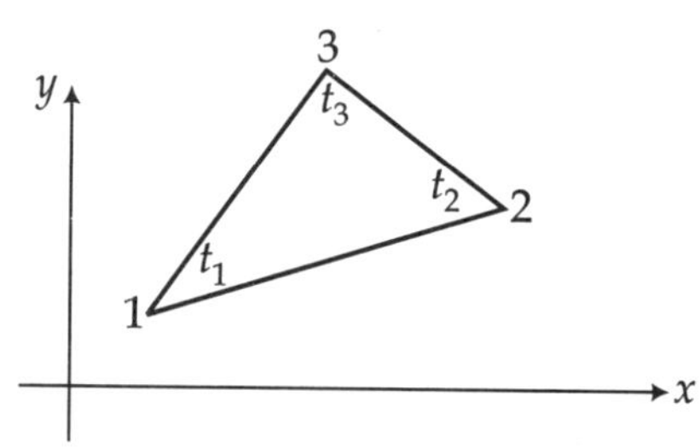

Fig. 4.13 A Triangular Element with Linearly Varying Thickness

The steps 1, 2 and 3 are the same but the strain energy expression from Eq. (4.9) becomes

$$U = \frac{1}{2}\int_A \{u\}^t [B]^t [D] [B] \{u\} (L_1t_1 + L_2t_2 + L_3t_3)\, dx\, dy \qquad (4.27)$$

and therefore, the element stiffness matrix becomes

$$[k] = \frac{A(t_1 + t_2 + t_3)}{3}[B]^t [D] [B] \qquad (4.28)$$

Naturally, the energy principle has used the average thickness of the element here. Continuing this procedure, one can develop a stiffness matrix for an triangular element whose modulus of elasticity varies from node to node linearly, in addition to the variation of its thicknesses. Here again, one can interpolate the modulus of elasticity and the thickness of the plate at any point within the element and substitute that in the strain energy function. The value of Et at any point (x,y) within the triangle is given as

$$Et = (L_1E_1 + L_2E_2 + L_3E_3)\,(L_1t_1 + L_2t_2 + L_3t_3).$$

Using the integration formula given in Eq. (4.18), the element stiffness matrix becomes

$$[k] = \frac{A}{12}\left(\sum_{i=1}^{3}\sum_{j=1}^{3} E_i\, t_j + \sum_{i=1}^{3} E_i\, t_i\right)[B]^t[D][B] \qquad (4.29)$$

and here the D matrix does not contain the value of E. The readers are expected to verify the above integration. The power of the finite element technique is easily noted here for its ability to accommodate complex material properties as encountered in materials having anisotropic or even extended to elastic-plastic properties. The only modification one has to do is to determine the coefficients in the D matrix and substitute in the stiffness matrix equation. The constant strain elements are often used in solving dynamic problems involving impact between bodies, stress wave propagation or earth-quake related stress analyses.

4.4 LINEAR STRAIN TRIANGLE (LST)

Having experienced that for a satisfactory finite element stress analysis one has to employ a large number of constant triangular elements, an alternative is to develop a linear strain triangular element just as the one developed for the case of the one dimensional linear strain bar element. The most important task here is the development of the element stiffness matrix for a linear strain condition.[2,4,8] Once the element stiffness matrix has been developed the rest of the finite element analysis where one assembles the element matrices, incorporating the prescribed boundary conditions, solution of the global equations, comes almost in the same fashion that was used for the previous elements and any discussions on that is only a repetition of what has been said earlier. One of the best ways for the development of the element stiffness matrix is to employ the four steps discussed earlier. This is done in the following sections.

Step 1. Assume Displacement Functions

In order to have a linear strain condition within the element, the assumed displacement function has to be at least a complete quadratic function in x and y. At this time, it may be interesting to examine the so-called Pascal triangle shown in Fig. 4.14 for the set of polynomial functions in x and y. The Pascal triangle is as shown below

$$1$$

$$x \;\ldots\ldots\; y$$

$$x^2 \;\ldots\ldots\; xy \;\ldots\ldots\; y^2$$

$$x^3 \;\ldots\ldots\; x^2y \;\ldots\ldots\; xy^2 \;\ldots\ldots\; y^3$$

$$x^4 \;\ldots\ldots\; x^3y \;\ldots\ldots\; x^2y^2 \;\ldots\ldots\; xy^3 \;\ldots\ldots\; y^4$$

Fig. 4.14 The Pascal Triangle

In each row of the Pascal triangle, the functions of x and y are homogeneous. A complete quadratic function in $u(x, y)$ has six parameters and so also for $v(x, y)$; thus the minimum number of independent parameters becomes twelve. Now by providing six nodes in the triangle as shown in the Fig. 4.15, and with two degrees of freedom per node, there are 12 d.o.f. that match with the parameters necessary for a complete quadratic in x and y, i.e.,

$$\begin{aligned} u(x, y) &= \alpha_1 + \alpha_2 x + \alpha_3 y + \alpha_4 x^2 + \alpha_5 xy + \alpha_6 y^2 \\ v(x, y) &= \alpha_7 + \alpha_8 x + \alpha_9 y + \alpha_{10} x^2 + \alpha_{11} xy + \alpha_{12} y^2 \end{aligned} \tag{4.30}$$

or in matrix form

$$\begin{Bmatrix} u \\ v \end{Bmatrix} = [A]\{\alpha\} \tag{4.31}$$

where $\{\alpha\}$ is a vector containing twelve parameters. This complete quadratic polynomial has the rigid body motions, constant strain conditions already embedded in addition to the linear strain condition. Next important criterion for the selection of the displacement function is to verify whether the assumed displacement functions satisfy compatibility of displacements between the element boundaries.

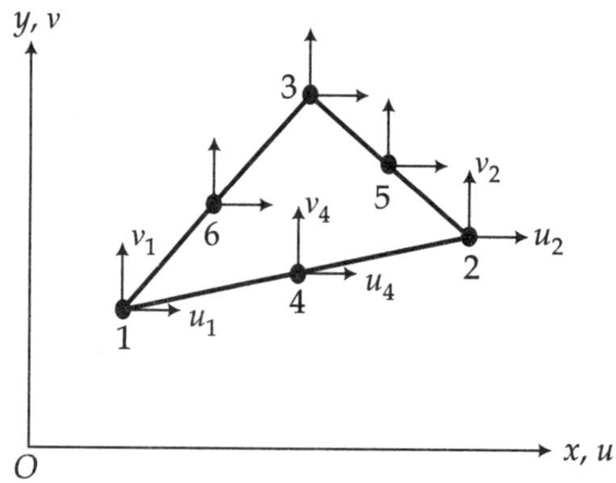

Fig. 4.15 A 2-D Linear Strain Triangular Element

To verify this, let us examine the side 2-3 of the triangle. There are three nodes on that side, and with three displacement parameters, a unique quadratic polynomial exists for each displacements along that side. Keeping the nodal displacements compatible between the interelement boundaries, it is seen that complete compatibility of u and v displacements exists on the interelement boundaries. Of course, note that only a C_0-continuity type compatibility exists between elements in this model. In other words, compatibility of strains between the elements is not guarenteed.

Having formed the displacement functions, the next step is to derive the stiffness matrix for the element. This can be done in three slightly different ways. Even though all three methods are similar, by studying or experimenting with them, new ideas on the computation of the stiffness matrix for finite elements are obtained. Simple exercises like these different procedures are very valuable and interesting.

Method 1

This is the simplest of the three. Following the procedure for the constant triangular element, the α-parameters in terms of the nodal displacements using the nodal coordinates are obtained, i.e.,

$$\{u\} = [G]\{\alpha\} \text{ or } \{\alpha\} = [G]^{-1}\{u\} \tag{4.32}$$

Substituting this into Eq. (4.31), the internal displacement functions are expressed in terms of the nodal displacement vector. The next step is to express the internal strain in terms of the nodal displacement vector. This is accomplished in the following way. The internal strain vector can be written as

$$\{\varepsilon\} = \begin{Bmatrix} \varepsilon_x \\ \varepsilon_y \\ \gamma_{xy} \end{Bmatrix} = \begin{Bmatrix} \dfrac{\partial u}{\partial x} \\ \dfrac{\partial v}{\partial y} \\ \dfrac{\partial u}{\partial y} + \dfrac{\partial v}{\partial x} \end{Bmatrix}$$

$$= \begin{bmatrix} 0 & 1 & 0 & 2x & y & 0 & 0 & 0 & 0 & 0 & 0 & 0 \\ 0 & 0 & 0 & 0 & 0 & 0 & 0 & 0 & 1 & 0 & x & 2y \\ 0 & 0 & 1 & 0 & x & 2y & 0 & 1 & 0 & 2x & y & 0 \end{bmatrix} \{\alpha\}$$

$$= [A]\{\alpha\} = [A][G]^{-1}\{u\} \tag{4.33}$$

$$= [B]\{u\}$$

Once the B matrix has been developed, the element stiffness matrix can be written using the formula

$$[k] = \int_{vol} [B]^t [D] [B] \, dvol \qquad (4.34)$$

$$= [G^{-1}]^t \int_{vol} [A]^t [D] [A] \, dvol \, [G^{-1}]$$

The integral can be explicitly calculated, and then multiplied by the inverse of G matrix as shown in the equation or the whole matrix can computed using a numerical integration technique. This technique used for triangular elements is discussed in the next chapter.

Method 2

Based on the theory and properties of the shape functions and some heuristic techniques, the shape functions are derived in terms of the area coordinates L_1, L_2, L_3. Remembering that the shape function N_i will have a value of one at node i, and will be zero at all other nodes, the following quadratic shape functions will indeed have that property (see Section on Lagrangian interpolation functions in Chapter 3, Eqs. (3.54), i.e.,

$$N_i = L_i(2L_i - 1) \text{ for } i = 1, 2, 3 \text{ and}$$

$$N_4 = 4L_1L_2,\ N_5 = 4L_2L_3 \text{ and } N_6 = 4L_1L_3 \qquad (4.35)$$

These shape functions are plotted in Fig. 4.16.

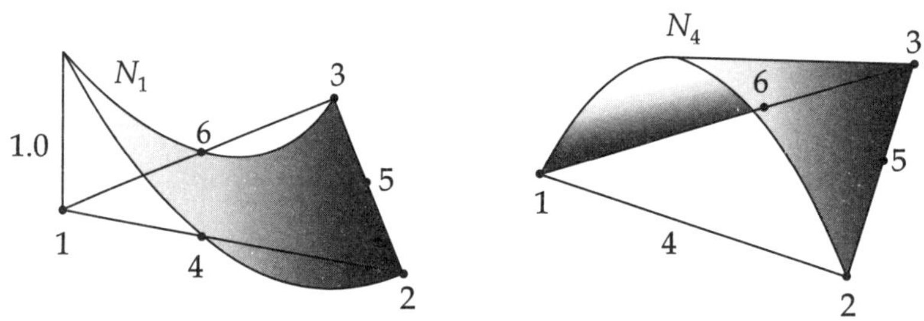

Fig. 4.16 Shape Functions N_1 and N_4

The displacement functions u and v are:

$$u = \sum_{i=1}^{6} N_i u_i \text{ and } v = \sum_{i=1}^{6} N_i v_i \qquad (4.36)$$

The internal strains can be expressed in terms of the nodal displacements, using Eqs. (4.36), (4.6) and (4.16), i.e.,

$$\varepsilon_x = \frac{\partial u}{\partial x} = \frac{\partial}{\partial x} \sum_{i=1}^{6} N_i u_i$$

$$= \sum_{i=1}^{6} \left(\frac{\partial N_i}{\partial L_1} \frac{\partial L_1}{\partial x} + \frac{\partial N_i}{\partial L_2} \frac{\partial L_2}{\partial x} + \frac{\partial N_i}{\partial L_3} \frac{\partial L_3}{\partial x} \right) u_i \qquad (4.37)$$

$$= \frac{1}{2A}[(4L_1 - 1)\, b_1u_1 + (4L_2 - 1)\, b_2u_2 + (4L_3 - 1)\, b_3u_3 +$$

$$4(L_2b_1 + L_1b_2)u_4 + 4(L_3b_2 + L_2b_3)u_5 + 4(L_1b_3 + L_3b_1)u_6]$$

$$= [B_1]\,\{u^e\}$$

In Eq. (4.37), the nodal displacement vector $\{u^e\}$ is

$$\{u_1 \quad u_2 \quad u_3 \quad u_4 \quad u_5 \quad u_6\}^t.$$

Similarly, differentiating v with respect to y,

$$\varepsilon_y = \frac{\partial v}{\partial y} = \frac{\partial}{\partial y}\sum_{i=1}^{6} N_i v_i$$

$$= \frac{1}{2A}[(4L_1 - 1)c_1\, v_1 + (4L_2 - 1)c_2\, v_2 + (4L_3 - 1)c_3\, v_3 +$$

$$4(L_2c_1 + L_1c_2)v_4 + 4(L_3c_2 + L_2c_3)v_5 + 4(L_1c_3 + L_3c_1)v_6]$$

$$= [B_2]\,\{v^e\} \tag{4.38}$$

The nodal displacement vector $\{v^e\}$ is similar to $\{u^e\}$ except v-components substituted for u-components. The shear strain can be expressed as

$$\gamma_{xy} = \frac{\partial u}{\partial y} + \frac{\partial v}{\partial x} = [B_2 \quad B_1]\begin{Bmatrix} u^e \\ v^e \end{Bmatrix} \tag{4.39}$$

and hence the internal strain vector is

$$\{\varepsilon\} = \begin{Bmatrix} \varepsilon_x \\ \varepsilon_y \\ \gamma_{xy} \end{Bmatrix} = \begin{bmatrix} B_1 & 0 \\ 0 & B_2 \\ B_2 & B_1 \end{bmatrix}\begin{Bmatrix} u^e \\ v^e \end{Bmatrix} = [B]\,\{u\} \tag{4.40}$$

Having obtained the matrix $[B]$ and using matrix $[D]$, the element stiffness matrix can be written as

$$[k] = \int_{vol} [B]^t\, [D][B]\, dvol$$

Special care is necessary to distinguish the nodal displacement vector $\{u\} = \{u_1\, v_1\, u_2\, v_2 \ldots u_6\, v_6\}^t$ from the vectors u^e and v^e, since the latter vectors are ordered in a different sequence.

Method 3

This is the most interesting of the three methods. It utilizes all the four steps of the formulation. Here it takes advantage of the fact that the internal strain components vary linearly within the element. Thus, instead of using directly the internal strains $\{\varepsilon\}$ in terms of its nodal displacement vector, their components are interpolated from the nodal strain components, represented by $\{\varepsilon_n\}$ using the area coordinates L_1, L_2 and L_3. Then $\{\varepsilon_n\}$ is written in terms of the nodal displacement vector using Eq. (4.40). It is very necessary to distinguish between $\{\varepsilon\}$ and $\{\varepsilon_n\}$ as shown in Eq. (4.41). The internal strain vector is expressed in terms of its nodal values using the area coordinates.

$$\{\varepsilon\} = \begin{Bmatrix} \varepsilon_x \\ \varepsilon_y \\ \gamma_{xy} \end{Bmatrix}$$

$$= \begin{bmatrix} L_1 & L_2 & L_3 & 0 & 0 & 0 & 0 & 0 & 0 \\ 0 & 0 & 0 & L_1 & L_2 & L_3 & 0 & 0 & 0 \\ 0 & 0 & 0 & 0 & 0 & 0 & L_1 & L_2 & L_3 \end{bmatrix} \begin{Bmatrix} \varepsilon_{x1} \\ \varepsilon_{x2} \\ \varepsilon_{x3} \\ \varepsilon_{y1} \\ \varepsilon_{y2} \\ \varepsilon_{y3} \\ \gamma_{xy1} \\ \gamma_{xy2} \\ \gamma_{xy3} \end{Bmatrix} \tag{4.41}$$

In matrix notation,

$$\{\varepsilon\} = [N_\varepsilon]\,\{\varepsilon_n\}$$

Now the nodal strain vector $\{\varepsilon_n\}$ is written in terms of the nodal displacement vector $\{u\}$ and this is done in parts using Eq. (4.35), (4.36) and (4.37), i.e., the normal strain components in the x-direction is obtained as:

$$\begin{Bmatrix} \varepsilon_{x1} \\ \varepsilon_{x2} \\ \varepsilon_{x3} \end{Bmatrix} = \frac{1}{2A} \begin{bmatrix} 3b_1 & -b_2 & -b_3 & 4b_2 & 0 & 4b_3 \\ -b_1 & 3b_2 & -b_3 & 4b_1 & 4b_3 & 0 \\ -b_1 & -b_2 & 3b_3 & 0 & 4b_2 & 4b_1 \end{bmatrix} \begin{Bmatrix} u_1 \\ u_2 \\ u_3 \\ u_4 \\ u_5 \\ u_6 \end{Bmatrix} \tag{4.42}$$

$$= [B_{nx}]\,\{u^e\}$$

Similarly, the normal strain components in the y-direction can be obtained from Eq. (4.43) or by replacing the b_i with c_i and u_i with corresponding v_i in the above equation. In other words,

$$\begin{Bmatrix} \varepsilon_{y1} \\ \varepsilon_{y2} \\ \varepsilon_{y3} \end{Bmatrix} = \frac{1}{2A}\begin{bmatrix} 3c_1 & -c_2 & -c_3 & 4c_2 & 0 & 4c_3 \\ -c_1 & 3c_2 & -c_3 & 4c_1 & 4c_3 & 0 \\ -c_1 & -c_2 & 3c_3 & 0 & 4c_2 & 4c_1 \end{bmatrix}\begin{Bmatrix} v_1 \\ v_2 \\ v_3 \\ v_4 \\ v_5 \\ v_6 \end{Bmatrix} \tag{4.43}$$

$$= [B_{ny}]\,\{v^e\}$$

Using the above two equations, the nodal shear strain components can be written as:

$$\begin{Bmatrix} \gamma_{xy1} \\ \gamma_{xy2} \\ \gamma_{xy3} \end{Bmatrix} = [B_{ny} \quad B_{nx}]\begin{Bmatrix} u^e \\ v^e \end{Bmatrix} \tag{4.44}$$

and therefore,

$$\{\varepsilon_n\} = \begin{bmatrix} B_{nx} & 0 \\ 0 & B_{ny} \\ B_{ny} & B_{nx} \end{bmatrix}\begin{Bmatrix} u^e \\ v^e \end{Bmatrix}$$

$$= [B_n]\,\{u\} \tag{4.45}$$

Again note the difference in the sequencing of the vectors $\{u^e\}$, $\{v^e\}$ and $\{u\}$. Rewriting $[B_n]$ according to the ordering of vector $\{u\}$ is left as an exercise to the reader. Now the internal strain vector can be written as:

$$\{\varepsilon\} = [N_\varepsilon]\,[B_n]\,\{u\} \tag{4.46}$$

Similarly the internal stresses also can be interpolated using the same technique from the corresponding nodal stress vector $\{\sigma_n\}$, i.e.,

$$\{\sigma\} = [N_\sigma]\,\{\sigma_n\} \tag{4.47}$$

Also it may be noted that the two interpolation matrices for strains and stresses are same, i.e.,

$$[N_\varepsilon] = [N_\sigma] \tag{4.48}$$

Instead of the stress strain relationships at a general point within the element as required in step 3 of the formulation, a 9×9 matrix that relates the nodal stress vector in terms of the corresponding nodal strain vector is developed, i.e.,

$$\{\sigma_n\} = [D_n]\,\{\varepsilon_n\} \tag{4.49}$$

where

$$[D_n] = \begin{bmatrix} d_{11}[I] & d_{12}[I] & d_{13}[I] \\ d_{12}[I] & d_{22}[I] & d_{23}[I] \\ d_{13}[I] & d_{23}[I] & d_{33}[I] \end{bmatrix}$$

and $[I]$ is a 3×3 identity matrix and d_{ij} are the components of the matrix $[D]$. The general components are used only to illustrate that the method is applicable to any orthotropic material with its axes in the plane of the coordinate axes x and y.

Now for the step 4, using the minimum potential energy theorem, the strain energy function is written as

$$\begin{aligned}\mathbf{U} &= \frac{1}{2}\int_{vol} \{\varepsilon\}^t \{\sigma\}\, dvol \\ &= \frac{1}{2}\int_{vol} \{\varepsilon_n\}^t [N_\varepsilon]^t [N_\sigma] \{\sigma_n\}\, dvol \\ &= \frac{1}{2}\int_{vol} \{u\}^t [B_n]^t [N_\varepsilon]^t [N_\sigma] [D_n] [B_n] \{u\}\, dvol \\ &= \frac{1}{2}\{u\}^t [B_n]^t \int_{vol} [N_\varepsilon]^t [N_\sigma]\, dvol\, [D_n] [B_n] \{u\}\end{aligned} \tag{4.50}$$

Noting that the integral can be explicitly evaluated using the formula for integration of polynomials in area coordinates and substituting

$$[C] = \int_{vol} [N_\varepsilon]^t [N_\sigma]\, dvol = \begin{bmatrix} [c] & [0] & [0] \\ [0] & [c] & [0] \\ [0] & [0] & [c] \end{bmatrix} \tag{4.51}$$

where
$$[c] = \frac{At}{12}\begin{bmatrix} 2 & 1 & 1 \\ 1 & 2 & 1 \\ 1 & 1 & 2 \end{bmatrix}$$

and $[0]$ is a 3×3 null matrix. Therefore, the element stiffness matrix can be written as

$$[k] = [B_n]^t [C] [D_n] [B_n] \tag{4.52}$$

In the computer code, the coefficients of the matrix $[C]$ can be efficiently stored in the data statement of the subroutine.

4.4.1 Equivalent Force Vector

It is observed earlier that for linear strain bar elements, in order to obtain best values of the components of the equivalent force vector, they have to be computed from the work potential function. Here also, the components of the equivalent force vector are computed by differentiating the work potential function **V** with respect to the nodal displacements.

Case 1. Prescribed Body Force Vector $\{\tilde{b}\}$ per unit volume

The work potential in this case is

$$\mathbf{V} = -\int_{vol} \begin{Bmatrix} \tilde{b}_x \\ \tilde{b}_y \end{Bmatrix}^t \begin{Bmatrix} u \\ v \end{Bmatrix} dvol \tag{4.53}$$

$$= -\int_{vol} (\tilde{b}_x u + \tilde{b}_y v)\ dvol$$

$$= -t \int \left(\tilde{b}_x \sum_{i=1}^{6} N_i u_i + \tilde{b}_y \sum_{i=1}^{6} N_i v_i \right) dx\ dy$$

and therefore,

$$\tilde{f}_{xi} = -\frac{\partial \mathbf{V}}{\partial u_i} = t\tilde{b}_x \int_A N_i\ dx\ dy$$

$$\tilde{f}_{yi} = -\frac{\partial \mathbf{V}}{\partial v_i} = t\tilde{b}_y \int_A N_i\ dx\ dy$$

and integrating the shape functions using Eq. (4.18) , the equivalent force vector is obtained as

$$\{\tilde{f}\}^t = \frac{At}{3}\{0 \quad 0 \quad 0 \quad 0 \quad 0 \quad 0 \quad \tilde{b}_x \quad \tilde{b}_y \quad \tilde{b}_x \quad \tilde{b}_y \quad \tilde{b}_x \quad \tilde{b}_y\}^t \tag{4.54}$$

It is interesting to note that the components of the equivalent force vector at the nodes 1, 2 and 3 are zeros.

Case 2. Equivalent Force Vector due to Surface Traction

The surface traction is assumed to be prescribed on side 2-3. There are two types for which the equivalent force expressions are derived. The first type deals with uniform traction and the second type deals with a linearly varying traction from node 2 to node 3 on the side 2-3. Both types of loading are shown in Figs. 4.17 and 4.18.

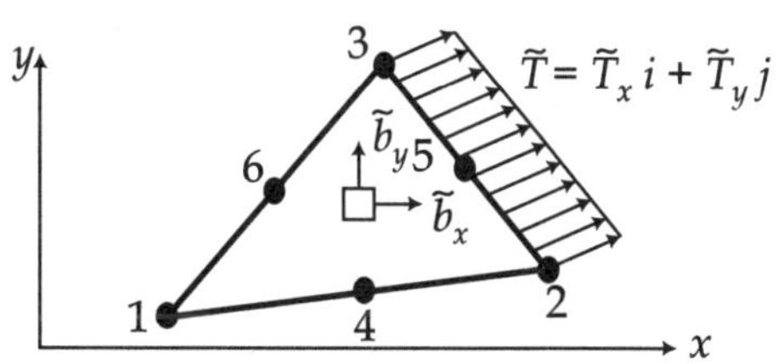

Fig. 4.17 Uniform Surface Traction on Side 2-5-3

Type 1: Uniform Traction on the Side 2-5-3

Here by definition, the work potential

$$\mathbf{V} = -t \int_{2-3} (\tilde{T}_x u + \tilde{T}_y v)\, dl \tag{4.55}$$

and differentiating the **V**-function with respect to the corresponding nodal displacements, the components of the force vector are:

$$\tilde{f}_{x2} = t\tilde{T}_x \int_{2-3} N_2 \, dl = \frac{t\,\tilde{T}_x}{6} l_{2\text{-}3} \tag{4.56}$$

$$= \tilde{f}_{x3}$$

and

$$\tilde{f}_{x5} = t\tilde{T}_x \int_{2-3} N_5 \, dl = \frac{4t\,\tilde{T}_x}{6} l_{2\text{-}3}$$

Similarly, the force vector in the y-direction is given as

$$\tilde{f}_{y2} = t\tilde{T}_y \int_{2-3} N_2 \, dl = \frac{t\,\tilde{T}_y}{6} l_{2-3} = \tilde{f}_{y3}$$

$$\tilde{f}_{y5} = t\tilde{T}_y \int_{2-3} N_5 \, dl = \frac{4t\,\tilde{T}_y}{6} l_{2-3} \tag{4.57}$$

Type 2: Surface Traction Linearly Varying from Nodes 2 to 3.

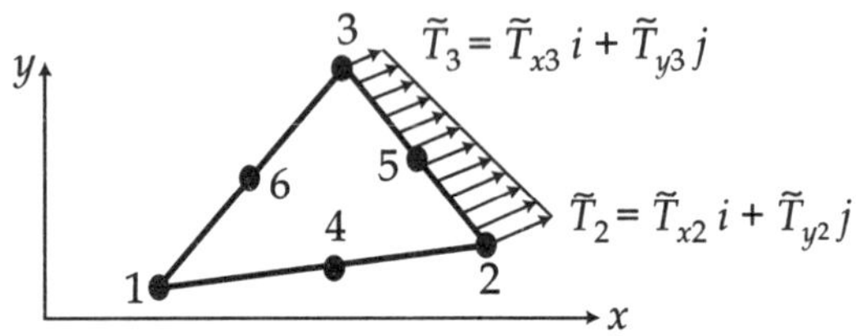

Fig. 4.18 Linearly Varying Surface Traction on Side 2-5-3

Here the surface traction varies from $\{\tilde{T}_{x2} \;\; \tilde{T}_{y2}\}^t$ at node 2 to $\{\tilde{T}_{x3} \;\; \tilde{T}_{y3}\}^t$ at node 3 linearly as shown in Fig. 4.18. Here the surface traction is interpolated using the area coordinates L_2 and L_3 respectively, i.e.,

$$\tilde{T}_x = L_2\tilde{T}_{x2} + L_3\tilde{T}_{x3}$$

and

$$\tilde{T}_y = L_2\tilde{T}_{y2} + L_3\tilde{T}_{y3} \tag{4.58}$$

Now

$$\tilde{f}_{x2} = t\int_{2-3} N_2\,(L_2\tilde{T}_{x2} + L_3\tilde{T}_{x3})\, dl \tag{4.59}$$

$$= t \int_{2-3} L_2(2L_2 - 1)\,(L_2\tilde{T}_{x2} + L_2\tilde{T}_{x3})\, dl$$

$$= \frac{1}{6}\, t\, l_{2-3}\tilde{T}_{x2}$$

Similarly,

$$\tilde{f}_{x3} = \frac{1}{6}\, t\, l_{2-3} \tilde{T}_{x3}$$

and

$$\tilde{f}_{x5} = t\int_{2-3} N_5 (L_2 \tilde{T}_{x2} + L_3 \tilde{T}_{x3})$$

$$= t\int_{2-3} 4L_2\, L_3\, (L_2 \tilde{T}_{x2} + L_3 \tilde{T}_{x3})\, dl$$

$$= \frac{1}{3}\, t\, l_{2-3}\, (\tilde{T}_{x2} + \tilde{T}_{x3})$$

Hence, part of the nodal force vector selected for the side 2-3 can be written as:

$$\begin{Bmatrix} \tilde{f}_{x2} \\ \tilde{f}_{y2} \\ \tilde{f}_{x3} \\ \tilde{f}_{y3} \\ \tilde{f}_{x5} \\ \tilde{f}_{y5} \end{Bmatrix} = \frac{t\, l_{2-3}}{6} \begin{Bmatrix} T_{x2} \\ T_{y2} \\ T_{x3} \\ T_{y3} \\ 2(T_{x2} + T_{x3}) \\ 2(T_{y2} + T_{y3}) \end{Bmatrix} \tag{4.60}$$

The rest of the procedures to be used in the finite element analysis is the same as discussed before in chapter three and hence not presented here.

In this chapter two types of triangular finite elements for two-dimensional elasticity problems are introduced: constant strain and linear strain. Both elements are very popular in solving two dimensional problems with complex geometry. With constant strain elements, a large number of elements are necessary for solids with complex geometry and high stress concentrations; the same problem can be solved with fewer number of elements using linear strain elements. However, the number of degrees of freedom and the half band width of the global stiffness matrix would increase. But such issues are not a major concern now-a-days since even desk top computers have large storage capacity. Most commercial finite element codes have big pre- and post- processors to create and massage the large data due to large number of nodes and elements in the domain.

A nice extension of the linear strain element is a curved triangular element with six degrees of freedom. But their derivation require numerical integration procedure and hence presented in the next chapter.

4.5 SUMMARY

Development of triangular finite elements is the beginning of the finite element method as it is known today. Assumed displacement functions, and shape functions using area coordinates, the steps needed to develop the element stiffness matrix and the equivalent prescribed load vector are presented. Techniques to compute the stiffness matrix for the higher order linear strain triangular elements with six nodes are also shown. Even though simple triangular elements are convenient, for most problems a large number of elements are necessary to get a satisfactory solution. The selection of the number of elements and mode of discretizing the domain depends on the type of the problem and the location of stress concentrations in the domain. A convergence study on the effect of the number of elements is recommended until the analyst have good experience to asses the possible associated errors in the use of finite element method. Use of linear strain triangular elements increases the accuracy of the solution, but increases the number of nodes and the half band width of the global stiffness matrix.

PROBLEMS

4.1 Calculate the integrals $\int x^2 y dA$ and $\int (x^2 + y^2) dA$ over the triangle as shown below by using area coordinates.

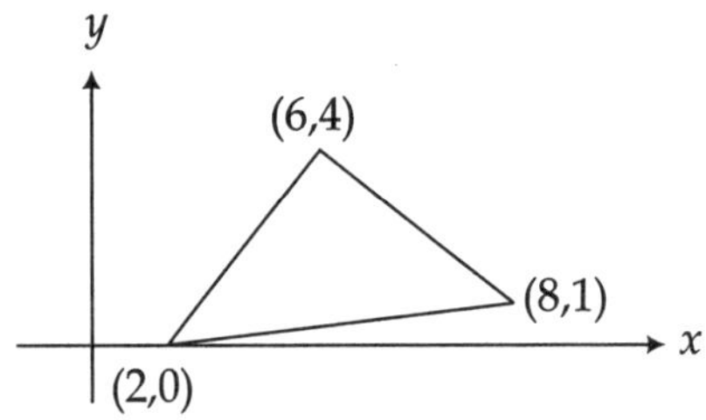

Fig. P4.1

4.2 Assume that the coordinates of nodes i, j and k for a constant strain triangle are given in the Fig. P4.2. Find the equivalent nodal loads due to a concentrated force $\widetilde{P}$ applied at point A.

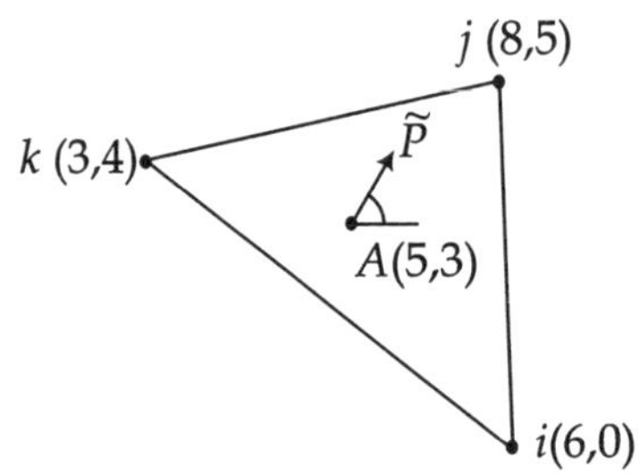

Fig. P4.2

4.3 Determine the element nodal force vector **Q** for distributed stress traction shown in the Figs. P4.3a and 4.3b.

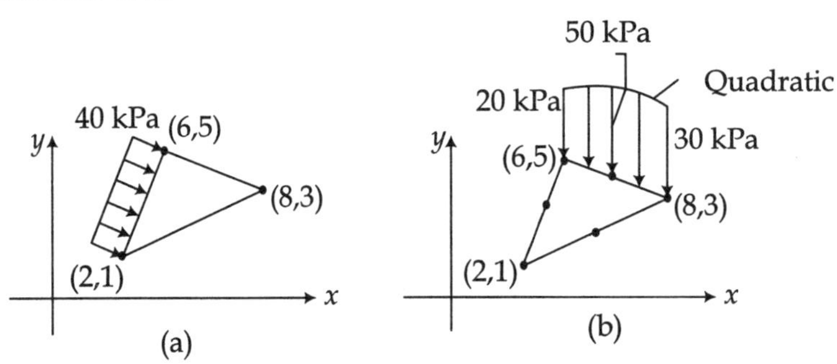

Fig. P4.3

4.4 Write the interpolation functions for the vertical displacement v along the boundary 123 of the element shown in Fig. P4.4, using the nodal displacements v_1, v_2, v_3. If the vertical load is uniform (force/unit length), determine the consistent loads at nodes 1,2 and 3.

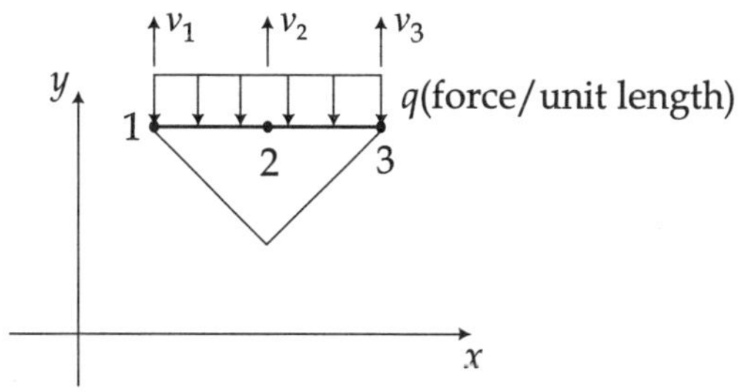

Fig. P4.4

4.5 A plane stress problem with two constant strain elements is given as shown below.

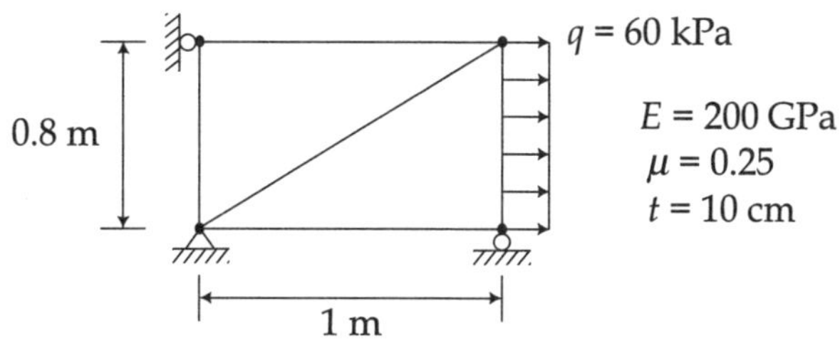

Fig. P4.5

i. Develop global stiffness matrix, K
ii. Solve for nodal displacements
iii. Solve for internal strains and stresses

4.6 Consider a plane stress problem with two linear strain elements as shown below in Fig. P4.6.

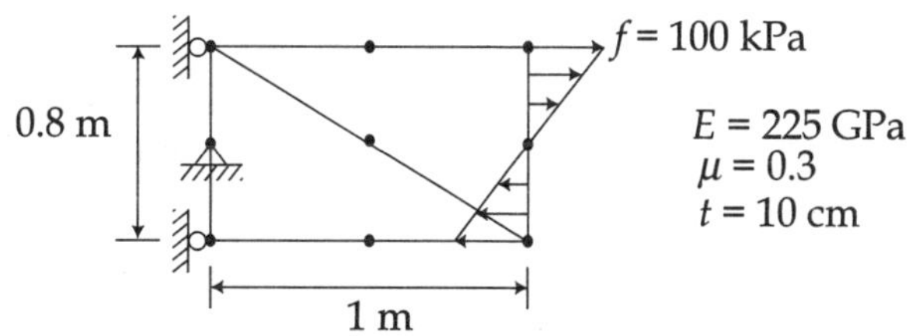

Fig. P4.6

i. Find the equivalent nodal loads
ii. Develop global stiffness matrix, K
iii. Solve for nodal displacements
iv. Solve for internal strains and stresses

4.7 A triangular element as shown has its thickness varying linearly from node to node and in the domain. Drive the element stiffness matrix for representing constant strain within the element for a plane stress problem.

$E = 200$ GPa

$\mu = 0.3$

$t_1 = 10$ cm

$t_2 = 5$ cm

$t_3 = 7$ cm

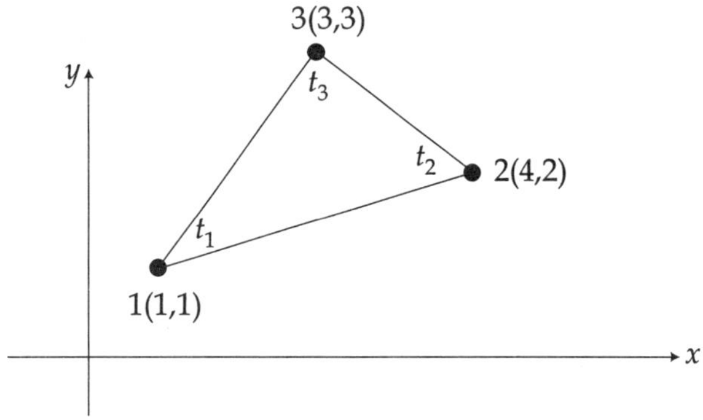

Fig. P4.7

4.8 Derive an element stiffness matrix for a constant strain triangular element having linearly varying modulus from node to node in the domain for plane stress condition as shown in Fig. P4.8.

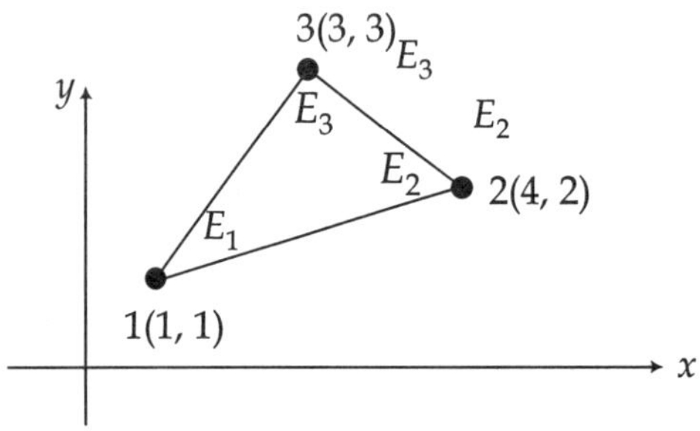

$E_1 = 200$ GPa, $E_2 = 100$ GPa, $E_3 = 150$ GPa, $\mu = 0.3$ and $t = 10$ cm.

Fig. P4.8

4.9 Construct a cubic polynomial displacement model for quadratic variation of strain inside a triangular element with 3 primary external node and one internal node. Use displacements and its derivatives at the primary node and displacement alone at the central node. Discuss how the completeness and compatibility requirements are satisfied.

4.10 A long plate with a hole of 10 cm radius is subjected to the uniform pressure $\sigma_0 = 100$ kPa at the sides vertical to the x-axis as shown below. Geometric and material properties are given on the figure. Analyze the portion of the plate with the constant strain and linear strain elements, as shown in the figure, by using an available finite element code. Plot the deformation graph and stress distribution along the line AB. Make your comments on the results.

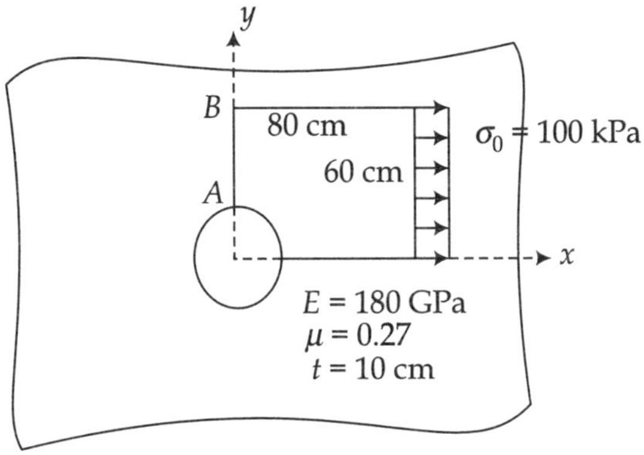

Fig. P4.9

4.11 A dam shown in the figure below is constructed for irrigation purpose. It is subjected to the water pressure on the upstream face. Calculate the displacements and critical stresses on the dam by using the linear strain elements. All necessary data is given on the figure. (Use available computer code.)

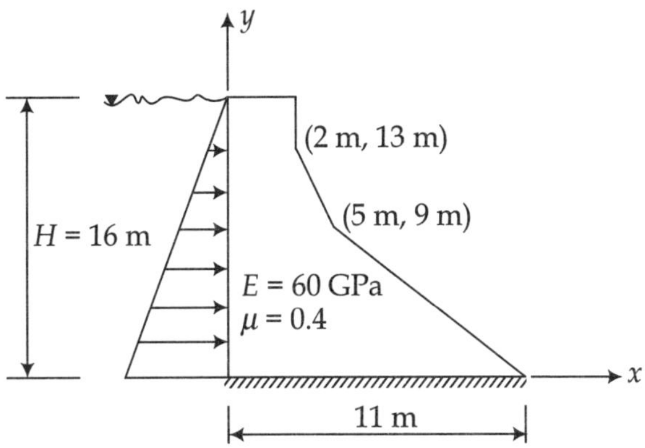

Fig. P4.10

REFERENCES

1. Bathe, K.J. and Wilson, E.L, "Numerical Methods in Finite Element Analysis", Prentice-Hall, Englewood Cliffs, New Jersey, 1976.
2. Chandrupatla, T.R. and Belegundu, A.D, "Introduction to Finite Elements in Engineering", Prentice Hall, New Jersey, 1997.
3. Clough, R.W, "The Finite Element in Plane Stress Analysis", Proc. 2nd ASCE Conference on Electronic Computation, Pittsburgh, Pa, Sept. 1960.
4. Desai, C.S. and Abel, J.F, "Introduction to Finite Element Method", Van Nostrand Reinhold, New York, 1972.
5. Krishnamoorthy, C.S, "Finite Element Analysis, Theory and Programming", Tata McGraw Hill Publishing Company, New Delhi, 1987.

6. Reddy, J.N, "An Introduction to the Finite Element Method",2nd Ed. McGraw Hill Book Company, New York, 1993.
7. Timoshenko, S.P. and Goodier, J.N, "Theory of Elasticity", McGraw Hill, New York, 1970.
8. Zienkiewicz, O.C. and Taylor, R.L, "The Finite Element Method", Vols. I & II, McGraw Hill Book Company, London, 1989.

5

CHAPTER

Two-Dimensional Problems - II: C_0-Quadrilateral and Curved Elements

5.1 INTRODUCTION

For solving two dimensional problems of elasticity, many finite element users prefer quadrilateral elements to constant triangular elements, since they yield slightly more accurate results for the same number of nodes with less number of elements in the discretization. Also they are quite convenient for finite element mesh generation and can be fitted to almost any geometry. For extremely complex geometry, quadrilateral elements can be combined with triangular elements in the overall discretization.

Compared to triangular and rectangular elements, the concept of quadrilateral element was formulated at a later time, since the development of the satisfactory displacement functions using x, y coordinates quadrilateral geometry was found to be very difficult. To circumvent this, researchers transformed the quadrilateral into a square and resulting equations were integrated using numerical quadrature formulae. To understand these in a simple manner, first the formulation of the stiffness matrix of a rectangular element is studied here.

5.2 RECTANGULAR FINITE ELEMENT

This element was developed[3,6,7,8] long before the formulation of the quadrilateral element and actually paved the way for the development of the quadrilateral element. Interestingly, the rectangular element is very useful in solving many stress problems involving laterally loaded shear walls, semi-infinite elastic continuum, layered medium as in the case of airport pavements, etc., i.e., wherever the domain can be discretized into rectangles. The steps for the development of the element stiffness matrix are as follows:

Step 1. Assume Displacement Functions

The simplest of these elements has four nodes and with two degrees of freedom per node; the total number of degrees of freedom becomes 8 per element. A local coordinate system is selected with origin at one corner of the rectangle as shown in Fig. 5.1.

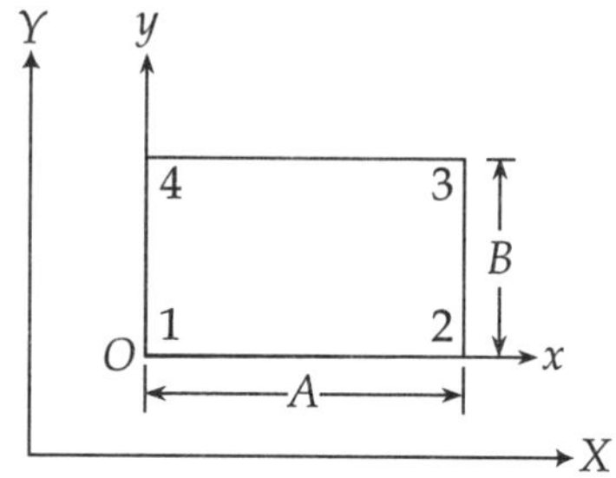

Fig. 5.1 A Rectangular Element with Origin of the Local Coordinates at a Corner

For the selection of a polynomial for the displacement functions, four parameters are chosen in each direction and revisiting the Pascal triangle given in Fig. 4.14, the assumed displacement functions are written as:

$$u = \alpha_1 + \alpha_2 x + \alpha_3 y + \alpha_4 xy$$
$$v = \alpha_5 + \alpha_6 x + \alpha_7 y + \alpha_8 xy \tag{5.1}$$

Here only one quadratic term xy is possible to make four parameters. In other words, if x^2 is selected, then y^2 also has to be selected to make it symmetric, and this will increase the number of functions that can be selected. The resulting functions given in Eq. (5.1) are bilinear in x and y. Just like the constant strain triangular element, these functions also allow the three rigid body motions by the coefficients α_1, α_5 and $\alpha_3 - \alpha_6$ and the three constant strains by α_2, α_7 and $\alpha_3 + \alpha_6$. Further examining the distributions of the strains, it can be found that the normal strain $\varepsilon_x = \alpha_2 + \alpha_4 y$, indicating that the normal strain is constant in x, but varies linearly in y within the element. Similarly, the normal strain $\varepsilon_y = \alpha_7 + \alpha_8 x$, is a constant in the y direction, but linearly varies in the x-direction. But, the shear strain $\gamma_{xy} = \alpha_3 + \alpha_4 x + \alpha_6 + \alpha_8 y$, varies linearly in both x and y directions. This is the reason why rectangular element formulation yields more accurate results compared to the constant strain triangular elements for same number of nodes, especially for problems where the shear distribution in the domain of the problem has relatively large gradients. At this point, it is necessary to examine whether the assumed displacement functions satisfy the interelement compatibility of displacements. Examining the variation of the u-displacement along the side 2-3 of the rectangular element shown in Fig. 5.1, it is seen that for $x = A$, the u- displacement is linear in y. Thus by keeping the two nodal displacements u_2 and u_3 compatible along the side 2-3 for the two adjacent elements, the interelement compatibility for u-displacement is satisfied. Same argument holds for the v-displacements and for all the other sides. It may be noted that if the other quadratic terms such as x^2 and y^2 had been included in the displacement functions, then the interelement displacement compatibility will not be satisfied for the same number of nodes in the element.

Former researchers[6, 7] have explicitly developed the element stiffness matrix for the rectangular element with origin of the coordinates as shown in the Fig. 5.1, using the four steps that are given in the earlier chapters. But the discussion is given here for an element where the origin of the axes is at the center of the element; a technique that could be used for the quadrilateral element, as shown in Fig. 5.2, with sides equal to $2a$ and $2b$ respectively.

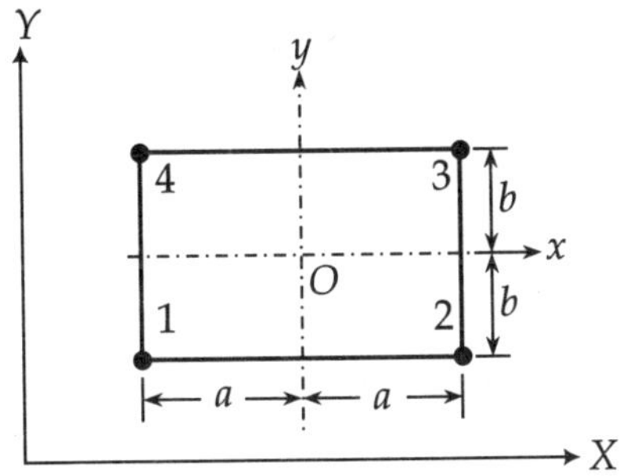

Fig. 5.2 A Rectangular Element with Origin of the Local Coordinates at the Center

Using the Lagrange interpolation formulae given in Eq. (3.53) along both axes, the shape functions for the four nodes of the rectangular element can be expressed as given below:

$$N_1 = \frac{1}{4ab}(a - x)(b - y)$$

$$N_2 = \frac{1}{4ab}(a + x)(b - y)$$

$$N_3 = \frac{1}{4ab}(a + x)(b + y)$$

$$N_4 = \frac{1}{4ab}(a - x)(b + y) \tag{5.2}$$

It is easy to verify that these shape functions have the properties that were given in the previous chapters.

Step 2. Internal Strains

Using the strain-displacement equations, the internal strain can be written as:

$$\{\varepsilon\} = \begin{Bmatrix} \varepsilon_x \\ \varepsilon_y \\ \gamma_{xy} \end{Bmatrix} = \begin{Bmatrix} \dfrac{\partial u}{\partial x} \\ \dfrac{\partial v}{\partial y} \\ \dfrac{\partial u}{\partial y} + \dfrac{\partial v}{\partial x} \end{Bmatrix}$$

$$= \frac{1}{4ab}\begin{bmatrix} -(b-y) & 0 & (b-y) & 0 & (b+y) & 0 & -(b+y) & 0 \\ 0 & -(a-x) & 0 & -(a+x) & 0 & (a+x) & 0 & (a-x) \\ -(a-x) & -(b-y) & -(a+x) & (b-y) & (a+x) & (b+y) & (a-x) & -(b+y) \end{bmatrix}\begin{Bmatrix} u_1 \\ v_1 \\ u_2 \\ v_2 \\ u_3 \\ v_3 \\ u_4 \\ v_4 \end{Bmatrix}$$

$$= [B]\{u\} \tag{5.3}$$

Step 3. Element Stiffness Matrix

Now, using the matrix D that relates the stress in terms of the internal strain in the element, the element stiffness matrix is:

$$[k] = \int_{-b}^{b}\int_{-a}^{a}[B]^t\,[D]\,[B]\,t\,dx\,dy$$

which can be explicitly integrated, if desired or it can be numerically integrated by a procedure discussed in the following section.

The equivalent force vector can be computed using the shape functions, and these values are found to be equal to simple lumping of the total loads for the case of uniform traction or body force. All the procedures for assembling the elements, forces, prescribing the displacements and the solution of the global equations to obtain the nodal displacements are the same as discussed before.

From the nodal displacement of each element, the calculation of internal strains are made using the B matrix as shown in Eq. (5.3) which is a function of x and y. As mentioned earlier, the internal strain ε_x varies linearly in y, while it is a constant in the x- direction. Similarly, the internal strain ε_y is a constant in the y-direction and varies linearly in the x-direction. The shear strain varies linearly in both x and y directions. Because of these variations within the element, it is a common practice to calculate the strains at the center of the element. The internal stresses are computed from the strains. Sometimes nodal stresses are computed for each element and then the average of the all nodal stresses in all the elements joining at each node are computed.

5.3 NUMERICAL INTEGRATION

Using the numerical integration, researchers have made many of the impossible calculations of the element stiffness matrix possible, many new sophisticated finite elements have been formulated. The quadrilateral element that is presented later in this chapter was made possible only because of this procedure. It is noteworthy to remember that early pioneers in the finite element research relied on closed-form integrations of the elements of the stiffness matrix, which resulted in long and tedious calculations. The rediscovery of numerical integration procedures[4] became a handy and interesting tool. There are many numerical integration procedures in the computational literature, but the Gauss quadrature is very well suited to the finite element formulation[1,2,9] because the assumed displacement functions are all polynomial functions.

5.3.1 Gauss Quadrature Formulae

Suppose a polynomial function $f(x)$ has to be integrated for $-1 \le x \le +1$. The integral is expressed as a sum of values of the function $f(x_i)$ at specified pivotal points called Gauss points such as x_i, for $i = 1,\ldots,n$ multiplied by corresponding weights, w_i, $i = 1,\ldots,n$, where n is the number of Gauss points. In other words, the integral,

$$I = \int_{-1}^{+1} f(x)\,dx = \sum_{i=1}^{n} w_i f(x_i) \tag{5.4}$$

Since there are n Gauss points and an equal number of weights, there are a total of $2n$ parameters involved in the procedure. A complete polynomial of the order of $(2n - 1)$ has $2n$ parameters, and by judiciously selecting these pivotal points and the weights, one can integrate this polynomial exactly. The Gauss points for a n-point integration formula are the roots of the Legendre polynomial of order n, and these Gauss points and the corresponding weights are given in the Table 5.1.

Table 5.1

No. of Points	*Gauss Points* x_i	*Weights* w_i
1	0.0	2.00000 00000 00000
2	± 0.57735 02691 89626	1.00000 00000 00000
3	± 0.77459 66692 41483	0.55555 55555 55555
	0.00000 00000 00000	0.88888 88888 88888
4	± 0.86113 63115 94053	0.34785 48451 37454
	± 0.33998 10435 84856	0.65214 51548 62546
5	± 0.90617 98459 38664	0.23692 68850 56189
	± 0.53846 93101 05683	0.47862 86704 99366
	0.00000 00000 00000	0.56888 88888 88889

Let the function to be integrated is a simple linear polynomial of order one, $f(x) = a_0 + a_1x$,

$$I = \int_{-1}^{+1} f(x)\,dx = \int_{-1}^{+1} (a_0 + a_1x)\,dx = 2a_0$$

In other words, no matter whatever be the value of a_1, the value of the integral is $2a_0$. If the Gauss point is taken as zero and the corresponding weight as 2, exact value of the integral is obtained by using a single Gauss point formula. For a two point integration formula, then there are four parameters, and thus a complete cubic polynomial can be integrated exactly. Here the pivotal points are $\pm 1/\sqrt{3}$ and the corresponding weights are both equal to 1.0.

$$\mathrm{I} = \int_{-1}^{+1} f(x)\,dx$$

$$= \int_{-1}^{+1} (a_0 + a_1x + a_2x^2 + a_3x^3)dx = 2a_0 + \frac{2}{3}a_2$$

The above expression is the exact value of the integral and is independent of the values of a_1 and a_3. If the integral is written as

$$I = w_1 f(x_1) + w_2 f(x_2)$$

and substituting the Gauss pivotal points x_1 and x_2 and the corresponding weights in the above expression, the integral becomes

$$I = 1\left[a_0 + a_1\left(\frac{-1}{\sqrt{3}}\right) + a_2\left(\frac{-1}{\sqrt{3}}\right)^2 + a_3\left(\frac{-1}{\sqrt{3}}\right)^3\right] + 1\left[a_0 + a_1\left(\frac{1}{\sqrt{3}}\right) + a_2\left(\frac{1}{\sqrt{3}}\right)^2 + a_3\left(\frac{1}{\sqrt{3}}\right)^3\right]$$

which will be equal to the exact value shown above. For integrating higher order polynomials, say a fifth order, a three point integration formula is necessary, and the use of a two point integration will result in some error depending upon the coefficients of the polynomial. Even though the technique is suited for integrating polynomials, they can also be used to integrate other functions as well. One has to consider the behavior of the function within the limits of integration. A convergence study is often very useful to establish the associated error in the final result. In the finite element procedure, one can always establish the order of polynomials to be integrated and thus, determine the exact number of pivotal points necessary for the integration. There are very few instances, such as in axisymmetric problems, singular functions may have to be integrated and sometimes even a reduced integration technique[9] is sufficient in such cases. Also there are other cases, where a reduced integration is used to integrate certain parts of the element stiffness matrix, and obtain a better and more accurate solution for certain finite element formulations used for plates and shell structures.

Now, in order to integrate a function $f(x)$ which has a domain varying from a to b, the variable x has to be transformed to ξ which varies from –1 to +1. From Eq. (3.56),

$$\xi = \frac{2(x-a)}{b-a} - 1, \text{ or } x = \frac{\xi(b-a)+(b+a)}{2}$$

and $$dx = \frac{b-a}{2}d\xi = |J|d\xi$$

Here $(b-a)/2$ is the Jacobian $|J|$ of the transformation of x to ξ. Goemetrically it is the ratio of the dx and $d\xi$. Now the following integral,

$$\int_a^b f(x)\,dx = \frac{b-a}{2}\int_{-1}^{+1} f\left(\frac{\xi(b-a)+(b+a)}{2}\right)d\xi$$

$$= \frac{b-a}{2}\sum_{i=1}^{n} w_i F(\xi_i) \tag{5.5}$$

Example Problem 5.1 A Numerical Example for Integration

Integrate: $$\int_1^2 \frac{dx}{x}$$

Since the limits are from 1 to 2, there is no singularity for the integrand between the limits, and thus it can be integrated exactly. The exact value of the integral is $\log_e 2 = 0.69314718$. Transforming $1 \le x \le 2$ to $-1 \le \xi \le +1$, using Eq. (3.56) or equation given above,

$$x = \frac{\xi + 3}{2} \text{ and } dx = \frac{d\xi}{2}$$

$$f(x) = \frac{1}{x} = \frac{2}{\xi + 3} = F(\xi) \text{ and therefore,}$$

$$\int_1^2 \frac{dx}{x} = \frac{1}{2}\int_{-1}^{1} \frac{2}{\xi + 3}\, d\xi = \sum_{i=1}^{n} w_i F(\xi_i)|J|$$

$$= 0.6666667 \text{ if } n = 1$$

$$= 0.6923076, \text{ if } n = 2$$

$$= 0.6931472, \text{ if } n = 5$$

This shows that even functions that are not polynomials can also be integrated using the Gauss formula. If the integrands are not polynomials, it is necessary to evaluate the behavior of the functions within the limits such as singularity, continuity and zero crossings and the integrals have to be verified for their convergence.

5.3.2 Integration in Two Dimensions

Extensions to integration of functions in two dimensions is very simple. For example, consider the following integral:

$$\int_{-1}^{+1}\int_{-1}^{+1} f(\xi, \eta)\, d\xi\, d\eta = \int_{-1}^{+1}\left[\sum_{i=1}^{n} w_i f(\xi_i, \eta)\right] d\eta$$

$$= \sum_{j=1}^{m} w_j \left[\sum_{i=1}^{n} w_i\, f(\xi_i, \eta_j)\right]$$

$$= \sum_{i=1}^{n}\sum_{j=1}^{m} w_i\, w_j\, f(\xi_i, \eta_j) \tag{5.6}$$

Here n, m are the number of Gauss points in the ξ *and* η directions respectively, in other words, different number of Gauss points can be used depending on the order of polynomials to be integrated in each direction.

An obvious extension of the procedure to integrations in three dimensions is self explanatory and the formula is given below.

$$\int_{-1}^{+1}\int_{-1}^{+1}\int_{-1}^{+1} f(\xi, \eta, \zeta)\, d\xi\, d\eta\, d\zeta = \sum_{i=1}^{l}\sum_{j=1}^{m}\sum_{k=1}^{n} w_i\, w_j\, w_k\, (\xi_i, \eta_j, \zeta_k) \tag{5.7}$$

Here l, m and n are the number of Gauss points required in each direction respectively.

Example Problem 5.2 A Numerical Example for Integration in 2-D

Integrate: $$\int_{y=1}^{5}\int_{x=2}^{6} 2x^3\, y^3\, dxdy$$

Solution:

Using Eq. (3.56),
$$x = \frac{\xi(6-2)+(6+2)}{2} = 2\xi + 4, \text{ and } dx = 2d\xi$$
$$y = \frac{\eta(5-1)+(5+1)}{2} = 2\eta + 3, \text{ and } dy = 2d\eta$$

$$\int_{y=1}^{5}\int_{x=2}^{6} 2x^3\, y^3\, dxdy = \int_{-1}^{1}\int_{-1}^{1} 2(2\xi+4)^3\,(2\eta+3)^2\,(2d\xi)\,(2d\eta)$$

$$= \sum_i\sum_j w_i\, w_j\, 8(2\xi+4)_i^3\,(2\eta+3)_j^2 \text{ for } i, j = 1, 2;\ \xi_i = \eta_i = \pm\frac{1}{\sqrt{3}},\ w_i = 1$$

$$= 26453.3$$

Since the polynomial was cubic, a two point integration yielded exact solution.

Example Problem 5.3 A Plane Stress Problem of a Shear Wall

An example of a simple classical plane stress problem of a shear wall shown in Fig. 5.3, is solved using both triangular and rectangular elements. See Figs. 5.4 (a) and (b). Shear walls are used in multistoried buildings to provide stiffness against lateral forces such as wind and earthquake forces. Majority of shear walls are constructed with rectangular shapes with holes inside to provide access from elevators, other rooms, stair cases etc. The wall is modeled using triangular and rectangular elements.[7] In the triangular finite element discretization of the example shear wall contained 1568 elements, while that using rectangular elements had 264

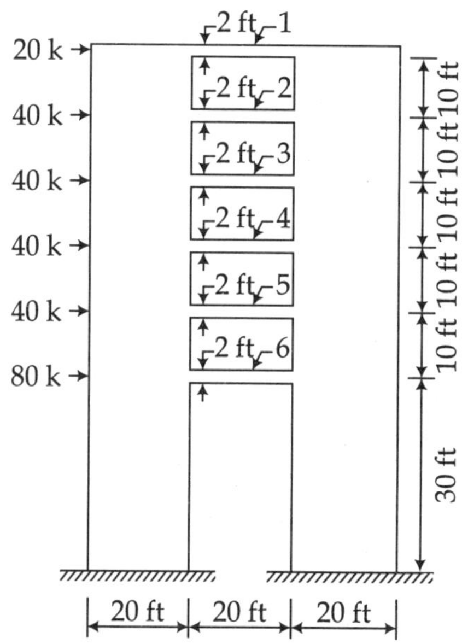

(1 ft = 0.3048 m, 1 k = 44.482 kN)

Fig. 5.3 Model Shear Wall Problem

elements. The distribution of vertical stresses at different storey level are shown in Fig. 5.5a and the overall displacement pattern of the entire shear wall is shown in Fig. 5.5b. The lateral displacement patterns of the left side of the wall obtained from using both triangular and rectangular elements are illustrated in Fig. 5.6. The results of the deflection of the left side of the wall from both idealizations are almost the same, indicating that rectangular elements yield good results even with less number of unknowns because rectangular elements are higher order elements compared to the three nodded triangular ones.

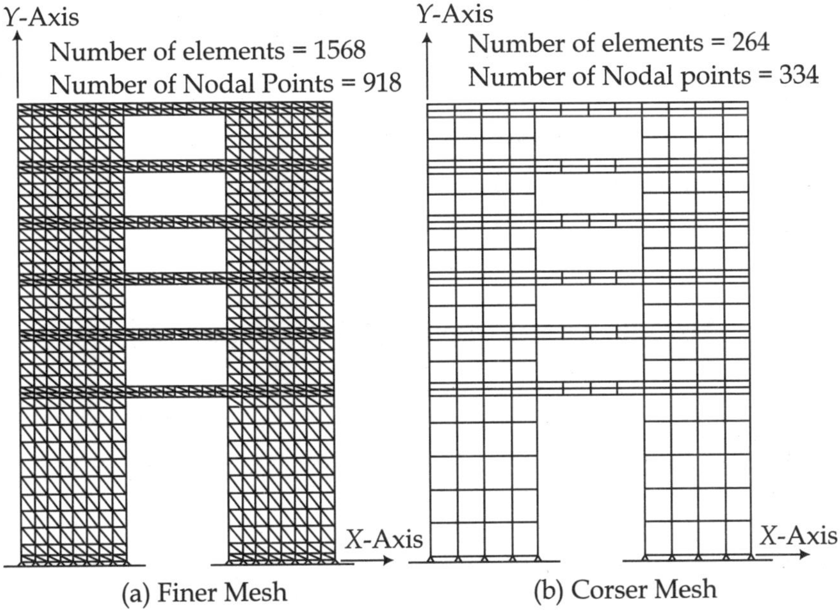

Fig. 5.4 Finite Element Representation

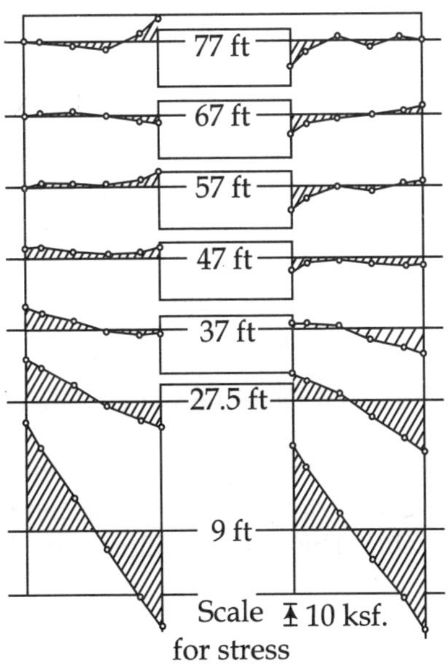

Fig. 5.5a Verical Stresses in Storey Levels

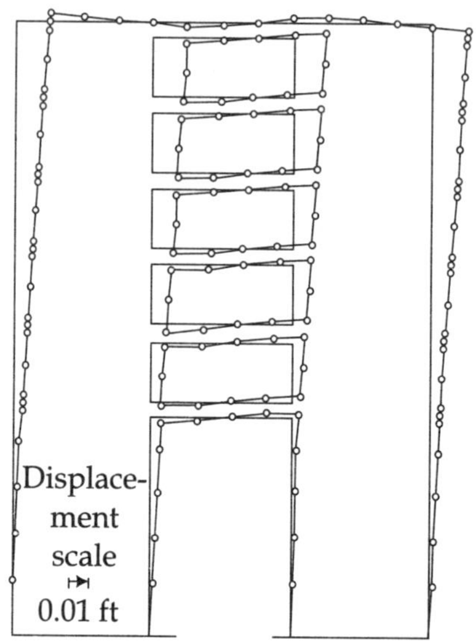

Fig. 5.5b Overall Displacement of the Shear Wall

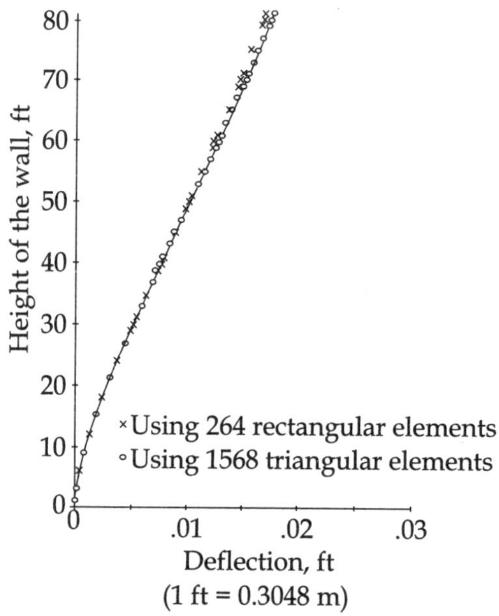

Fig. 5.6 Lateral Deflection - Left Side of Wall

If there is complicated geometry for the shear wall (plane stress condition) with curved openings, both elements can be used; triangular elements can be used to fill up the complex geometry where ever it is applicable. This example illustrates the power of the finite element method to solve a practical structural engineering problem. When shear walls have geometries with walls connected in other directions, the problem may no longer belong to the plane stress condition, in such instances a three dimensional finite element analysis including shell finite elements have to be used, and more accurate analysis is obtained. The sample shown here is just for a simple illustration of the use of triangular elements.

Example Problem 5.4 A Plane Strain Problem of a Long Footing

Another simple problem is that of a long continuous footing resting on a layered soil medium. The footing can be flexible or rigid and can be placed at any level as required in the analysis; in other words the depth of the footing can also be modeled. Problems of this type have no exact solutions in the literature. For a simple illustration here, the problem of a flexible footing resting on a semi-infinite continuum is selected so that the results obtained can be compared to the available exact solutions. Semi-infinite medium can be modeled only as a finite medium with depths and widths taken sufficiently away from the loaded region. These dimensions are taken based on some experience on similar stress analysis problems. The problem and the finite element idealization are shown in Figs. 5.7 and 5.8. The vertical stresses σ_z (z-axis is taken as vertical) is compared to the Boussinesq solution with semi-infinite medium.[3] This solution is independent of the value of the Poisson's ratio of the medium. But in the finite element model, one has to input the value of the Poisson's ratio. Results for different values of Poisson's ratio are computed at a depth of 2.0 feet below the surface and plotted and compared with the closed form solution. Good comparison for stresses are obtained even though in the finite element model the depth and the width of the soil layer is limited.

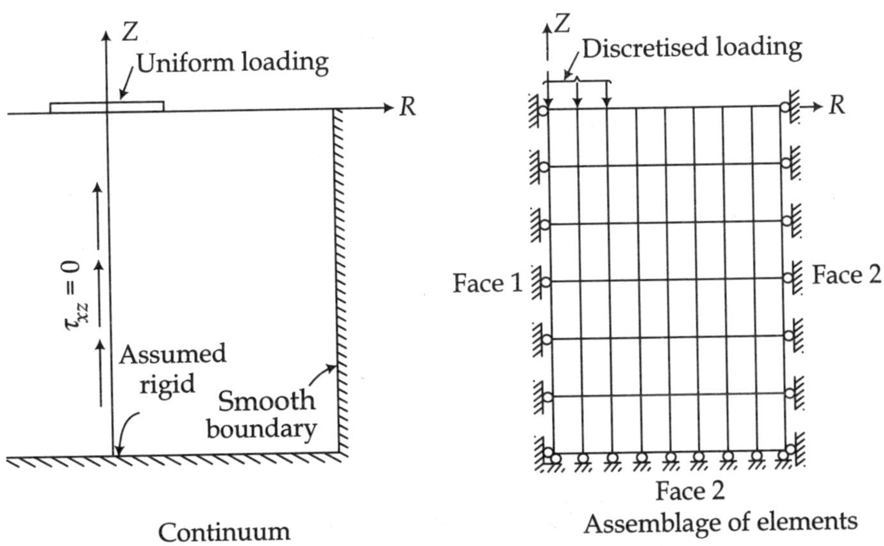

Fig. 5.7 Assumed and Specified Boundary Conditions

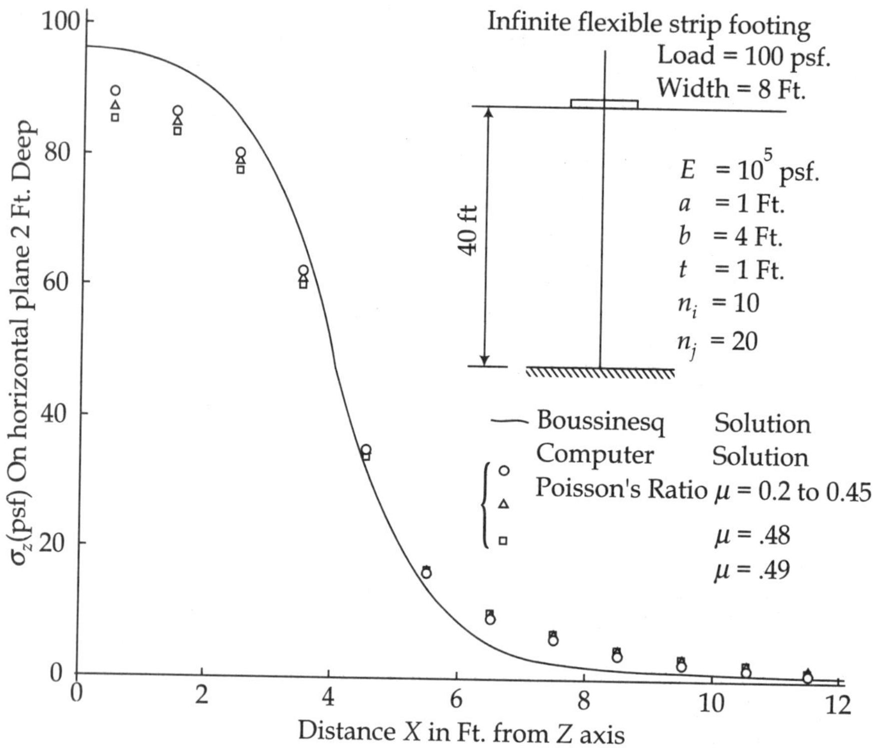

Fig. 5.8 Distribution of σ_z in a Horizontal Plane at a Depth of 2 ft.

If the footing is more rigid than the soil medium as usually is the case, the stiffness of the footing can be incorporated into the model very easily. Also if the footing is resting below the surface inside the soil medium, and if the soil properties vary within the medium, such geometries and variable material properties can be easily incorporated in the finite element model.

Other practical application of finite element model is to analyze long dams and navigation lock structures. In a navigation lock structure there is water inside and the walls support soil on the outside. This is a classical example of a practical plane strain problem where the behavior of the structure clearly depends on the interaction between the structure and the soil surrounding it. The complex geometry of the structure and the variable soil properties as it exists in different layers, etc. can be modeled relatively easily.

5.4 QUADRILATERAL FINITE ELEMENT

Step 1. Assumed Displacement Functions

A four noded quadrilateral[2,5,6] is shown in Fig. 5.9(a). Connecting the midpoints of the opposite sides, a new coordinate system (ξ, η) is developed as shown in Fig. 5.9(b), such that the quadrilateral element could be transformed into a square with side equal to 2, and the corresponding coordinates vary from –1 to +1 as shown in the Fig. 5.9(b). Noting that substituting $\xi = \dfrac{x}{a}$ and $\eta = \dfrac{y}{b}$ in the Eq. (5.2), the new shape functions for the quadrilateral element in terms of the ξ and η coordinates, can be written as:

$$N_1 = \frac{1}{4}(1 - \xi)(1 - \eta)$$

$$N_2 = \frac{1}{4}(1 + \xi)(1 - \eta)$$

$$N_3 = \frac{1}{4}(1 + \xi)(1 + \eta)$$

$$N_4 = \frac{1}{4}(1 - \xi)(1 + \eta) \tag{5.9}$$

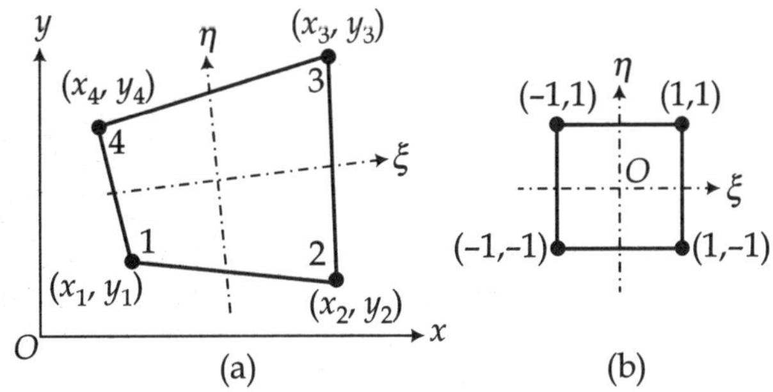

Fig. 5.9

These functions can be written in a compact form

$$N_i = \frac{1}{4}(1 + \xi_i\xi)(1 - \eta_i\eta)$$

where $\xi_i \Rightarrow (-1\ \ 1\ \ 1\ \ -1)$ and $\eta_i \Rightarrow (-1\ \ -1\ \ 1\ \ 1)$ for $i = 1, 2, 3, 4$.

This kind of representation can be useful in computer programming. The displacement functions are therefore interpolated as:

$$u = \sum_{i=1}^{4} N_i(\xi, \eta)\, u_i \text{ and } v = \sum_{i=1}^{4} N_i(\xi, \eta)\, v_i \tag{5.10}$$

Step 2. Internal Strains

In order to obtain the components of strains, the displacement functions u and v (which are functions of ξ and η) have to be differentiated with respect to x and y. Thus, the relationships between (x, y) coordinate system with the corresponding (ξ, η) coordinate system have to be determined. Observing that the coordinates x and y within the element can also be interpolated using the same shape functions given in Eq. (5.8),

$$x = \sum_{i=1}^{4} N_i(\xi, \eta)\, x_i$$

$$y = \sum_{i=1}^{4} N_i(\xi, \eta) y_i \tag{5.11}$$

Because the shape functions for the u, v displacements and x, y coordinates are the same, this quadrilateral element is called an 'isoparametric element'. Expressions for the internal strain components are obtained by differentiating the u and v functions with respect to x and y. Using the chain rule in partial differentiation, a Jacobian for the transformation is developed.

5.4.1 The Jacobian of Transformation

Since ξ and η are also functions x and y, using the chain rule for partial differentiation,

$$\frac{\partial}{\partial \xi} = \frac{\partial}{\partial x}\frac{\partial x}{\partial \xi} + \frac{\partial}{\partial y}\frac{\partial y}{\partial \xi}$$

$$\frac{\partial}{\partial \eta} = \frac{\partial}{\partial x}\frac{\partial x}{\partial \eta} + \frac{\partial}{\partial y}\frac{\partial y}{\partial \eta}$$

In other words,

$$\begin{Bmatrix} \dfrac{\partial}{\partial \xi} \\ \dfrac{\partial}{\partial \eta} \end{Bmatrix} = \begin{bmatrix} \dfrac{\partial x}{\partial \xi} & \dfrac{\partial y}{\partial \xi} \\ \dfrac{\partial x}{\partial \eta} & \dfrac{\partial y}{\partial \eta} \end{bmatrix} \begin{Bmatrix} \dfrac{\partial}{\partial x} \\ \dfrac{\partial}{\partial y} \end{Bmatrix} = [J] \begin{Bmatrix} \dfrac{\partial}{\partial x} \\ \dfrac{\partial}{\partial y} \end{Bmatrix}$$

or

$$\begin{Bmatrix} \dfrac{\partial}{\partial x} \\ \dfrac{\partial}{\partial y} \end{Bmatrix} = [J]^{-1} \begin{Bmatrix} \dfrac{\partial}{\partial \xi} \\ \dfrac{\partial}{\partial \eta} \end{Bmatrix} \tag{5.12}$$

The elements of the Jacobian matrix $[J]$ are:

$$J_{11} = \frac{\partial x}{\partial \xi} = \sum_{i=1}^{4} \frac{\partial N_i}{\partial \xi} x_i,$$

$$J_{12} = \frac{\partial y}{\partial \xi} = \sum_{i=1}^{4} \frac{\partial N_i}{\partial \xi} y_i$$

$$J_{21} = \frac{\partial x}{\partial \eta} = \sum_{i=1}^{4} \frac{\partial N_i}{\partial \eta} x_i \quad \text{and}$$

$$J_{22} = \frac{\partial y}{\partial \eta} = \sum_{i=1}^{4} \frac{\partial N_i}{\partial \eta} y_i \tag{5.13}$$

i.e.,
$$[J] = \begin{bmatrix} N_{1,\xi} & N_{2,\xi} & N_{3,\xi} & N_{4,\xi} \\ N_{1,\eta} & N_{2,\eta} & N_{3,\eta} & N_{4,\eta} \end{bmatrix} \begin{bmatrix} x_1 & y_1 \\ x_2 & y_2 \\ x_3 & y_3 \\ x_4 & y_4 \end{bmatrix} \quad (5.14)$$

where the comma in the subscripts denotes differentiation with respect to the variable following the comma, and therefore

$$\begin{aligned} N_{1,\xi} &= -(1-\eta)/4, & N_{2,\xi} &= (1-\eta)/4, \\ N_{3,\xi} &= (1+\eta)/4, & N_{4,\xi} &= -(1+\eta)/4, \\ N_{1,\eta} &= -(1-\xi)/4, & N_{2,\eta} &= -(1+\xi)/4, \\ N_{3,\eta} &= (1+\xi)/4, & N_{4,\eta} &= (1-\xi)/4, \end{aligned} \quad (5.15)$$

Now the inverse of the $[J]$ matrix can be written as:

$$[J]^{-1} = \frac{1}{\det|J|}\begin{bmatrix} J_{22} & -J_{12} \\ -J_{21} & J_{11} \end{bmatrix} = \begin{bmatrix} J^*_{11} & J^*_{12} \\ J^*_{21} & J^*_{22} \end{bmatrix} \quad (5.16)$$

where $\det|J| = J_{11}J_{22} - J_{12}J_{21}$ and the determinant of the Jacobian is the ratio of the product of the differentials in the (x, y) and (ξ, η) coordinate system. In other words,

$$dxdy = \det|J|\,d\xi\,d\eta \quad (5.17)$$

The terms J^*_{ij} are the components of the inverse of the Jacobian matrix.

Example Problem 5.5

Calculate the following for: (a) the rectangle (b) the quadrilateral shown in Figs. 5.10(a) and (b)

1. the Jacobian transformation matrix $[J]$ and its determinant,
2. integrate $\iint_A dxdy$ and $\iint_A xdxdy$.

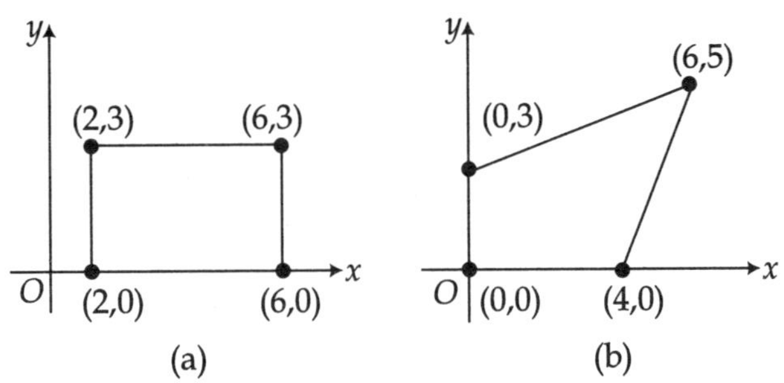

Fig. 5.10

Solution:

(a) for the rectangle:

Using Eq. (5.14), the Jacobain matrix is,

1. $$[J] = \begin{bmatrix} N_{1,\xi} & N_{2,\xi} & N_{3,\xi} & N_{4,\xi} \\ N_{1,\eta} & N_{2,\eta} & N_{3,\eta} & N_{4,\eta} \end{bmatrix} \begin{bmatrix} x_1 & y_1 \\ x_2 & y_2 \\ x_3 & y_3 \\ x_4 & y_4 \end{bmatrix}$$

$$= \begin{bmatrix} \frac{-(1-\eta)}{4} & \frac{(1-\eta)}{4} & \frac{(1+\eta)}{4} & \frac{-(1+\eta)}{4} \\ \frac{-(1-\xi)}{4} & \frac{-(1+\xi)}{4} & \frac{(1+\xi)}{4} & \frac{(1-\xi)}{4} \end{bmatrix} \begin{bmatrix} 2 & 0 \\ 6 & 0 \\ 6 & 3 \\ 2 & 3 \end{bmatrix}$$

$$= \frac{1}{16}\begin{bmatrix} 8 & 0 \\ 0 & 6 \end{bmatrix}$$

Therfore the Determinant of the Jacobian Matrix = $|J| = 3.0$

(This is the ratio of the area of the rectangle in x, y axes and the transformed rectangle in ξ, η axes)

2. Using one point integration, at $\xi = \eta = 0$, $w = 2$,

$$\iint_A dxdy = \int_{-1}^{+1}\int_{-1}^{+1} |J| d\xi \, d\eta$$

$$= \sum_i \sum_j w_i w_j |J|_{i,j} = (2)(2)(3)$$

$$= 12.0$$

$$\iint_A x \, dxdy = \int_{-1}^{+1}\int_{-1}^{+1}\sum_{i=1}^{4} N_i x_i |J| d\xi d\eta$$

$$= \sum_i \sum_j W_i W_j (N_1 x_1 + N_2 x_2 + N_3 x_3 + N_4 x_4)_{i,j}, \text{ at } \xi = n = 0$$

$$= (2)\ (2)\left(\frac{1}{4}\right)(2 + 6 + 6 + 2)\ (3)$$

$$= 48.0$$

Both the integrals yield exact values for one point integration in this case a.

(b) for the quadrilateral.

Using Eq. (5.14), the matrix for the Jacobian is,

1. $$[J] = \begin{bmatrix} \frac{-(1-\eta)}{4} & \frac{(1-\eta)}{4} & \frac{(1+\eta)}{4} & \frac{-(1+\eta)}{4} \\ \frac{-(1-\xi)}{4} & \frac{-(1+\xi)}{4} & \frac{(1+\xi)}{4} & \frac{(1-\xi)}{4} \end{bmatrix} \begin{bmatrix} 0 & 0 \\ 4 & 0 \\ 6 & 5 \\ 0 & 3 \end{bmatrix}$$

$$= \frac{1}{16}\begin{bmatrix} (10+2\eta) & (2+2\eta) \\ (2+2\xi) & (8+2\xi) \end{bmatrix}$$

Jacobian $[J] = \frac{1}{4}[(5 + \eta)\ (4 + \xi) - (1 + \eta)\ (1 + \xi)]$.

2. Using one point integration, at $\xi = \eta = 0$, $w = 2$,

$$\iint_A dxdy = \int_{-1}^{+1}\int_{-1}^{+1} |J| d\xi\, d\eta$$

$$= \sum_i \sum_j w_i w_j |J|_{i,j} = (2)(2)\left(\frac{19}{4}\right)$$

$$= 19.0$$

which is the exact area of the quadrilateral. Now for the second integral, using one point integration, at $\xi = \eta = 0$, $w = 2$,

$$\iint_A x\; dxdy = \int_{-1}^{+1}\int_{-1}^{+1} \sum_{i=1}^{4} N_i x_i |J| d\xi\, d\eta$$

$$= \sum_{i=1}\sum_{j=1} w_i w_j \,(N_1 x_1 + N_2 x_2 + N_3 x_3 + N_4 x_4\,)_{i,j}, \quad \text{at } \xi = \eta = 0$$

$$= (2)(2)\left(\frac{1}{4}\right)(0 + 4 + 6 + 0)\left(\frac{19}{4}\right)$$

$$= 47.5$$

With two point integration, $\xi_i = \eta_i = \pm \dfrac{1}{\sqrt{3}}$, and $w_i = 1.0$, for $i = 1, 2$, there are four points within the transformed square,

$$\iint_A x\; dxdy = \int_{-1}^{+1}\int_{-1}^{+1} \sum_{i=1}^{4} N_i x_i |J| d\xi\, d\eta$$

$$= \sum_{i=1}^{2}\sum_{j=1}^{2}\left(w_i w_j \sum_{k=1}^{4} N_k x_k |J| \right)_{i,j}$$

$$= 51.33$$

This is the exact value of the integral.

Step 2. Strain-displacement Equations

The internal strains are written as derivatives of the displacement vector using Eq. (2.2),

$$\{\varepsilon\} = \begin{Bmatrix} \varepsilon_x \\ \varepsilon_y \\ \gamma_{xy} \end{Bmatrix} = \begin{Bmatrix} \dfrac{\partial u}{\partial x} \\ \dfrac{\partial v}{\partial y} \\ \dfrac{\partial u}{\partial y} + \dfrac{\partial v}{\partial x} \end{Bmatrix} = \begin{bmatrix} 1 & 0 & 0 & 0 \\ 0 & 0 & 0 & 1 \\ 0 & 1 & 1 & 0 \end{bmatrix} \begin{Bmatrix} \dfrac{\partial u}{\partial x} \\ \dfrac{\partial u}{\partial y} \\ \dfrac{\partial v}{\partial x} \\ \dfrac{\partial v}{\partial y} \end{Bmatrix} \tag{5.18}$$

and the vector on the extreme right end of the above equation can be written as:

$$\begin{Bmatrix} \dfrac{\partial u}{\partial x} \\ \dfrac{\partial u}{\partial y} \\ \dfrac{\partial v}{\partial x} \\ \dfrac{\partial v}{\partial y} \end{Bmatrix} = \begin{bmatrix} J^{-1} & 0 \\ 0 & J^{-1} \end{bmatrix} \begin{Bmatrix} \dfrac{\partial u}{\partial \xi} \\ \dfrac{\partial u}{\partial \eta} \\ \dfrac{\partial v}{\partial \xi} \\ \dfrac{\partial v}{\partial \eta} \end{Bmatrix} \tag{5.19}$$

Combining Eqs. (5.12), (5.13), (5.14) and (5.16), the internal strain vector is written as

$$\{\varepsilon\} = \begin{Bmatrix} \varepsilon_x \\ \varepsilon_y \\ \gamma_{xy} \end{Bmatrix} \begin{bmatrix} J^*_{11} & J^*_{12} & 0 & 0 \\ 0 & 0 & J^*_{21} & J^*_{22} \\ J^*_{21} & J^*_{22} & J^*_{11} & J^*_{12} \end{bmatrix} \begin{Bmatrix} \dfrac{\partial u}{\partial \xi} \\ \dfrac{\partial u}{\partial \eta} \\ \dfrac{\partial v}{\partial \xi} \\ \dfrac{\partial v}{\partial \eta} \end{Bmatrix} = [A] \begin{Bmatrix} \dfrac{\partial u}{\partial \xi} \\ \dfrac{\partial u}{\partial \eta} \\ \dfrac{\partial v}{\partial \xi} \\ \dfrac{\partial v}{\partial \eta} \end{Bmatrix} \tag{5.20}$$

Now the vector on the right end of the above equation is expressed in terms of nodal displacements as,

$$\begin{Bmatrix} \dfrac{\partial u}{\partial \xi} \\ \dfrac{\partial u}{\partial \eta} \\ \dfrac{\partial v}{\partial \xi} \\ \dfrac{\partial v}{\partial \eta} \end{Bmatrix} = \begin{bmatrix} N_{1,\xi} & 0 & N_{2,\xi} & 0 & N_{3,\xi} & 0 & N_{4,\xi} & 0 \\ N_{1,\eta} & 0 & N_{2,\eta} & 0 & N_{3,\eta} & 0 & N_{4,\eta} & 0 \\ 0 & N_{1,\xi} & 0 & N_{2,\xi} & 0 & N_{3,\xi} & 0 & N_{4,\xi} \\ 0 & N_{1,\eta} & 0 & N_{2,\eta} & 0 & N_{3,\eta} & 0 & N_{4,\eta} \end{bmatrix} \begin{Bmatrix} u_1 \\ v_1 \\ u_2 \\ v_2 \\ u_3 \\ v_3 \\ u_4 \\ v_4 \end{Bmatrix}$$

$$= [G]\,\{u\} \tag{5.21}$$

Since $\{\varepsilon\} = [B]\{u\}$, and using Eqs. (5.20) and (5.21), the matrix $[B]$ can be expressed as:

$$[B] = [A][G] \tag{5.22}$$

$$\begin{bmatrix} J^*_{11} N_{1,\xi} + J^*_{12} N_{1,\eta} & 0 & J^*_{11} N_{2,\xi} + J^*_{12} N_{2,\eta} & 0 & J^*_{11} N_{3,\xi} + J^*_{12} N_{3,\eta} & 0 & J^*_{11} N_{4,\xi} + J^*_{12} N_{4,\eta} & 0 \\ 0 & J^*_{21} N_{1,\xi} + J^*_{22} N_{1,\eta} & 0 & J^*_{21} N_{2,\xi} + J^*_{22} N_{2,\eta} & 0 & J^*_{21} N_{3,\xi} + J^*_{22} N_{3,\eta} & 0 & J^*_{21} N_{4,\xi} + J^*_{22} N_{4,\eta} \\ J^*_{21} N_{1,\xi} + J^*_{22} N_{1,\eta} & J^*_{11} N_{1,\xi} + J^*_{12} N_{1,\eta} & J^*_{21} N_{2,\xi} + J^*_{22} N_{2,\eta} & J^*_{11} N_{2,\xi} + J^*_{12} N_{2,\eta} & J^*_{21} N_{3,\xi} + J^*_{22} N_{3,\eta} & J^*_{11} N_{3,\xi} + J^*_{12} N_{3,\eta} & J^*_{21} N_{4,\xi} + J^*_{22} N_{4,\eta} & J^*_{11} N_{4,\xi} + J^*_{12} N_{4,\eta} \end{bmatrix}$$

Step 3 and 4.

Using the stress-strain matrix and the B matrix, the element stiffness matrix now is written as:

$$[k] = \iint_A [B]^t\,[D]\,[B]\;dx\;dy$$

$$= \int_{-1}^{+1}\int_{-1}^{+1} [B]^t\,[D]\,[B]\;\det|J|\;d\xi\,d\eta$$

$$= \sum_{i=1}^{n}\sum_{j=1}^{n} w_i w_j \left([B]^t\,[D]\,[B]\;\det|J| \right)_{i,j} \tag{5.23}$$

Since the polynomials in both directions are of the same order, equal number of integration points is used for integration.

5.4.2 Subroutines for the Quadrilateral Element Stiffness Matrix

A sample subroutine called QUAD for the computation of a quadrilateral finite element is given below. The subroutine QUAD calls another subroutine BMAT which computes the coefficients of the B matrix at each Gauss point. Both subroutines are given here. These codes are only to help beginners to understand how the finite element program using quadrilateral elements is developed. The variables employed within the codes are self explanatory.

```
      SUBROUTINE QUAD (NN,SM,NOD,X,Y,D,THICK)
      DIMENSION D(3,3),B(3,8),XE(4),YE(4),SM(8,8),
     *NOD(100,4),X(100),Y(100)
      DIMENSION XG(2),YG(2),WGT(2),DB(3)
      DATA XG/ -0.5773502691,0.5773502691
      DATA YG/ -0.5773502691,0.5773502691
      DATA WGT/1.0,1.0
      DO 10 I = 1,8
      DO 10 J = 1,8
   10 SM(I,J) = 0.0
```

```
      DO 15 I = 1,4
      NE = NOD(NN,I)
      XE(I) = X(NE)
   15 YE(I) = Y(NE)
      DO 100 IG = 1,2
      XP = XG(IG)
      DO 100 JG = 1,2
      YP = YG(JG)
      CALL BMAT(XE,YE,XP,YP,B,DJAC)
      WT = WGT(IG)*WGT(JG) * THICK * DJAC
      DO 100 J = 1,8
      DO 60 K = 1,3
      DB(K) = .0
      DO 60 L = 1,3
   60 DB(K) = DB(K) + D(K,L)*B(L,J)
      DO 70 I=J,8
      SUM=0.
      DO 65 L=1,3
      65 SUM = SUM + B(L,I)*DB(L)
   70 SM(I,J) = SM(I,J) + SUM * WT
  100 CONTINUE
      DO 110 J = 1,8
      DO 110 I = J,8
  110 SM(J,I) = SM(I,J)
      RETURN
      END

      SUBROUTINE BMAT(XE,YE,XP,YP,B,DJAC)
      DIMENSION DS(2,4),B(3,8),XE(4),YE(4),XT(4),ET(4)
      DATA XT/-1,1,1,-1
      DATA ET/ -1,-1,1,1
      DO 10 I = 1,4
      DS(1,I) = XT(I)*(1+ET(I)*YP)/4.0
   10 DS(2,I) = ET(I)*(1+XT(I)*XP)/4.0
      DO 20 I = 1,2
      SUM1 = 0
      SUM2 = 0
      DO 15 K = 1,4
```

```
      SUM1 = SUM1 + DS(I,K)*XE(K)
   15 SUM2 = SUM2 + DS(I,K)*YE(K)
      AJ(I,1) = SUM1
   20 AJ(I,2) = SUM2
      DJAC = AJ(1,1)*AJ(2,2) - AJ(1,2)*AJ(2,1)
      TEMP = AJ(1,1)
      AJ(1,1) = AJ(2,2)/DJAC
      AJ(1,2) = -AJ(1,2)/DJAC
      AJ(2,1) = -AJ(2,1)/DJAC
      AJ(2,2) = TEMP/DJAC
      DO 30 I = 1,4
      J = 2* I
      K = J-1
      B(1,K) = AJ(1,1)*DS(1,I) + AJ(1,2)*DS(2,I)
      B(2,J) = AJ(2,1)*DS(1,I) + AJ(2,2)*DS(2,I)
      B(3,K) = B(2,J)
   30 B(3,J) = B(1,K)
      RETURN
      END
```

5.4.3 Equivalent Force Vector

As in the case of the plane elasticity problems using triangular elements, the equivalent nodal forces due to surface traction and body forces have to be determined for quadrilateral elements also. Prescribed nodal forces are directly read into the global force vector, while the distributed forces have to be transformed into equivalent nodal forces using the minimum potential energy principle. Two simple cases of the distributed forces are described here.

Case 1. Uniform Surface Traction on Side 2-3 (from node 2 to 3)

A surface traction $\tilde{T}_x$ is prescribed in the x-direction on side 2-3.

Since the work potential function

$$\mathbf{V} = -\int_{2-3} \tilde{T}_x u\, dS = -\int_{2-3} \tilde{T}_x \sum_i N_i u_i\, dS \tag{5.24}$$

and therefore the equivalent force in the x-direction at node 2 is

$$f_{x2} = -\frac{\partial \mathbf{V}}{\partial u_2} = \int_{2-3} \tilde{T}_x N_2\big|_{\xi=1} dS$$

$$= \int_{2-3} \tilde{T}_x \frac{1}{2}(1-\eta)\, dS \tag{5.25}$$

Similarly,

$$f_{x3} = \int_{2-3} \tilde{T}_x \frac{1}{2}(1 + \eta)\, dS \tag{5.26}$$

Observing that $\tilde{T}_x$ is a constant and the functions $(1 - \eta)/2$ and $(1 + \eta)/2$ are linear interpolation functions along the side 2-3, the components of the force vector can be integrated as

$$f_{x2} = f_{x3} = \frac{\tilde{T}_x}{2}(l_{2-3})\, t \tag{5.27}$$

The components of the equivalent force vector $\tilde{T}_y$ in the y-direction on side 2-3 can be expressed in the same way as

$$f_{y2} = f_{y3} = \frac{\tilde{T}_y}{2}(l_{2-3})\, t \tag{5.28}$$

where $\tilde{T}_y$ is the component of the prescribed traction in the y-direction on side 2-3.

Case 2. Prescribed Body Force per unit volume

For this case, the work potential function

$$\mathbf{V} = -\int_{vol} \{\tilde{b}_x \quad \tilde{b}_y\}^t \begin{Bmatrix} u \\ v \end{Bmatrix} dvol$$

$$= -\int_{vol} (\tilde{b}_x u + \tilde{b}_y v)\, dvol$$

$$= \int_{vol} \left(\tilde{b}_x \sum_i N_i u_i + \tilde{b}_y \sum_i N_i v_i \right) dvol \tag{5.29}$$

and therefore the equivalent nodal force vector

$$f_{xi} = \int_{vol} \tilde{b}_x N_i \, dvol = t \iint_A \tilde{b}_x N_i \, dx\, dy$$

$$= t \int_{-1}^{+1}\int_{-1}^{+1} \tilde{b}_x N_i \, |J| \, d\xi d\eta$$

$$= t\, \tilde{b}_x \sum_{k=1}^{n}\sum_{l=1}^{n} w_k w_l (N_i \,|J|)_{k,l} \quad \text{for } i = 1,2,3. \tag{5.30}$$

Similarly,

$$f_{yi} = t \tilde{b}_y \sum_{k=1}^{n}\sum_{l=1}^{n} w_k w_l (N_i \,|J|)_{k,l} \quad \text{for } i = 1,2,3. \tag{5.31}$$

5.4.4 Calculation of Internal Strains and Stresses

Calculation of internal strains in a quadrilateral element is more complicated than for the rectangular element, because the B matrix that connects the internal strain and the nodal displacement vector is a function of ξ and η as shown in Eq. (5.22). The matrix also contains elements of the inverse of the Jacobian matrix. However, one can calculate the values of the internal strains at any ξ and η. Because of the complex variation of the internal strains within the element, it is customary to determine the internal strains at Gauss points that are used for integration of the stiffness matrix. These values would indicate some distribution of the strains within the element. If 2×2 Gauss integration scheme is used, there will be four points where the strains are computed. Another technique is to use the center of the quadrilateral, i.e., $\xi = \eta = 0$. In either case, the parts of the subroutines QUAD and BMAT is used to compute the Jacobian and the elements of the B matrix for a given value of ξ and η. With this procedure the methodology of the 4-noded quadrilateral finite element for solving plane elasticity problem is complete.

5.5 CURVED TRIANGULAR ELEMENTS

Numerical integration schemes allowed more complex triangular elements with curved boundaries as shown in Fig. 5.11. In the following sections, only the theory involved in developing the element stiffness matrix is shown. The rest of the procedures are similar to the overall finite element formulation. However the steps involved for numerical integration is shown with some examples because the understanding the development of these elements depends on the clear understanding of the numerical integration procedures.

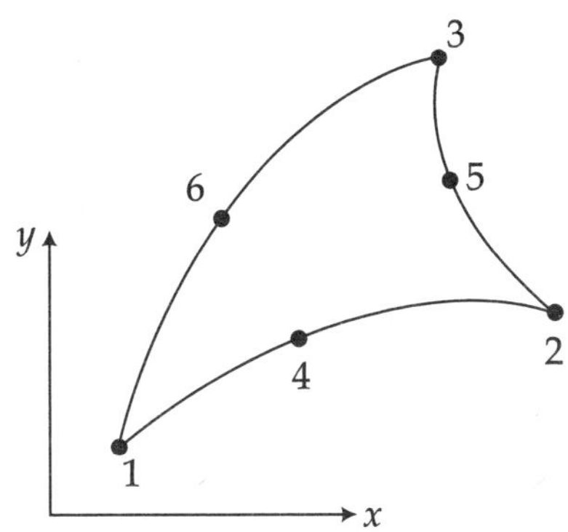

Fig. 5.11 Curved Triangular Element and Transformation

5.5.1 Element Stiffness Matrix

Step 1. Assumed Displacement Functions

In Chapter 4, all the steps required for the development of a higher order linear strain triangular element were presented. The procedure is almost the same with minor changes. Same interpolation equations can be used to formulate a curved triangular finite element with some modifications, see Eqs. (4.35),

$$N_i = L_i(2L_i - 1) \quad \text{for} \quad i = 1, 2, 3 \text{ and}$$
$$N_4 = 4L_1L_2, \quad N_5 = 4L_2L_3 \quad \text{and} \quad N_6 = 4L_1L_3 \tag{5.32}$$

Since the area coordinates L_1, L_2, L_3 are dependent, two new coordinates ξ, η that are independent from the set L_1, L_2, L_3 are used to develop the Jacobian, i.e.,

$$L_1 = \xi, \quad L_2 = \eta \text{ and } L_3 = 1 - \xi - \eta \tag{5.33}$$

Now the shape functions given in Eq. (5.32) can be rewritten in the new coordinates as

$$\begin{aligned} N_1 &= \xi(2\xi - 1) \\ N_2 &= \eta(2\eta - 1) \\ N_3 &= 1 + \xi(2\xi - 3) + \eta(2\eta - 3) + 4\xi\eta \\ N_4 &= 4\xi\eta \\ N_5 &= 4\eta(1 - \xi - \eta) \\ N_6 &= 4\xi(1 - \xi - \eta) \end{aligned} \tag{5.34}$$

In other words, all shape functions are written as functions of ξ, η only.

Using these quadratic shape functions, the internal displacement functions are written as

$$u = \sum_{i=1}^{6} N_i u_i \text{ and } v = \sum_{i=1}^{6} N_i v_i \tag{5.35}$$

Just as the displacement functions are expressed using quadratic shape functions, the six nodal coordinates of a curved triangular element are also expressed using the same shape functions. i.e.,

$$x = \sum_{i=1}^{6} N_i x_i \text{ and } y = \sum_{i=1}^{6} N_i y_i \tag{5.36}$$

Since both displacements and coordinates use the same shape functions, this curved triangluar element is also an isoparametric element. The most important derivation required is that of the B matrix. But, first the Jacobian of the transformation, from (x, y) to (ξ, η) has to be established.

Jacobian of the Transformation for Curved Elements

Using the chain rule,

$$\begin{Bmatrix} \dfrac{\partial u}{\partial \xi} \\ \dfrac{\partial u}{\partial \eta} \end{Bmatrix} = \begin{bmatrix} \dfrac{\partial x}{\partial \xi} & \dfrac{\partial y}{\partial \xi} \\ \dfrac{\partial x}{\partial \eta} & \dfrac{\partial y}{\partial \eta} \end{bmatrix} \begin{Bmatrix} \dfrac{\partial u}{\partial x} \\ \dfrac{\partial u}{\partial y} \end{Bmatrix} = [J] \begin{Bmatrix} \dfrac{\partial u}{\partial x} \\ \dfrac{\partial u}{\partial y} \end{Bmatrix}$$

where $[J]$ is the Jacobain matrix. Taking inverse of the above expression,

$$\begin{Bmatrix} \dfrac{\partial u}{\partial x} \\ \dfrac{\partial u}{\partial y} \end{Bmatrix} = \frac{1}{|J|} \begin{bmatrix} \dfrac{\partial y}{\partial \eta} & -\dfrac{\partial y}{\partial \xi} \\ -\dfrac{\partial x}{\partial \eta} & \dfrac{\partial x}{\partial \xi} \end{bmatrix} \begin{Bmatrix} \dfrac{\partial u}{\partial \xi} \\ \dfrac{\partial u}{\partial \eta} \end{Bmatrix}$$

where the determinant of the Jacobian is

$$|J| = \frac{\partial x}{\partial \xi}\frac{\partial y}{\partial \eta} - \frac{\partial x}{\partial \eta}\frac{\partial y}{\partial \xi}$$

$$= \sum_i N_{i,\xi} x_i \sum_j N_{j,\eta} y_j - \sum_i N_{i,\eta} x_i \sum_j N_{j,\xi} y_j$$

$$= \sum_i \sum_j x_i (N_{i,\xi} N_{j,\eta} - N_{i,\eta} N_{j,\xi}) y_j$$

$$= \sum_i \sum_j x_i g_{ij} y_j = \{X\}^t\,[G]\,\{Y\} \tag{5.37}$$

where $g_{ij} = N_{i,\xi} N_{j,\eta} - N_{i,\eta} N_{j,\xi}$ and

$$\{X\}^t = \{x_1 \quad x_2 \quad x_3 \quad x_4 \quad x_5 \quad x_6\} \text{ and}$$

$$\{Y\}^t = \{y_1 \quad y_2 \quad y_3 \quad y_4 \quad y_5 \quad y_6\}$$

In all the above equations, both i and j have values from 1 to 6.

Step 2. Internal Strain-Displacement Matrix

The components of the internal strains are:

$$\varepsilon_x = \frac{\partial u}{\partial x} = \frac{1}{|J|}\left(\frac{\partial y}{\partial \eta}\frac{\partial u}{\partial \xi} - \frac{\partial y}{\partial \xi}\frac{\partial u}{\partial \eta}\right)$$

$$= \frac{1}{|J|}\sum_i \sum_j (y_i\, N_{i,\eta}\, N_{j,\xi}\, u_j - y_i\, N_{i,\xi}\, N_{j,\eta}\, u_j)$$

$$= -\frac{1}{|J|}\{Y\}^t\,[G]\,\{u^e\} \tag{5.38}$$

Similarly,

$$\varepsilon_y = \frac{\partial v}{\partial y} = \frac{1}{|J|}\,\{X\}^t\,[G]\,\{v^e\}$$

and

$$\gamma_{xy} = \frac{\partial u}{\partial y} + \frac{\partial v}{\partial x} = \frac{1}{|J|}(\{X\}^t\,[G]\,\{u^e\} - \{Y\}^t\,[G]\,\{v^e\})$$

In the above equations,

$$\{u^e\}^t = \{u_1 \quad u_2 \quad u_3 \quad u_4 \quad u_5 \quad u_6\}$$

and

$$\{v^e\}^t = \{v_1 \quad v_2 \quad v_3 \quad v_4 \quad v_5 \quad v_6\}.$$

Rearranging the coefficients in equations for strain-displacement equations, the B matrix must be developed so that the total element nodal displacement vector is as given in Eq. 5.39.

$$\{\varepsilon\} = [B]\begin{Bmatrix} u_1 \\ v_1 \\ u_2 \\ v_2 \\ . \\ v_6 \end{Bmatrix} = [B]\{u\} \tag{5.39}$$

Step 3 and 4. The stiffness matrix is obtained as

$$[k] = t\iint_A [B]^t\,[D]\,[B]\,dxdy = t\iint_A [B]^t\,[D]\,[B]|\,\mathrm{J}\,|\,d\xi\,d\eta \tag{5.40}$$

$$= t\sum_{i=1}^{n}\sum_{j=1}^{n} w_i w_j ([B]^t[D][B]|\,\mathrm{J}\,|)_{i,j}$$

The matrix expression inside the bracket in Eq. (5.40) above should be expressed in terms of the area coordinates L_1, L_2, and L_3. This numerical integration formula is discussed in the following section.

5.5.2 Numerical Integration for Triangular Elements

One of the great landmark in finite element method was the introduction of numerical integration techniques. Researchers started to develop complex elements as all the tedious integrations were accomplished numerically. An integral of a function $f(x, y)$ within a triangle just as given in Eq. (5.40) is expressed as a sum of the weighted values of the function at some pivotal points[1,5,9] just like the Guass Quadrature formula discussed earlier in one, two and three dimensions.

Integration within a Triangle

The integration formulae are made using area coordinates L_1, L_2 and L_3, and the integration formulas use values of all the three area coordinates at some pivotal points. In other words, the function $f(x, y)$ to be integrated has to be transformed into a function of area coordinates L_1, L_2, and L_3. However, since the three area coordinates are not independent, only two of them have to be used to compute the Jacobian. To avoid any confusion, new notations are introduced where $L_1 = \xi$, $L_2 = \eta$, and $L_3 = 1 - \xi - \eta$. The same equations were used in Eq. (5.33) with two independent coordinates ξ, η. In general, the nodal coordinates are interpolated using the shape functions and they have to be expressed in terms of ξ, η coordinates. Similar to the Eqs. (5.11),

$$x = \sum_i N_i(\xi, \eta)\, x_i$$

$$y = \sum_i N_i(\xi, \eta)\, y_i, \quad \text{where } i = \text{the number of nodes}$$

Table 5.2 Numerical Integration for a Triangle

Number of integration points	*Degree of polynomial and order of the error*	*Location of integration points*		*Weights and integration points* W	L_1	L_2	L_3
1	1 $O(h^2)$	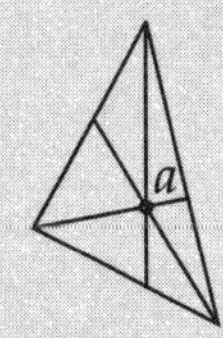	a	$\frac{1}{2}$	$\frac{1}{3}$	$\frac{1}{3}$	$\frac{1}{3}$
3	2 $O(h^3)$	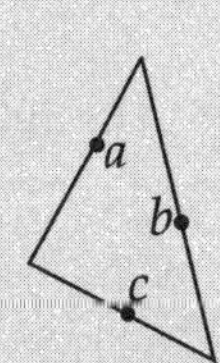	a	$\frac{1}{6}$	$\frac{1}{2}$	0	$\frac{1}{2}$
			b	$\frac{1}{6}$	$\frac{1}{2}$	$\frac{1}{2}$	0
			c	$\frac{1}{6}$	0	$\frac{1}{2}$	$\frac{1}{2}$
4	3 $O(h^4)$	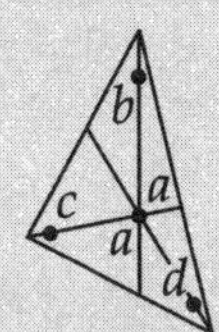	a	$-\frac{27}{96}$	$\frac{1}{3}$	$\frac{1}{3}$	$\frac{1}{3}$
			b	$\frac{25}{96}$	0.6	0.2	0.6
			c	$\frac{25}{96}$	0.2	0.6	0.2
			d	$\frac{25}{96}$	0.2	0.2	0.6

and the Jacobian matrix $[J] = \begin{bmatrix} \dfrac{\partial x}{\partial \xi} & \dfrac{\partial y}{\partial \xi} \\ \dfrac{\partial x}{\partial \eta} & \dfrac{\partial y}{\partial \eta} \end{bmatrix}$ and its determinant have to be evaluated using the derivatives of the shape functions and the nodal coordinates similar to the expression given in Eq. (5.37). Now the integral,

$$\iint_A F(x, y)\, dx\, dy = \int_0^1 \int_0^{1-\eta} F(L_1, L_2, L_3)|J|\, d\xi\, d\eta \tag{5.41}$$

$$= \sum_{i=1}^{n} w_i F_i(L_1, L_2, L_3)|J|_i$$

In other words, the integral is the sum of the values of the function evaluated at some pivotal points multiplied by corresponding weights and the value of the Jacobian of transformation evaluated at the pivotal point. The pivotal points and the corresponding weights are given in the following table.

The best way to understand the procedure is to go through a simple example.

Example Problem 5.6

Evaluate the following three integrals in a triangle given in Fig. 5.12.

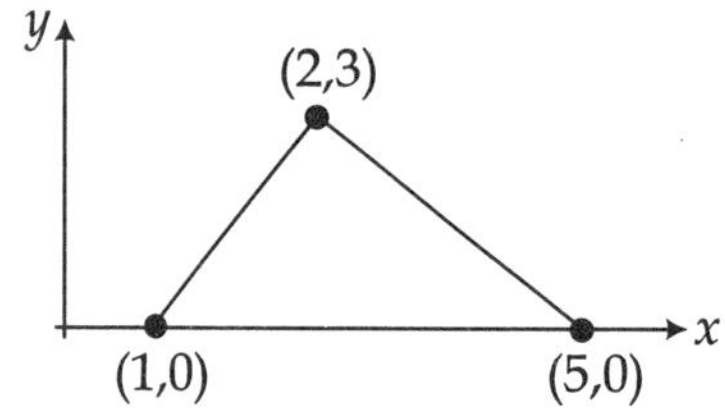

Fig. 5.12

$$I_1 = \iint_A dx\, dy,\ I_2 = \iint_A (x + y)dx\, dy, \text{ and } I_3 = \iint_A (x^2 + y^2)dx\, dy$$

Solution:

Since the given triangle consists of straight sides, using the nodal coordinates,

$$x = L_1x_1 + L_2x_2 + L_3x_3 \text{ and } y = L_1y_1 + L_2y_2 + L_3y_3 \text{ and by Eq. 5.33,}$$

$$x = 1\xi + 5\eta + 2(1 - \xi - \eta) = 2 - \xi + 3\eta$$

and $$y = +3(1 - \xi - \eta)$$

Hence the Jacobian $|J| = \begin{vmatrix} \frac{\partial x}{\partial \xi} & \frac{\partial y}{\partial \xi} \\ \frac{\partial x}{\partial \eta} & \frac{\partial y}{\partial \eta} \end{vmatrix} = \begin{vmatrix} -1 & +3 \\ -3 & -3 \end{vmatrix} = 12$

Using one point integration,

$$I_1 = \iint_A dx\, dy = \iint_A |J|\, d\xi\, d\eta = \sum_1 w_1 |J| = \frac{1}{2}(12) = 6$$

$$I_2 = \iint_A (x + y)dx\, dy = \iint_A (L_1 + 5L_2 + 5L_3)d\xi\, d\eta\ (12) = 0.5\left(\frac{11}{3}\right)12 = 22$$

$$I_3 = \iint_A (x^2 + y^2)dx\, dy = \iint_A \{(L_1 + 5L_2 + 2L_3)^2 + (3L_3)^2\}d\xi\, d\eta\ (12)$$

$$= 48.67,$$

where as the exact value is equal to 56.0. Here the first two integrals were computed with one point integration and they yielded correct values, where as the third integral where the integrand was a quadratic, the one point integration resulted with some error. Repeating the integration with three points which is good for integrating quadratic functions,

$$I_3 = \iint_A (x^2 + y^2)dx\, dy = 12\left(\frac{1}{6}\right)\left[\left(\frac{1}{2} + 0 + \frac{2}{2}\right)^2 + \left(\frac{3}{2}\right)^2 + \right.$$

$$\left.\left(\frac{1}{2} + \frac{5}{2} + 0\right)^2 + (0)^2 + \left(0 + \frac{5}{2} + \frac{2}{2}\right)^2 + \left(\frac{3}{2}\right)^2\right] = 56.0$$

which is the exact value. When the function to be integrated becomes complex, more pivotal points are to be taken.

5.6 HIGHER ORDER QUADRILATERAL ELEMENTS

The numerical integration schemes have revolutionized the finite element procedures. One can create elements which have even curved boundaries with the addition of extra nodes on the boundary. New elements can also allow the internal strains to vary linearly or with higher order polynomial. There are two types of higher order linear strain quadrilateral elements. For the first type, the element has nine nodes with an internal node with in the element and the second type has only eight nodes. The latter is popular among engineers because all the nodes are on the boundary of the element. The nine-noded quadrilateral element and the transformed square are shown in Figs. 5.13(a) and (b) and the quadrilateral element can be developed with or without curved boundaries.

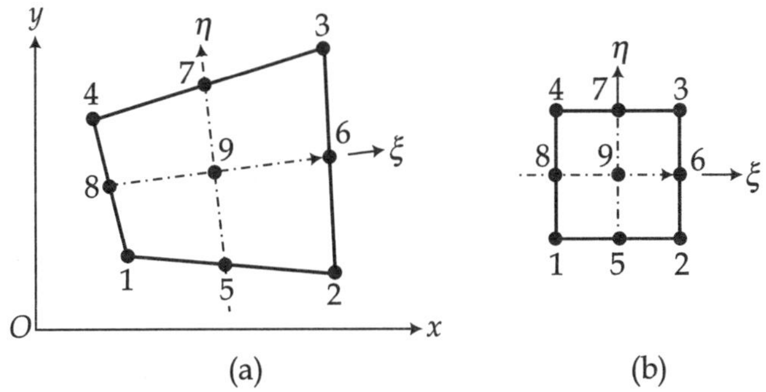

Fig. 5.13 Nine Noded Quadrilateral Element

5.6.1 Nine-Noded Quadrilateral Element

Here the quadrilateral has four nodes at the corners, four nodes at the midpoints of the sides (will give better results, if they are at the midpoints) and one at the center of the quadrilateral and altogether it has nine nodes. It is not necessary that the quadrilateral should have straight sides; even curved sides can be accommodated in this formulation. In other words, the shape functions developed to interpolate the displacements can also be used to interpolate the coordinates. This quadrilateral is transformed into a square with a side having length equal to 2 units as shown in Fig. 5.13(b). The shape functions for this element are developed using the Lagrange polynomials for the one-dimensional elements that are described in Chapter three. One-dimensional shape functions are first developed for the ξ and η directions individually. For example, for the ξ-direction, the one-dimensional shape functions are:

$$\begin{aligned} n_1(\xi) &= \frac{1}{2}\xi(\xi - 1) \\ n_2(\xi) &= (1 + \xi)(1 - \xi) \\ n_3(\xi) &= \frac{1}{2}\xi(\xi + 1) \end{aligned} \tag{5.41}$$

and similarly, the one-dimensional shape functions in the η-direction are:

$$
\begin{aligned}
n_1(\eta) &= \frac{1}{2}\eta(\eta - 1) \\
n_2(\eta) &= (1 + \eta)(1 - \eta) \\
n_3(\eta) &= \frac{1}{2}\eta(\eta + 1)
\end{aligned} \tag{5.42}
$$

These one-dimensional shape functions are multiplied in pairs in a logical manner such that the fundamental properties of the required new shape functions in the two-dimensional domain of the square are satisfied. For example, the following are the shape functions for the nine-noded quadrilateral element. These shape functions are developed first for the corner nodes, then for the mid-side nodes and lastly for the center node. The reader is expected to verify the properties of these shape functions.

For corner nodes:

$$
\begin{aligned}
N_1 &= n_1(\xi)n_1(\eta) = \frac{1}{4}\xi\,\eta(\xi - 1)(\eta - 1) \\
N_2 &= n_3(\xi)n_1(\eta) = \frac{1}{4}\xi\,\eta(\xi + 1)(\eta - 1) \\
N_3 &= n_3(\xi)n_3(\eta) = \frac{1}{4}\xi\,\eta(\xi + 1)(\eta + 1) \\
N_4 &= n_1(\xi)n_3(\eta) = \frac{1}{4}\xi\,\eta(\xi - 1)(\eta + 1)
\end{aligned} \tag{5.43}
$$

and for the mid-side nodes,

$$
\begin{aligned}
N_5 &= n_2(\xi)n_1(\eta) = \frac{1}{2}\eta(1 - \xi^2)(\eta - 1) \\
N_6 &= n_3(\xi)n_2(\eta) = \frac{1}{2}\xi(\xi + 1)(1 - \eta^2) \\
N_7 &= n_2(\xi)n_3(\eta) = \frac{1}{2}\eta(1 - \xi^2)(\eta + 1) \\
N_8 &= n_1(\xi)n_2(\eta) = \frac{1}{2}\xi(\xi - 1)(1 - \eta^2)
\end{aligned} \tag{5.44}
$$

and for the center node,

$$N_9 = n_2(\xi)n_2(\eta) = (1 - \xi^2)(1 - \eta^2) \tag{5.45}$$

Once the shape functions are determined, one can employ the other three steps required to compute the element stiffness matrix. The necessary equations for the development of the element stiffness matrix are left as an exercise for the reader.

5.6.2 Eight-Noded Quadrilateral Element

The internal node, represented by the ninth node in the nine-noded element has been considered as an unwanted node not from a mathematical point of view, but from a practical sense. Engineers have been trying to get rid of the extra equations in the stiffness matrix due to this node by simple Gauss elimination process, a technique which came to be known as 'static condensation' of the stiffness matrix. The reason for the elimination of the node was that the ninth node did not have any connection with the nodes of the neighboring elements except those on the inter-element boundaries. While some of the engineers were trying to remove the effect of the extra node by elimination process, others developed shape functions for an eight - noded element heuristically. The basic requirement was that the new shape functions must have the necessary properties. A quadrilateral element with four corner nodes and four mid-side nodes along with the corresponding transformed square element are shown in Figs. 5.14(a) and 5.14(b) respectively. The shape functions for the mid-side nodes are developed using the shape functions for the one-dimensional, linear and quadratic Lagrange shape functions in the ξ and η directions respectively. For example, the following are the shape functions for the mid-side nodes.

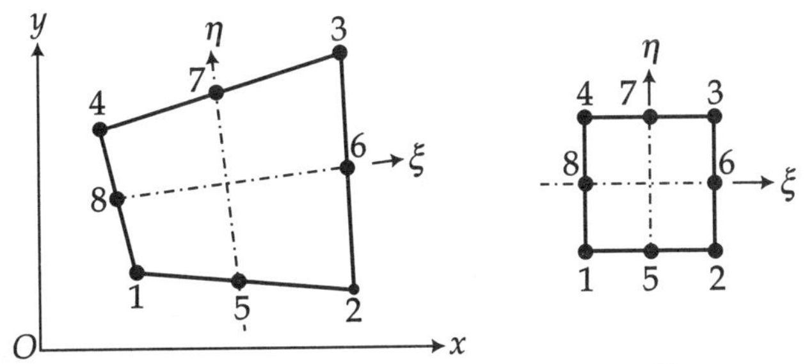

Fig. 5.14 Eight Noded Quadrilateral Element

$$
\begin{aligned}
\tilde{N}_5 &= \frac{1}{2}(1 - \xi^2)(1 - \eta), \\
\tilde{N}_6 &= \frac{1}{2}(1 + \xi)(1 - \eta^2), \\
\tilde{N}_7 &= \frac{1}{2}(1 - \xi^2)(1 + \eta), \\
\tilde{N}_8 &= \frac{1}{2}(1 - \xi)(1 - \eta^2)
\end{aligned} \tag{5.46}
$$

Upon examination, it can be verified that these shape functions satisfy the required properties. The shape functions for the corner nodes are developed using a linear combination of the shape functions for the four-noded quadrilaterals and the above shape functions at the mid-side nodes. The derivation of the shape function for the node 1 is as follows. First consider the shape function for node 1 for a four -noded quadrilateral, i.e.,

$$n_1 = \frac{1}{4}(1 - \xi)(1 - \eta)$$

It is observed that this function has values equal to 1.0 at node 1, 0.5 at nodes 5 and 8 and zeros at all other nodes as shown in Fig. 5.14. Combining this function with half of the shape functions N_5 and N_8, such that a new function is formed as

$$\tilde{N}_1 = n_1 - \frac{1}{2}N_5 - \frac{1}{2}N_8 \tag{5.47}$$

It is easily verified that the above shape function has value equal to one at node 1 and equal to zero at all other nodes. Similar expressions for the other three shape functions can be developed by proper combination of shape functions.

It is interesting to note that for the case of a finite element discretization with small number of elements representing a rectangular simply supported beam problem (2-D plane stress), the nine-noded finite element will yield a more accurate solution than the corresponding eight-noded element discretization. However, when a large number of finite elements are employed in a domain, the difference becomes small.

5.7 ERROR ANALYSIS OF FINITE ELEMENT SOLUTIONS

So far much has been discussed on the formulation of the finite elements and their solutions. Results of very few problems are also presented. A finite element user in the beginning is always confronted with the type and number of elements to be used for a given problem. An experienced person can always foretell approximately how many elements are needed to get some reasonable solutions to practical problems. Poor shapes and high aspect ratios (width to height ratio) of elements can result in poor solutions. The experience of the author[8] while solving a standard beam problem using four nodded and eight nodded rectangular elements with different aspect ratios yield slightly different results as they converge differently. It is always better to use a good shape for the element, such as having a shape similar to equilateral triangles, in other words, trying to avoid obtuse angles for triangles is a good strategy. And for rectangles and quadrilaterals, make the element look similar to a square. High aspect ratios for the elements in rectangular or quadrilateral elements always yield poor solutions. Normally, it is observed that the convergence of the solution has some definite relation with the number of elements, shapes and the aspect ratios. While using curved elements, too much curvature can yield poor solutions. It is recommended therefore to have a convergence study for important practical engineering problems before the results are utilized for design purposes.

During early days of finite element computing, when the same beam problems mentioned above were solved on computers with 7 digit word length, results diverged after converging to some value. This is due to the round-off errors in numerical computation. The error associated with the above solution disappeared when a double precision was used in computing. Now-a-days, computers are made with 64 bits word processors and for such a computer, perhaps round off problem may not be an issue. Currently, engineers are used to solving very large number of equations on the computer, as many as the computer can handle, such as 10 million equations or even more. A double precision solution is highly recommended for solving such large number of equations, because, it is possible that round-off errors can creep into the solution exactly as it was experienced in the beam problem.[8]

Use of the finite element methodology in modern engineering is currently directed toward automatic discretization, re-meshing and seeking convergence in the solution. These techniques would eventually be very helpful to the users when they are not sure of the selection of the element, number of elements, accuracy and convergence.

5.8 SUMMARY

Transformation of a quadrilateral into a square with a Jacobian for transformation of derivatives is shown in this chapter. Numerical integration using Guass quadrature formula are employed to integrate the otherwise impossible expressions evolved during the transformations. This new popular quadrilateral element is called isoparametric and is for solving two-dimensional plane elasticity problems. New higher order elements with nine and eight noded quadrilaterals are also developed. The derivation for the stiffness matrix of a new higher order curved triangular element is also shown here. Special numerical integration procedure required for integration of functions in a triangular element is also shown here with examples. The theory for the higher order elements are shown only up to the derivation of the element stiffness matrices and the rest of the operations are left to the reader as an exercise.

The effect of the number of elements, shape of elements and the aspect ratio of the rectangular elements are discussed only qualitatively. The best method is to do perform a convergence study at least with a limited number of elements so as to get a feel for the solution obtained.

PROBLEMS

5.1 Using three point Gauss Legendre Quadrature formula, integrate the following:

$$\int_0^4 (x^4 - 3x^3 + 5)dx$$

Do you expect exact result? Why?

5.2 Calculate the integrals $\int x^2 y \, dA$ and $\int (x^2 + y^2) \, dA$ over the quadrilateral as shown below.

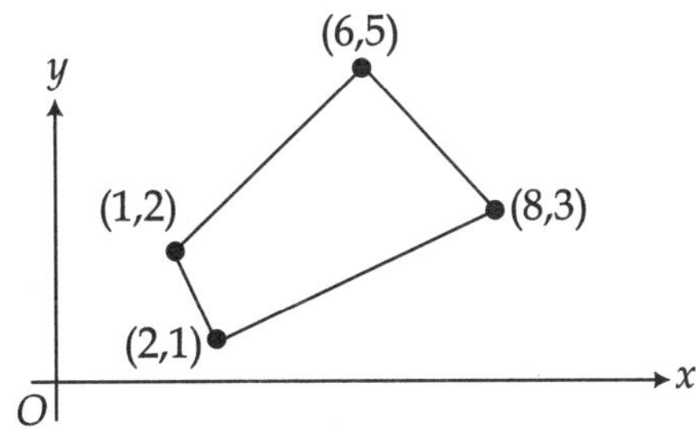

Fig. P5.1

5.3 Write down a subroutine to form a global stiffness matrix using element stiffness matrices of a quadrilateral element with 3 d.o.f per node.

5.4 A rectangular finite element for a plane stress problem has an assumed displacement function

$$u = \alpha_1 + \alpha_2 x + \alpha_3 y + \alpha_4 xy$$
$$v = \alpha_5 + \alpha_6 x + \alpha_7 y + \alpha_8 xy$$

Explain how these functions are satisfying the necessary conditions for the convergence of the solution, such as (i) rigid body motion, (ii) constant strain condition, (iii) interelement compatibility.

5.5 Laplace equation $\nabla^2 T = 0$ has to be solved by using a rectangular finite element. The functional to be minimized is

$$\iint_{\Omega} \frac{1}{2}\left\{\left(\frac{\partial T}{\partial x}\right)^2 + \left(\frac{\partial T}{\partial y}\right)^2\right\} dxdy$$

using $T = \sum N_i T_i$ where N_i are

$$N_1 = \frac{1}{4ab}(a - x)(b - y)$$
$$N_2 = \frac{1}{4ab}(a + x)(b - y)$$
$$N_3 = \frac{1}{4ab}(a + x)(b + y)$$
$$N_4 = \frac{1}{4ab}(a - x)(b + y)$$

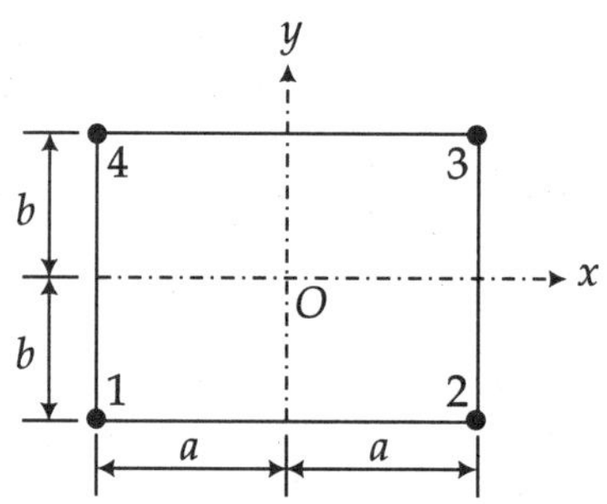

Fig. P5.2

T_i' s are nodal parameters. Derive the system equations for one element.

5.6 A one dimensional bar element with quadratic variation of displacement u can be obtained with

$$u = N_1 u_1 + N_2 u_2 + N_3 u_3$$

Where $N_1 = 1 - \frac{3x}{a} + \frac{2x^2}{a^2}$ $N_2 = 4\left(\frac{x}{a} - \frac{2x^2}{a^2}\right)$ $N_3 = -\frac{x}{a}\left(1 - \frac{2x}{a}\right)$

Using this idea, develop shape function for a nine noded rectangular element to solve plane elasticity problems with internal strains varying linearly within the element.

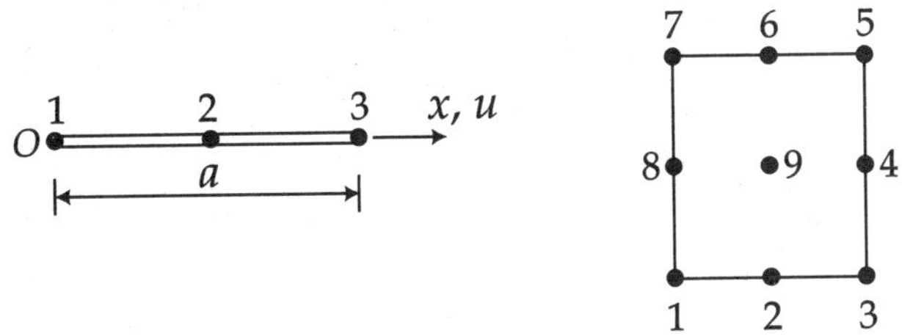

Fig. P5.3

5.7 Solve a beam shown below by using rectangular finite elements as a plane stress problem. Show convergence of the results and compare with elementary beam theory. Take E = 200GPa, b = 25 cm, h = 50 cm, l = 2.6 m, $\tilde{q}$ = 20 kN/m, t = 10 cm.

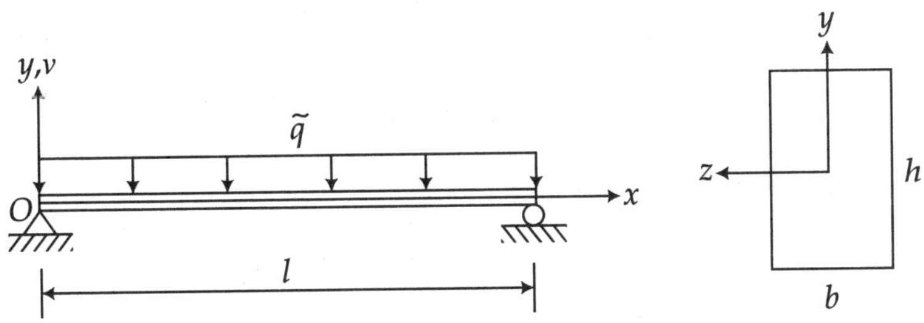

Fig. P5.4

REFERENCES

1. Bathe, K.J. and Wilson, E.L, "Numerical Methods in Finite Element Analysis", Prentice Hall, Englewood Cliffs, New Jersey, 1976.
2. Bathe, K.J, "Finite Element Procedures", Prentice Hall, Englewood Cliffs, New Jersey, 1996.
3. Clough, R.W, "The Finite Element Method in Structural Mechanics", Chapter 7, "Stress Analysis", O.C. Zienkiewicz and G.G. Hollister, Eds, John Wiley & Sons, Inc., NY 1965.
4. Irons, B.M., "Engineering Applications of Numerical Integrations in Stiffness Method", J.A.I.AA., Vol. 14, 1966.
5. Reddy, J.N, "An Introduction to the Finite Element Method", 2nd Ed. McGraw Hill Book Company, New York, 1993.
6. Vallabhan, Girija. C.V. and Reese, L.C., "Finite Element Method for Problems in Soil Mechanics", Proc. ASCE, Vol. 94, SM2, Mar. 1968.
7. Vallabhan, Girija. C.V, "Analysis of Shear Walls with Openings", Proc. ASCE, Vol. 95, ST10, Oct. 1969.
8. Vallabhan, Girija, C.V. and Narayanaswami, R., "Error Analysis for Rectangular Finite Elements", Proc. ASCE, Vol. 98, ST1, Jan. 1972.
9. Zienkiewicz, O.C. and Taylor, R.L., "The Finite Element Method", Vols. I & II, McGraw Hill Book Company, London, 1989.

6

CHAPTER

Axisymmetric Problems

6.1 INTRODUCTION

As discussed in Chapter 2, that certain three-dimensional problems can be simplified as two-dimensional ones, and some of these simplified ones as axisymmetric problems. To be categorized as an axisymmetric problem, the geometry of the solid, the prescribed loads and the boundary conditions should be axially symmetric with respect to an axis within the body.[4] The cylindrical coordinates r, θ and z are employed for the analysis of these problems, and it can be easily verified that the problem is independent of θ. Most analysts compare the axisymmetric finite element model with the plane elasticity finite element models, because of the resemblance of the equations involved in the two formulations. Naturally, the elements used for this analysis also belong to the C_0-continuity type. It is better to study the theory and formulation separately, since the equations of elasticity are written in terms of cylindrical coordinates and the introduction of the normal strain in the θ-direction makes a small difference in the formulation of the strain-displacement matrix, the B matrix. Also, the determination of the element stiffness matrix of which have nodal coordinates on the axis of symmetry requires numerical integration because of a singularity of the equations along the axis of symmetry.

Once the equations for the stiffness matrix and the equivalent force vectors for axisymmetric finite element are formulated, it can be seen that the resulting equations[6,7] are similar to those obtained for plane elasticity models. With minor modifications in the computer codes for plane elasticity models, axisymmetric equations can be developed. All types of finite element models discussed earlier for solving two-dimensional plane elasticity problems are applicable here also. Only the theory for the formulations of the stiffness matrices and the corresponding force vectors for the different finite elements are given in this chapter.

6.2 TRIANGULAR RING FINITE ELEMENT

The simplest of the element for an axisymmetric problem is triangular one as illustrated in Fig. 6.1. First of all, the physical representation of this triangular ring element has to be visualized, in other words, this triangular ring element denoted as 1-2-3 in the figure represents a solid body created by a circular ring of revolution of the triangular element 1-2-3 with respect to the *z*-axis. Clough and Rashid[1] were the first to develop this type of axisymmetric finite element. The nodal displacements and forces are to be visualized with components in the *r*- and *z*-directions only. The four steps discussed earlier for the derivation of the element stiffness matrix are repeated here also.

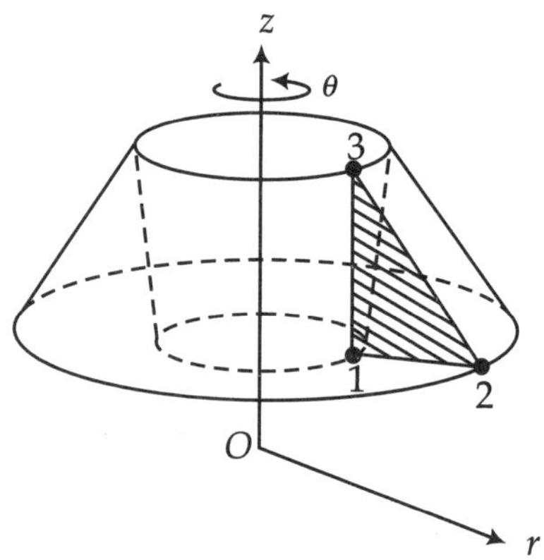

Fig. 6.1 An Axisymmetric Triangular Ring Element

Step 1. Assume Displacement Functions

Since there are three nodes and two displacement functions u and w, one can use a minimum of two degrees of freedom per node, thus producing a six d.o.f. element. Following the assumptions used for the formulation of constant triangular elements, the internal displacement functions are written as:

$$\begin{aligned} u(r, z) &= \alpha_1 + \alpha_2 r + \alpha_3 z \\ w(r, z) &= \alpha_4 + \alpha_5 r + \alpha_6 z \end{aligned} \tag{6.1}$$

It can be easily verified that these functions satisfy all the necessary requirements of rigid body motions, constant strain conditions (except ε_θ) and the interelement compatibility conditions. Shape functions that are given in Eqs. (4.6) can again be employed here, i.e.,

$$u(r, z) = \sum_{i=1}^{3} N_i \, u_i \tag{6.2}$$

$$w(r, z) = \sum_{i=1}^{3} N_i \, w_i$$

where

$$N_i = \frac{1}{2A}(a_i + b_i \, r + c_i \, z) \text{ for } i = 1, 2, 3 \text{ and}$$

$$a_i = r_j z_k - r_k z_j, \; b_i = z_j - z_k \text{ and } c_i = r_k - r_j$$

where i, j, k vary in a cyclic order.

Step 2. Internal Strains

For the development of the internal strains from the displacements u and w, the strain-displacement equations from theory of elasticity in cylindrical coordinates as given in Eq. (2.29) of Chapter 2 are used. Using these equations and differentiating the displacement functions accordingly, the internal strains are expressed in terms of the nodal displacement vector from which the B matrix is obtained.

$$\{\varepsilon\} = \begin{Bmatrix} \varepsilon_r \\ \varepsilon_\theta \\ \varepsilon_z \\ \gamma_{rz} \end{Bmatrix} = \begin{Bmatrix} \dfrac{\partial u}{\partial r} \\ \dfrac{u}{r} \\ \dfrac{\partial w}{\partial z} \\ \dfrac{\partial u}{\partial z} + \dfrac{\partial w}{\partial r} \end{Bmatrix} = \frac{1}{2A} \begin{bmatrix} b_1 & 0 & b_2 & 0 & b_3 & 0 \\ \dfrac{2AN_1}{r} & 0 & \dfrac{2AN_2}{r} & 0 & \dfrac{2AN_3}{r} & 0 \\ 0 & c_1 & 0 & c_2 & 0 & c_3 \\ c_1 & b_1 & c_2 & b_2 & c_3 & b_3 \end{bmatrix} \begin{Bmatrix} u_1 \\ w_1 \\ u_2 \\ w_2 \\ u_3 \\ w_3 \end{Bmatrix}$$

$$= [B]\{u\} \tag{6.3}$$

It is to be noted that all strain components are constants except ε_θ, which is a function of r and z. Hence theoretically this element is not a constant strain element, but it is close to one. Furthermore, there is a singularity for ε_θ at $r = 0$, which requires special considerations.

Step 3. Internal Stresses

The stress-strain relations as given in Eq. (2.30) for the axisymmetric elasticity problem is repeated here for clarity,

$$\begin{Bmatrix} \sigma_r \\ \sigma_z \\ \sigma_\theta \\ \tau_{rz} \end{Bmatrix} = \frac{E(1-\mu)}{(1+\mu)(1-2\mu)} \begin{bmatrix} 1 & \dfrac{\mu}{1-\mu} & \dfrac{\mu}{1-\mu} & 0 \\ \dfrac{\mu}{1-\mu} & 1 & \dfrac{\mu}{1-\mu} & 0 \\ \dfrac{\mu}{1-\mu} & \dfrac{\mu}{1-\mu} & 1 & 0 \\ 0 & 0 & 0 & \dfrac{1-2\mu}{2(1-\mu)} \end{bmatrix} \begin{Bmatrix} \varepsilon_r \\ \varepsilon_z \\ \varepsilon_\theta \\ \gamma_{rz} \end{Bmatrix} \tag{2.30}$$

Step 4. Energy Theorem and Element Stiffness Matrix

Having obtained the matrices B and D, the element stiffness matrix is written as

$$[k] = \iiint\limits_{vol} [B]^t\,[D]\,[B]\, r\, dr\, d\theta\, dz$$

$$= 2\pi \iint_A [B]^t \, [D] \, [B] \, r \, dr \, dz \qquad (6.4)$$

Early investigators[1] have explicitly evaluated and integrated the elements of the stiffness matrix. However while integrating these functions, there were a few terms in the stiffness matrix with a singularity at the axis of symmetry and hence they adopted a large negative value for the functions involving log r when the value of r became equal to zero. If the axisymmetric solid under investigation had a hole in the axis of symmetry, then this problem does not arise. Others[8] have found that these volume integrations can be approximated as a value calculated at the centroid of the triangle multiplied by the circumference of the triangle at its centroid. This technique is equivalent to a one-point Gauss quadrature formula for integrating in a triangular domain. Experience shows that use of higher order integration formulae does not substantially improve the accuracy of the solution. Hence the element stiffness matrix is evaluated as

$$[k] = 2\pi \, \bar{r} A \, [\bar{B}]^t \, [D] [\bar{B}] \qquad (6.5)$$

where the matrix $\bar{B}$ is evaluated with r and z substituted with $\bar{r}$ and $\bar{z}$ and, i.e.,

$$\bar{r} = \frac{1}{3}(r_1 + r_2 + r_3) \text{ and } \bar{z} = \frac{1}{3}(z_1 + z_2 + z_3) \qquad (6.6)$$

6.2.1 Equivalent Nodal Force Vector

As discussed earlier, there are three types of prescribed forces. The components of a prescribed nodal concentrated force which represents a circular uniform line load, are directly added on to the global force vector. The other two forces to be considered are the prescribed body forces and surface tractions.

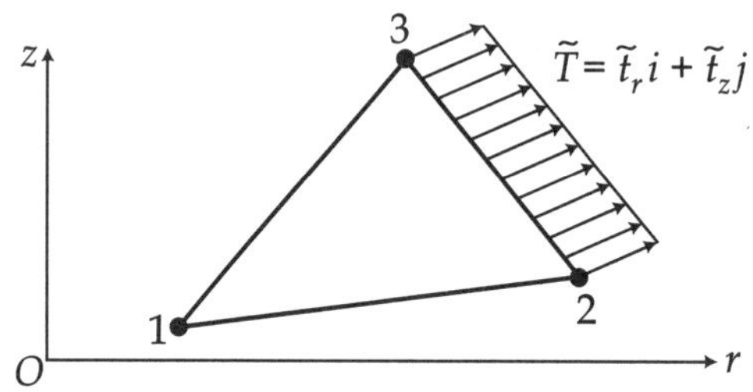

Fig. 6.2 Prescribed Body Forces and Surface Tractions

Type 1. Equivalent Force Vector for Prescribed Body Forces

Here the components of the body forces per unit volume are represented by a vector $\{\tilde{b}_r \; \tilde{b}_z\}^t$. The components of the equivalent force vector for this case are computed by differentiating the work potential function V with respect to the corresponding nodal displacements.

$$\mathbf{V} = -\int_{vol} \{\tilde{b}_r \; \tilde{b}_z\} \begin{Bmatrix} u \\ w \end{Bmatrix} r \, dr \, d\theta \, dz$$

$$= -2\pi \iint_A \{\tilde{b}_r\, u + \tilde{b}_z\, w\} r\, dr\, dz \tag{6.7}$$

$$= -2\pi \iint_A \left\{ \tilde{b}_r \sum_i N_i\, u_i + \tilde{b}_z \sum_i N_i\, w_i \right\} r\, dr\, dz$$

Therefore the component of the equivalent force vector due to $\tilde{b}$ in the r-direction is given as

$$f_{ri} = -\frac{\partial \mathbf{V}}{\partial u_i} = 2\pi \tilde{b}_r \iint_A N_i r\, dr\, dz \tag{6.8}$$

Note that the shape function $N_i = L_i$, the area coordinates, for i = 1, 2, 3, see Eq. (4.16). Therefore the above expression can be integrated in two ways. One way is to assume the value of r equal to $\bar{r}$ as given in Eq. (6.6) and then integrating the rest,

$$f_{ri} = \frac{2}{3} \pi \bar{r}\, \tilde{b}_r\, A \quad \text{for } i = 1, 2, 3. \tag{6.9}$$

Similarly, the equivalent force vector for the z-direction can be derived as

$$f_{zi} = \frac{2}{3} \pi \bar{r}\, \tilde{b}_z\, A \quad \text{for } i = 1, 2, 3. \tag{6.10}$$

This is equivalent to lumping the total force equally among the nodes. The second way is to integrate the whole expression explicitly, without making the approximation, i.e.,

$$f_{r1} = 2\pi \tilde{b}_r \iint_A N_1 \left(L_1\, r_1 + L_2\, r_2 + L_3\, r_3\right) dr\, dz \tag{6.11}$$

$$= \frac{\pi}{6} \tilde{b}_r\, A(2r_1 + r_2 + r_3)$$

and similar expressions can be written for the other components. The difference in the solution using these two equivalent force vectors are negligible, if the body force vector is not the predominant prescribed force.

Type 2. Equivalent Force Vector for Surface Traction

For convenience, let us assume that side 2-3 of the triangular element has a uniform prescribed force with components $\tilde{T}_r$ and $\tilde{T}_z$ per unit surface area respectively as shown in Fig. 6.2. The work potential function is

$$\mathbf{V} = -2\pi \int_{l_{2-3}} (\tilde{T}_r u + \tilde{T}_z w) r\, dl \tag{6.12}$$

$$= -2\pi \int_{l_{2-3}} \left\{ \tilde{T}_r (L_2 u_2 + L_3 u_3) + \tilde{T}_z (L_2 w_2 + L_3 w_3) \right\} (L_2 r_2 + L_3 r_3)\, dl$$

Therefore

$$f_{r2} = \frac{\pi}{3} l_{2\text{-}3} \tilde{T}_r (2r_2 + r_3) \text{ and } f_{r3} = \frac{\pi}{3} l_{2\text{-}3} \tilde{T}_r (r_2 + 2r_3)$$

$$f_{z2} = \frac{\pi}{3} l_{2\text{-}3} \tilde{T}_z (2r_2 + r_3) \text{ and } f_{z3} = \frac{\pi}{3} l_{2\text{-}3} \tilde{T}_z (r_2 + 2r_3) \tag{6.13}$$

where $$l_{2\text{-}3} = \sqrt{(r_3 - r_2)^2 + (z_3 - z_2)^2}$$

If the side 2-3 of the element is vertical, i.e., $r_2 = r_3$ = a constant, then by substituting this in the above equation, the component equivalent force vector f_{r2} will be equal to f_{r3} and each will be equal to half of the total force the side 2-3 is subjected to in the r-direction. If the external applied force varies linearly from node 2 to node 3, as in the case of a hydrostatic pressure, the integral for the work potential contains three functions such as $\tilde{T}_r$, u and r to be linearly interpolated using the area coordinates L_2 and L_3 on the side 2-3. These integrations are left as an exercise to the reader.

6.3 FOUR-NODED RECTANGULAR RING ELEMENT

Vallabhan and Reese[5] developed four nodded rectangular elements in order to model geotechnical stress problems involving footings, pavements, circular caisson foundations, etc. The formulation of the element stiffness matrix is very similar to the four noded plane stress element explained in Chapter 5. As in the case of the plane stress problem, the stiffness matrix of this element can be explicitly integrated. Only a brief explanation is given here, the rest of the formulation is as an exercise for the student. In general, the four noded quadrilateral element has become versatile and used extensively in practice.

Step 1. Assume Displacement Functions.

Using a similar figure for the rectangular element as given in Fig. 5.2, the displacement functions can be written as:

$$u(r, z) = \sum_{i=1}^{4} N_i u_i, \quad \text{and}$$

$$w(r, z) = \sum_{i=1}^{4} N_i w_i \tag{6.14}$$

using the same shape functions given in Eq. (5.2) where x and y are replaced with r and z.

Step 2. Internal Strains

The internal strain vector is developed similar to Eqs. (6.3) and (5.3).

$$\{\varepsilon\} = \begin{Bmatrix} \dfrac{\partial u}{\partial r} \\ \dfrac{u}{r} \\ \dfrac{\partial w}{\partial z} \\ \dfrac{\partial u}{\partial z} + \dfrac{\partial w}{\partial r} \end{Bmatrix}$$

$$= \frac{1}{4ab}\begin{bmatrix} -(b-z) & 0 & (b-z) & 0 & (b+z) & 0 & -(b+z) & 0 \\ \dfrac{N_1}{r} & \dfrac{N_1}{r} & \dfrac{N_2}{r} & \dfrac{N_2}{r} & \dfrac{N_3}{r} & \dfrac{N_3}{r} & \dfrac{N_4}{r} & \dfrac{N_4}{r} \\ 0 & -(a-r) & 0 & -(a+r) & 0 & (a+r) & 0 & (a-r) \\ -(a-r) & -(b-z) & -(a+r) & (b-z) & (a+r) & (b+z) & (a-r) & -(b+z) \end{bmatrix} \begin{Bmatrix} u_1 \\ w_1 \\ u_2 \\ w_2 \\ u_3 \\ w_3 \\ u_4 \\ w_4 \end{Bmatrix}$$

$$= [B]\{u\} \tag{6.15}$$

Step 4. Element Stiffness Matrix

Using the matrix D for the stress-strain relations as given in Eq. (2.30), the element stiffness matrix is calculated using the Eq. (6.4). This equation can be integrated explicitly, however if there is a node in the axis of symmetry, then the integrated functions will have a function containing *log r* which becomes minus infinity at the origin, a large negative value of that function is used in the stiffness matrix. Finite element analyses of problems with solutions given later in this chapter are obtained by employing this type of elements.

6.4 FOUR-NODED QUADRILATERAL RING ELEMENT

Similar to the four-noded quadrilateral element for the plane stress or strain problems, one can develop a quadrilateral ring element for solving axisymmetric problems.[2,6,7] The derivations are comparable to the previous case, therefore only a brief outline of the formulation using the four steps is presented here.

Step 1. Assumed Displacement Functions

The displacement functions are expressed by the same equations given in Eqs. (5.9) and (5.10), i.e.,

$$u = \sum_{i=1}^{4} N_i(\xi, \eta)u_i$$

$$w = \sum_{i=1}^{4} N_i(\xi, \eta) w_i \tag{6.16}$$

using the same shape functions given in Eq. (5.9) which can be expressed more concisely as

$$N_i = \frac{1}{4}(1 + \xi_i\xi)(1 + \eta_i\eta) \quad \text{for } i = 1, 2, 3, 4$$

where ξ_i, η_i are the respective nodal coordinates such that their values are:

$$\xi_i \Rightarrow (-1 \quad 1 \quad 1 \ -1)$$
$$\eta_i \Rightarrow (-1 - 1 \quad 1 \quad 1)$$

Representation of the shape function in the above form has some advantage in computer coding especially for higher order elements.

Step 2. Internal Strains

The components of the internal strains are derivatives of the displacements with respect to r and z, while the displacements are expressed as functions of ξ, η and the Jacobian transformation matrix and its inverse are necessary here also. The coordinates of an internal point within the element are:

$$r = \sum_{i=1}^{4} N_i(\xi, \eta)\, r_i$$

$$z = \sum_{i=1}^{4} N_i(\xi, \eta)\, z_i \tag{6.17}$$

The Jacobian matrix is expressed in the same way as given in Eqs. (5.11) through (5.14), the only difference being that the nodal coordinates r_i, z_i are to be substituted instead of x_i, y_i in these equations. The components of the strain and their relationship with displacements are given previously in Eq. (6.3), but they are expressed in a different form so as to facilitate the transformation of the components into the ξ, η coordinate system..

$$\{\varepsilon\} = \begin{Bmatrix} \varepsilon_r \\ \varepsilon_\theta \\ \varepsilon_z \\ \gamma_{rz} \end{Bmatrix} = \begin{Bmatrix} \dfrac{\partial u}{\partial r} \\ \dfrac{u}{r} \\ \dfrac{\partial w}{\partial z} \\ \dfrac{\partial u}{\partial z} + \dfrac{\partial w}{\partial r} \end{Bmatrix} = \begin{bmatrix} 1 & 0 & 0 & 0 & 0 \\ 0 & 0 & 0 & 0 & \dfrac{1}{r} \\ 0 & 0 & 0 & 1 & 0 \\ 0 & 1 & 1 & 0 & 0 \end{bmatrix} \begin{Bmatrix} \dfrac{\partial u}{\partial r} \\ \dfrac{\partial u}{\partial z} \\ \dfrac{\partial w}{\partial r} \\ \dfrac{\partial w}{\partial z} \\ u \end{Bmatrix} \tag{6.18}$$

Now the vector on the right hand side of Eq. (6.18) is expressed as:

$$\begin{Bmatrix} \dfrac{\partial u}{\partial r} \\ \dfrac{\partial u}{\partial z} \\ \dfrac{\partial w}{\partial r} \\ \dfrac{\partial w}{\partial z} \\ u \end{Bmatrix} = \begin{bmatrix} J_{11}^* & J_{12}^* & 0 & 0 & 0 \\ J_{21}^* & J_{22}^* & 0 & 0 & 0 \\ 0 & 0 & J_{11}^* & J_{12}^* & 0 \\ 0 & 0 & J_{21}^* & J_{22}^* & 0 \\ 0 & 0 & 0 & 0 & 1 \end{bmatrix} \begin{Bmatrix} \dfrac{\partial u}{\partial \xi} \\ \dfrac{\partial u}{\partial \eta} \\ \dfrac{\partial w}{\partial \xi} \\ \dfrac{\partial w}{\partial \eta} \\ u \end{Bmatrix} \tag{6.19}$$

Here the matrix $[J^*]$ is a 2×2 matrix which is the inverse of $[J]$,

$$[J]^{-1} = \frac{1}{\det|J|} \begin{bmatrix} J_{22} & -J_{12} \\ -J_{21} & J_{11} \end{bmatrix} = \begin{bmatrix} J_{11}^* & J_{12}^* \\ J_{21}^* & J_{22}^* \end{bmatrix}$$

the same as given in Eqs. 5.14 though 5.16.

Thus,

$$\{\varepsilon\} = \begin{bmatrix} J_{11}^* & J_{12}^* & 0 & 0 & 0 \\ 0 & 0 & 0 & 0 & \dfrac{1}{r} \\ 0 & 0 & J_{21}^* & J_{22}^* & 0 \\ J_{21}^* & J_{22}^* & J_{11}^* & J_{12}^* & 0 \end{bmatrix} \begin{Bmatrix} \dfrac{\partial u}{\partial \xi} \\ \dfrac{\partial u}{\partial \eta} \\ \dfrac{\partial w}{\partial \xi} \\ \dfrac{\partial w}{\partial \eta} \\ u \end{Bmatrix} \tag{6.20}$$

and

$$\begin{Bmatrix} \dfrac{\partial u}{\partial \xi} \\ \dfrac{\partial u}{\partial \eta} \\ \dfrac{\partial w}{\partial \xi} \\ \dfrac{\partial w}{\partial \eta} \\ u \end{Bmatrix} = \begin{Bmatrix} \sum N_{i,\xi} u_i \\ \sum N_{i,\eta} u_i \\ \sum N_{i,\xi} w_i \\ \sum N_{i,\eta} w_i \\ \sum N_i u_i \end{Bmatrix}$$

$$
= \begin{bmatrix} N_{1,\xi} & 0 & N_{2,\xi} & 0 & N_{3,\xi} & 0 & N_{4,\xi} & 0 \\ N_{1,\eta} & 0 & N_{2,\eta} & 0 & N_{3,\eta} & 0 & N_{4,\eta} & 0 \\ 0 & N_{1,\xi} & 0 & N_{2,\xi} & 0 & N_{3,\xi} & 0 & N_{4,\xi} \\ 0 & N_{1,\eta} & 0 & N_{2,\eta} & 0 & N_{3,\eta} & 0 & N_{4,\eta} \\ N_1 & 0 & N_2 & 0 & N_3 & 0 & N_4 & 0 \end{bmatrix} \begin{Bmatrix} u_1 \\ w_1 \\ u_2 \\ w_2 \\ u_3 \\ w_3 \\ u_4 \\ w_4 \end{Bmatrix} \tag{6.21}
$$

Substituting all the equations in correct order, the B matrix can be obtained as

$$
[B] = \begin{bmatrix}
J^*_{11} N_{1,\xi} + J^*_{12} N_{1,\eta} & 0 & J^*_{11} N_{2,\xi} + J^*_{12} N_{2,\eta} & 0 & J^*_{11} N_{3,\xi} + J^*_{12} N_{3,\eta} & 0 & J^*_{11} N_{4,\xi} + J^*_{12} N_{4,\eta} & 0 \\
\frac{N_1}{r} & 0 & \frac{N_2}{r} & 0 & \frac{N_3}{r} & 0 & \frac{N_4}{r} & 0 \\
0 & J^*_{21} N_{1,\xi} + J^*_{22} N_{1,\eta} & 0 & J^*_{21} N_{2,\xi} + J^*_{22} N_{2,\eta} & 0 & J^*_{21} N_{3,\xi} + J^*_{22} N_{3,\eta} & 0 & J^*_{21} N_{4,\xi} + J^*_{22} N_{4,\eta} \\
J^*_{21} N_{1,\xi} + J^*_{22} N_{1,\eta} & J^*_{11} N_{1,\xi} + J^*_{12} N_{1,\eta} & J^*_{21} N_{2,\xi} + J^*_{22} N_{2,\eta} & J^*_{11} N_{2,\xi} + J^*_{12} N_{2,\eta} & J^*_{21} N_{3,\xi} + J^*_{22} N_{3,\eta} & J^*_{11} N_{3,\xi} + J^*_{12} N_{3,\eta} & J^*_{21} N_{4,\xi} + J^*_{22} N_{4,\eta} & J^*_{11} N_{4,\xi} + J^*_{12} N_{4,\eta}
\end{bmatrix}
$$

The element stiffness matrix is obtained by the standard formula

$$
[k] = 2\pi \iint_{area} [B]^t \, [D][B] r \, dr \, dz \tag{6.22}
$$

$$
= 2\pi \sum_{i=1}^{N} \sum_{j=1}^{N} W_i \, W_j \, (r[B]^t \, [D][B] \, || \, J \, ||)_{i,j}
$$

where N is the number of pivotal points in the Gauss quadrature formulae in one direction.

Example Problem 6.1 A Circular Footing on a Layered Soil Medium

There are many axisymmetric problems in foundation engineering such as circular footing, circular pile, piers, caissons etc. For most of these problems classical elasticity solutions are non-existent. Finite element methods become very handy for solving such problems where the soil continuum is layered with different material properties. Here a circular footing is assumed to be resting on the top of a layered soil medium so that the finite element solutions can be compared with the classical Boussinesq solutions.[3] Rectangular axisymmetric four nodded elements are employed. The load is assumed to be flexible, in other words, a uniform pressure is applied on a circle on the top of the soil. The surface displacements are compared with the classical solution as shown in the Fig. 6.3. Steinbrenner[3] having realized that there will be difference in the surface displacements between the Boussinesq solution and the layered solution, made an approximate solution for a layered soil medium where he subtracted the computed displacement at the bottom of the layered medium using Boussinesq equations. The solution given by Stein Brenner matched with the computed displacements well. The vertical stress distribution is compared with the classical solution in Fig. 6.4. Also the vertical along the axis of the symmetry is also compared with the classical solution.

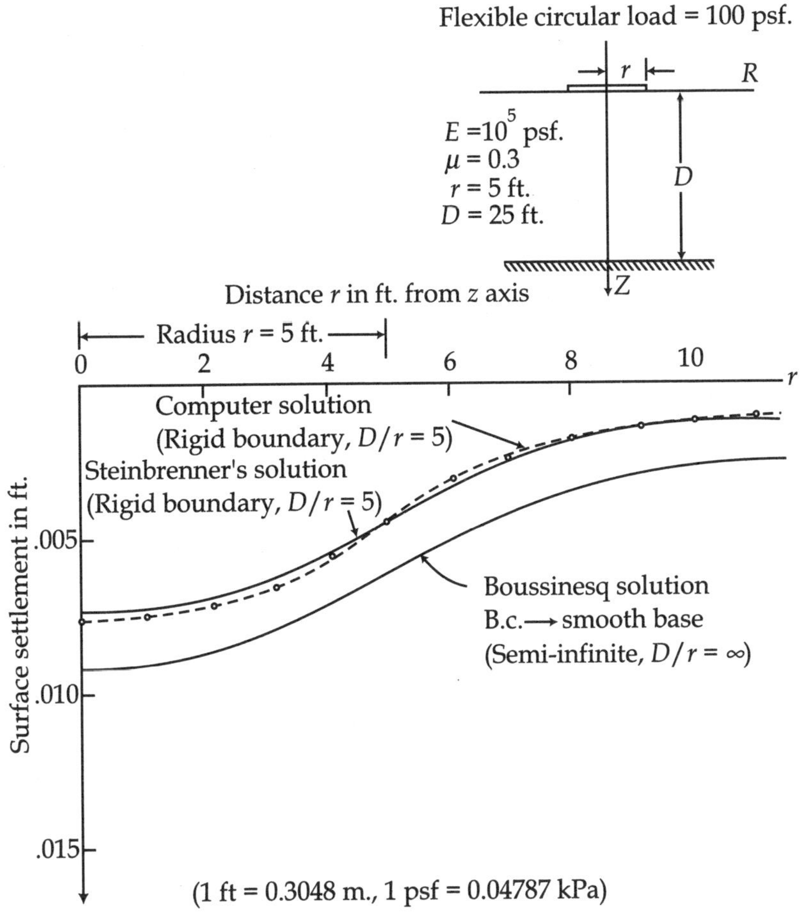

Fig. 6.3 Comparison of Surface Settlement of a Flexible Circular Footing in Semi-infinite and Rigid Boundary Cases

6.5 SUMMARY

Derivation of the element stiffness matrix of a triangular ring element for solving axisymmetric problems is shown here. Equations for computing equivalent nodal force vector for prescribed distributed forces are given. The technique is extended to four noded rectangular and quadrilateral ring elements. The quadrilateral element is transformed into a square. Using the Jacobian of transformation and numerical integrations, the steps for the derivation of the stiffness matrix are shown. An example classical Boussinesq problem of a circular flexible load on a semi infinite continuum is solved with finite dimensions and the solutions are compared with the classical ones.

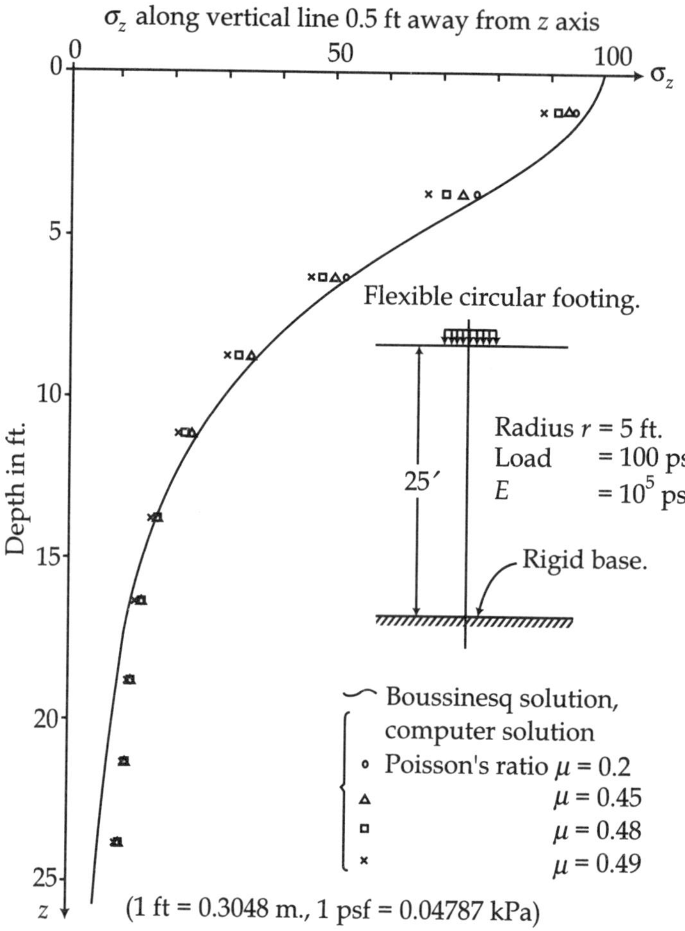

Fig. 6.4 Distribution of σ_z along a Vertical Line 0.5 ft. away from z- Axis

PROBLEMS

6.1 The triangle shown in the figure represents an axisymmetric triangular element. The side 23 is subjected to a uniform traction q as shown. Determine the equivalent nodal forces if the displacement varies linearly within the element and $\tilde{q}$ = 50 kPa.

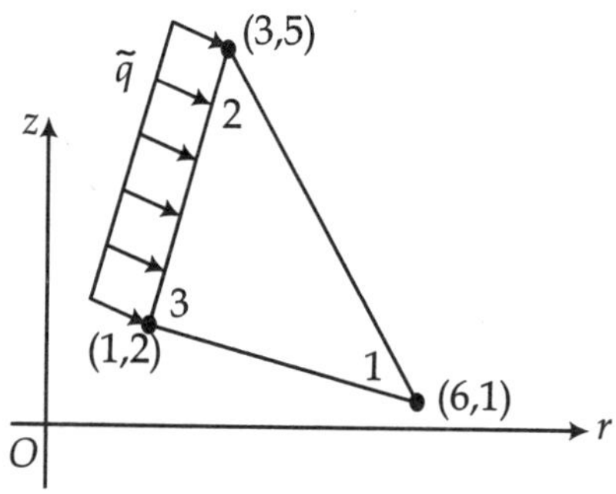

Fig. P6.1

6.2 For the rectangular element given in Fig. P6.2, calculate the element stiffness matrix in terms of modulus of elasticity E and Poisson's ratio μ. (Hint: Jacobian and Shape functions are given in chapter 5 by equations Eqs. 5.14 and 5.15 respectively)

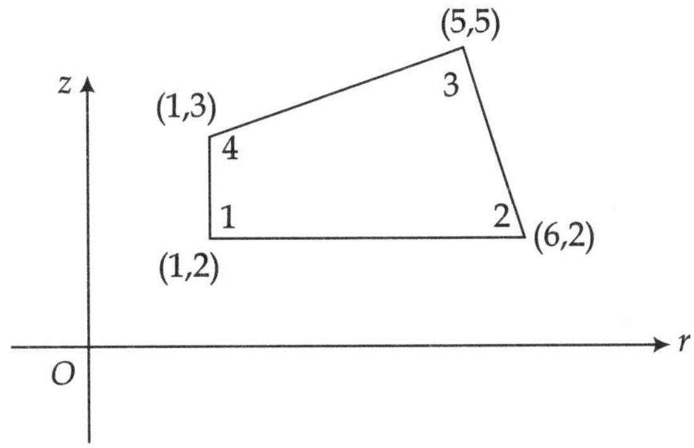

Fig. P6.2

6.3 Set up all the matrices required for the evaluation of the element stiffness matrix of a triangular element for solving axisymmetric problems using quadratic variation of displacements.

6.4 Develop the necessary equations for the stiffness matrix of a quadrilateral axisymmetric finite element. (Use isoparametric element formulation)

6.5 The following axisymmetric problem is given.

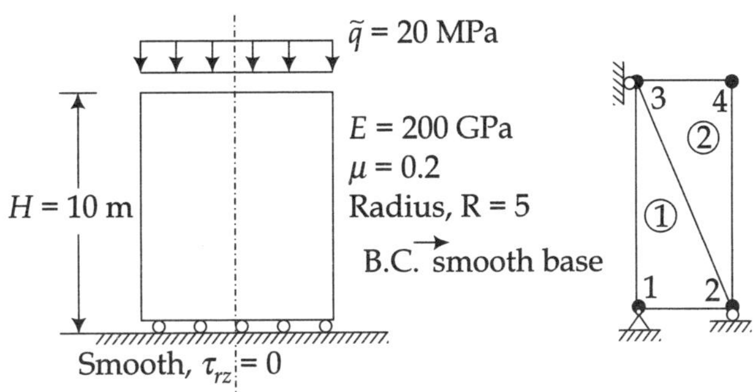

Fig. P6.3

(i) Develop element stiffness matrices
(ii) Global stiffness matrix
(iii) Apply boundary conditions
(iv) Prescribe nodal forces
(v) Calculate nodal displacements
(vi) Calculate internal strain and stresses

6.6 A storage tank with the cross section shown in the figure is constructed to store fluid material. Pressure distribution due to the fluid material on the wall is linearly increasing with the depth and it is 200kPa at the bottom of the tank. Calculate displacements and stresses for the tank by using an available computer code, plot the deformed shape and stresses. Tank material has a modulus of elasticity equal to 200GPa and Poisson's ratio 0.25.

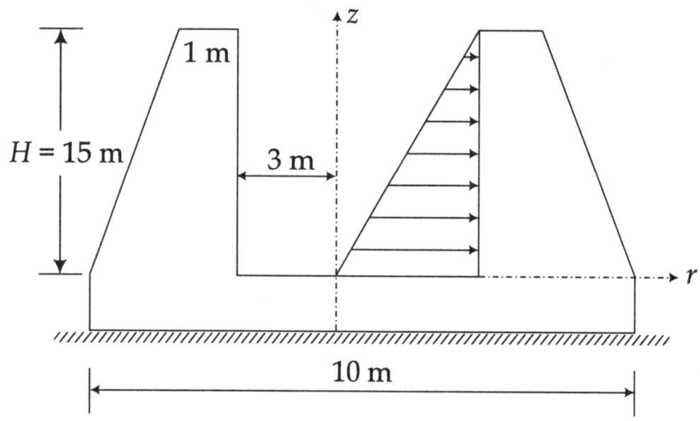

Fig. P6.4

6.7 Consider a semi-infinite solid subjected to a normal force acting on the boundary as shown in the figure. This problem can be modeled as an axisymmetric problem to simulate the behaviour and, to predict the settlement and stresses due to the normal force P distributed over a circle of diameter 0.8 m. To find the deformed shape and stresses, solve the problem in the Fig. P6.5c with E = 65GPa and ν = 0.4. Assume the solid is a hemi-sphere with zero displacements on its lateral surface. The radius of the hemi-sphere can be taken as 20.0 m.

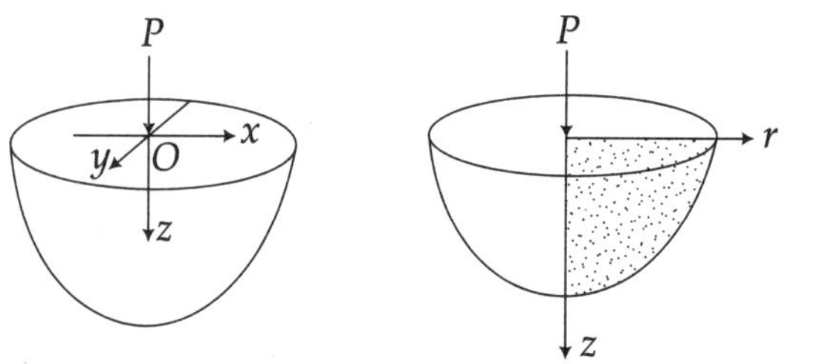

(a) Semi-infinite solid (b) Hemi-sphere model

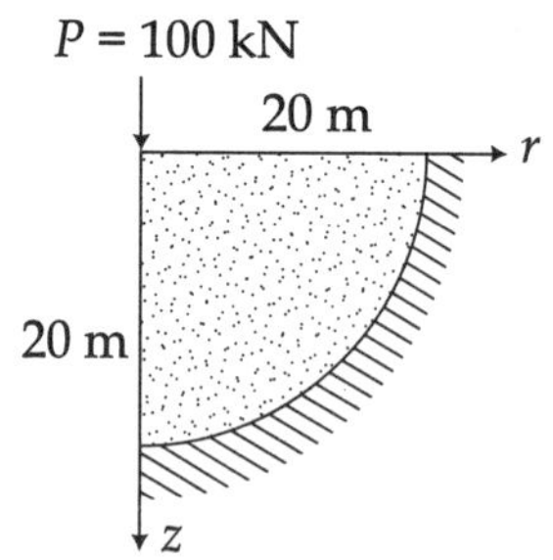

(c) Axisymmetric model

Fig. P6.5

REFERENCES

1. Clough, R.W. and Rashid, Y.R, "Finite Element Analysis of Axisymmetric Solids", Proc. ASCE, Vol.91, EM, 1965.
2. Cook, R.D, Malkus, D.S. and Plesha, M.E, "Concepts and Applications of Finite Element Analysis", John Wiley & Sons, New York, 1989.
3. Terzhaghi, K, "Theoretical Soil Mechanics" John Wiley & Sons, Inc., New York, 1943.
4. Timoshenko, S.P., and Goodier, J.N, "Theory of Elasticity", McGraw Hill Book Company, New York, 1970.
5. Vallabhan, C.V. Girija, and Lymon C. Reese, "Application of Finite Element Methods in Soil Mechanics", Proc. ASCE,Vol. 94, SM2,1968.
6. Weaver, W. and Johnston, P.R, "Finite Elements for Structural Analysis", Prentice Hall, Englewood Cliffs, New Jersey, 1984.
7. Yang, T.Y, "Finite Element Structural Analysis", Prentice Hall, Englewood Cliffs, New Jersey, 1986.
8. Zienkiewicz, O.C. and Taylor, R.L, "The Finite Element Method", Vols. I & II, McGraw Hill Book Company, London, 1989.

7

CHAPTER

Three Dimensional Stress Analysis

7.1 INTRODUCTION

There are only very few closed form solutions for three dimensional stress analysis of solids even with simple geometry. Finite element method provides a relatively simple means of handling the stress analysis of solids having complicated geometry and material properties. The only restriction is that one needs to employ a fairly large number of elements and for that one needs a high speed computer with big storage space to solve the large set of simultaneous equations representing the global system. There are several three dimensional finite elements having different geometry and nodes available in the literature; however, only two types of elements are discussed in this chapter. The most elementary 3-D element is four-noded tetrahedron[1,2,] the theory of which can be studied as an extension of the three-noded triangular plane stress element. This element is not a very convenient one because of the complexity in the preparation of input data; but rapid development of preprocessors for creating input data, three dimensional finite element stress analysis has become easy. Now-a-days, many users prefer the next 3-D element, an eight-noded brick element which is relatively easier for the preparation of input data and they yield slightly better stresses compared to tetrahedron elements with the same number of nodes.

Because the finite element methodology for getting global equations and solution are the same, only derivation of the finite element stiffness matrices are presented.

7.2 FOUR-NODED TETRAHEDRON

For the three dimensional stress analysis, this tetrahedron element was the first element developed by several researchers almost at the same time in different places. The derivation of the element stiffness matrix and equivalent nodal forces is given in the following steps.

Step 1. Assume Displacement Functions

A four-noded tetrahedron is shown in Fig. 7.1. Three dimensional stress analysis requires three internal displacements such as u, v and w, and therefore there are three displacement parameters for each node, thus requiring a 12 d.o.f. element.

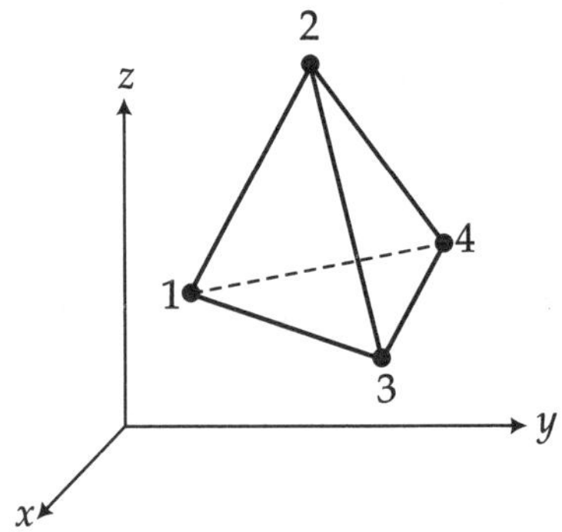

Fig. 7.1 A Tetrahedron Element

Following the techniques used for the constant strain triangular element, 4 parameters can be used for each displacement function so that,

$$
\begin{aligned}
u(x, y, z) &= \alpha_1 + \alpha_2 x + \alpha_3 y + \alpha_4 z \\
v(x, y, z) &= \alpha_5 + \alpha_6 x + \alpha_7 y + \alpha_8 z \\
w(x, y, z) &= \alpha_9 + \alpha_{10} x + \alpha_{11} y + \alpha_{12} z
\end{aligned}
\tag{7.1}
$$

In the above expressions, the rigid body translations are given by α_1, α_5 and α_9, while the rotations with respect to the coordinate axes are:

$$
\begin{aligned}
\theta_x &= \frac{1}{2}\left(\frac{\partial w}{\partial y} - \frac{\partial v}{\partial z}\right) = \frac{1}{2}(\alpha_{11} - \alpha_8) \\
\theta_y &= \frac{1}{2}\left(\frac{\partial u}{\partial z} - \frac{\partial w}{\partial x}\right) = \frac{1}{2}(\alpha_4 - \alpha_{10}) \\
\theta_z &= \frac{1}{2}\left(\frac{\partial v}{\partial x} - \frac{\partial u}{\partial y}\right) = \frac{1}{2}(\alpha_6 - \alpha_3)
\end{aligned}
\tag{7.2}
$$

Constant strain conditions are achieved by the α-parameters except α_1, α_5 and α_9. Since the displacements vary linearly within the element, and by specifying common values for the nodal displacements for neighboring elements, the displacements on the interelement surfaces vary linearly at every point on the surface, and thus the interelement displacement compatibility is satisfied. The next step is to express the internal displacements in terms of the nodal displacements using the shape functions. By extending the technique of area coordinates used in the constant strain triangular elements, 'volume coordinates' (also the natural coordinates for this element) are developed in a tetrahedron. Let us consider an arbitrary point x, y, z within the element. With four triangular faces, this interior point can create four small tetrahedrons with volumes V_1, V_2 V_3 and V_4, where the volume V_i corresponds to the tetrahedron opposite to node i, for $i = 1, 2, 3, 4$. If V is the volume of the tetrahedron 1234, then

$$V_1 + V_2 + V_3 + V_4 = V \tag{7.3}$$

and by defining a new set of volume coordinates L_i, such that

$$L_i = \frac{V_i}{V} \text{ for } i = 1, 2, 3, 4 \tag{7.4}$$

and $\sum_{i=1}^{4} L_i = 1.0$ and the value of $L_i = 1.0$ at node i and equal to zero at all other nodes.

This way the new set of functions defined by L_i has properties of the shape functions and hence used in the development of the stiffness matrix, i.e.,

$$N_i = L_i \text{ and this can be written as}$$

$$= \frac{1}{6V}(a_i + b_i x + c_i y + d_i z) \text{ for } i = 1, 2, 3, 4 \tag{7.5}$$

where the coefficients a_i, b_i, c_i and d_i can be computed from the following equations.

$$a_i = \det \begin{vmatrix} x_j & y_j & z_j \\ x_k & y_k & z_k \\ x_l & y_l & z_l \end{vmatrix}, \qquad b_i = -\det \begin{vmatrix} 1 & y_j & z_j \\ 1 & y_k & z_k \\ 1 & y_l & z_l \end{vmatrix}$$

$$c_i = \det \begin{vmatrix} x_j & 1 & z_j \\ x_k & 1 & z_k \\ x_l & 1 & z_l \end{vmatrix} \quad \text{and} \quad d_i = -\det \begin{vmatrix} x_j & y_j & 1 \\ x_k & y_k & 1 \\ x_l & y_l & 1 \end{vmatrix} \tag{7.6}$$

The subscripts i, j, k, l vary in a cyclic order and the volume of the tetrahedron, V is given as

$$6V = \det \begin{vmatrix} 1 & x_i & y_i & z_i \\ 1 & x_j & y_j & z_j \\ 1 & x_k & y_k & z_k \\ 1 & x_l & y_l & z_l \end{vmatrix} \tag{7.7}$$

The numbering of the nodes in the tetrahedron should be such that, if one views the solid through the node 1, the nodes on the other triangular face must be numbered in an anti-clockwise manner 2, 3, 4 as shown in the Fig. 7.1. The above equations are valid only if one numbers the tetrahedron this way. Also any polynomials of these shape functions can be integrated using the following formula:

$$\int_{vol} L_1^a \, L_2^b \, L_3^c \, L_4^d \, dvol = \frac{a!\,b!\,c!\,d!}{(a+b+c+d+3)!} 6V \tag{7.8}$$

Now the three displacements within the element are interpolated as:

$$u = \sum_{i=1}^{4} N_i u_i, \quad v = \sum_{i=1}^{4} N_i v_i \text{ and } w = \sum_{i=1}^{4} N_i w_i \tag{7.9}$$

and thus
$$\begin{Bmatrix} u \\ v \\ w \end{Bmatrix} = [N]\{u\} \tag{7.10}$$

where $[N]$ is a matrix containing the shape functions and $\{u\}$ is the nodal displacement vector equal to

$$\{u_1 \quad v_1 \quad w_1 \quad u_2 \quad v_2 \quad w_2 \quad \dots \quad \dots \quad \dots \quad u_4 \quad v_4 \quad w_4\}^t$$

Step 2. Internal Strains

The internal strains can be expressed using the strain-displacement equations given in Eq. (2.2). Thus,

$$\{\varepsilon\} = \begin{Bmatrix} \varepsilon_x \\ \varepsilon_y \\ \varepsilon_z \\ \gamma_{xy} \\ \gamma_{yz} \\ \gamma_{zx} \end{Bmatrix} = \begin{Bmatrix} \dfrac{\partial u}{\partial x} \\ \dfrac{\partial v}{\partial y} \\ \dfrac{\partial w}{\partial z} \\ \dfrac{\partial u}{\partial y} + \dfrac{\partial v}{\partial x} \\ \dfrac{\partial w}{\partial y} + \dfrac{\partial v}{\partial z} \\ \dfrac{\partial u}{\partial z} + \dfrac{\partial w}{\partial x} \end{Bmatrix} \tag{7.11}$$

$$= \frac{1}{6V_t} \begin{bmatrix} b_1 & 0 & 0 & b_2 & 0 & 0 & b_3 & 0 & 0 & b_4 & 0 & 0 \\ 0 & c_1 & 0 & 0 & c_2 & 0 & 0 & c_3 & 0 & 0 & c_4 & 0 \\ 0 & 0 & d_1 & 0 & 0 & d_2 & 0 & 0 & d_3 & 0 & 0 & d_4 \\ c_1 & b_1 & 0 & c_2 & b_2 & 0 & c_3 & b_3 & 0 & c_4 & b_4 & 0 \\ 0 & d_1 & c_1 & 0 & d_2 & c_2 & 0 & d_3 & c_3 & 0 & d_4 & c_4 \\ d_1 & 0 & b_1 & d_2 & 0 & b_2 & d_3 & 0 & b_3 & d_4 & 0 & b_4 \end{bmatrix} \begin{Bmatrix} u_1 \\ v_1 \\ w_1 \\ u_2 \\ v_2 \\ w_3 \\ u_3 \\ v_3 \\ w_3 \\ u_4 \\ v_4 \\ w_4 \end{Bmatrix}$$

$$= [B]\{u\}$$

Step 3. Internal Stresses

Using the stress strain relationships given in Eq. (2.14) for an ideal linearly elastic, homogeneous and isotropic material,

$$\{\sigma\} = [D]\{\varepsilon\}$$

Step 4. Development of Stiffness Matrix and Force Vector

Having obtained the matrices B and D, the element stiffness matrix can be written as

$$[k] = \int_{vol} [B]^t [D] [B] \, dvol$$

$$= [B]^t [D] [B] V \tag{7.12}$$

because the matrices B and D are independent of (x, y, z). V is the volume of the element.

At this point, a few more ideas and expressions may be found to be advantageous for the development of the element stiffness matrix. The coefficients a_i, $i = 1,\ldots, 4$ are not necessary for the computation of the stiffness matrix, hence not calculated. Upon examination of the determinants given in Eq. (7.6), the coefficients b_i, c_i, d_i can be computed by some simple formulae, For example, the coefficient b_1 is computed as

$$b_1 = (y_2 - y_4)(z_3 - z_4) - (y_3 - y_4)(z_2 - z_4)$$

Recognizing the similarity of expressions for these coefficients, the nodal coordinates are transformed with node 4 as the origin, such that

$$x_{i4} = x_i - x_4$$

$$y_{i4} = y_i - y_4$$

$$z_{i4} = z_i - z_4 \tag{7.13}$$

Using these new coordinates, the coefficients for $i = 1, 2, 3$ are:

$$b_i = y_{j4}\, z_{k4} - y_{k4}\, z_{j4}$$

$$c_i = x_{k4}\, z_{j4} - x_{j4}\, z_{k4}$$

$$d_i = x_{j4}\, y_{k4} - x_{k4}\, y_{j4} \tag{7.14}$$

where i, j, k vary in a cyclic order. The respective fourth coefficients are

$$b_4 = -(b_1 + b_2 + b_3)$$

$$c_4 = -(c_1 + c_2 + c_3)$$

$$d_4 = -(d_1 + d_2 + d_3) \tag{7.15}$$

The volume of the tetrahedron, V is computed as:

$$V = \frac{1}{6}(x_{14}b_1 + y_{14}c_1 + z_{14}d_1) \tag{7.16}$$

7.2.1 The Equivalent Force Vectors

The prescribed nodal force vectors are read directly from the input data. The equivalent surface traction vector is

$$\{f\}_T = \int_A [N]^t \begin{Bmatrix} \tilde{T}_x \\ \tilde{T}_y \\ \tilde{T}_z \end{Bmatrix} dA \tag{7.17}$$

where A is the area of the surface of the tetrahedron. This integral can be computed using the formula for the area coordinates on the triangular surface. The body force vector

$$\{f\}_b = \int_V [N]^t \begin{Bmatrix} \tilde{b}_x \\ \tilde{b}_y \\ \tilde{b}_z \end{Bmatrix} dvol \tag{7.18}$$

can be computed using the integration formula given in Eq. (7.8).

7.2.2 Assembly of Element Stiffness Matrices

This is accomplished by a procedure similar to what has been discussed in Chapter 3. Because of the three displacement parameters at each node i, the global displacement vector $\{Q\}$ numbered according to global numbering as

$$\begin{aligned} u_i &= Q_{3i-2} \\ v_i &= Q_{3i-1} \\ w_i &= Q_{3i} \end{aligned} \tag{7.19}$$

Because of the difficulty of visualizing each tetrahedron in an assemblage of several such elements to form a solid continuum, finite element analysts have been discretizing the solid as eight-noded brick elements and each eight-noded brick element is considered to be composed of 5 tetrahedra. In other words, the input data is prepared for these eight-noded brick elements and the computer code divides each eight-noded element into five internal tetrahedra. The element stiffness matrices of all tetrahedra are further assembled to form the global stiffness matrix and the associated force vector.

Tetrahedron elements are employed in dynamic finite element analysis involving elastic-plastic behavior during impact between bodies, but in general eight-noded brick elements has become more popular among stress analysts.

7.3 EIGHT-NODED BRICK ELEMENT

The use of numerical integration procedures using Gauss quadrature formulas led to the development of more complex finite element formulations. One of the application is in the development of eight-noded brick elements also called eight noded hexahedron[1.2.3]. The development of element stiffness matrix is as follows.

Step 1. Assumed Displacement Functions

With eight nodes, the basic element has 24 degrees of freedom as shown in Fig. 7.2a. Selecting the appropriate terms from the Pascal pyramid (here it is a pyramid instead of a triangle used for two dimensional problems), such that they satisfy the displacement compatibility between adjacent elements, each displacement at any point inside the element in the x, y and z axes is expressed as

$$\begin{aligned} u(x, y, z) &= \alpha_1 + \alpha_2 x + \alpha_3 y + \alpha_4 z + \alpha_5 xy + \alpha_6 yz + \alpha_7 xz + \alpha_8 xyz \\ v(x, y, z) &= \alpha_9 + \alpha_{10} x + \alpha_{11} y + \alpha_{12} z + \alpha_{13} xy + \alpha_{14} yz + \alpha_{15} xz + \alpha_{16} xyz \\ w(x, y, z) &= \alpha_{17} + \alpha_{18} x + \alpha_{19} y + \alpha_{20} z + \alpha_{21} xy + \alpha_{22} yz + \alpha_{23} xz + \alpha_{24} xyz \end{aligned} \tag{7.20}$$

At this point it is easy to understand the basic principles to be employed in the derivation of the element stiffness matrix. However, the integrations

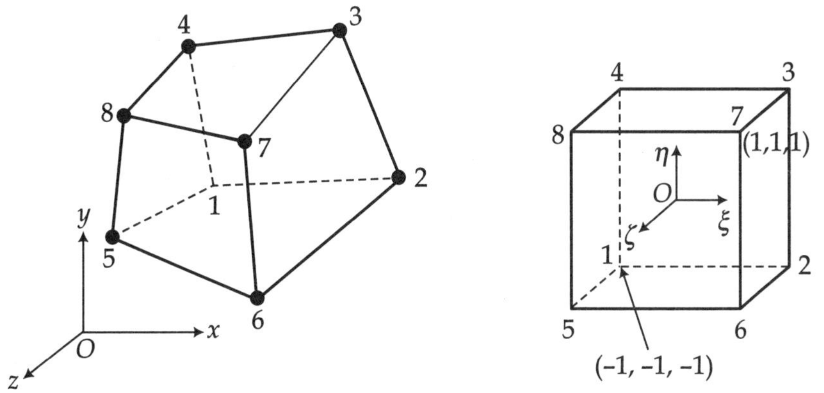

(a) A Hexahedron Element (b) Transformed Element

Fig. 7.2 A Hexahedron Element

involved to develop the element stiffness matrix are too complicated and hence a numerical integration technique has to be employed. To do this one has to transform the basic eight noded element into a cube with side 2 using a coordinate transformation similar to what was done in the basic four-noded quadrilateral element. The origin of the new coordinate axes is at the center of the cube as illustrated in the Fig. 7.2 b. It is important to identify the correspondence of nodal numbering and axes as given in the figure. The internal displacements can be written in terms of the shape functions in the new coordinate system as follows:

$$\begin{aligned} u(x, y, z) &= \sum_{i=1}^{8} N_i(\xi, \eta, \zeta) u_i \\ v(x, y, z) &= \sum_{i=1}^{8} N_i(\xi, \eta, \zeta) v_i \\ w(x, y, z) &= \sum_{i=1}^{8} N_i(\xi, \eta, \zeta) w_i \end{aligned} \tag{7.21}$$

Extending the concepts used to write the shape functions for the quadrilateral element, one can write the required eight shape functions N_i, $i = 1,\ldots,8$. The verification of the shape functions can be made by examining the value of the function by the following formula.

$$N_i(j) = \delta_{ij} \quad \text{where } \delta_{ij} \text{ is the Kronecker delta.}$$

As an example the first shape function at node 1 is written as

$$N_1 = (1 - \xi)(1 - \eta)(1 - \zeta)$$

All shape functions can be represented by one equation using three new terms such as ξ_i, η_i, ζ_i, for $i = 1,\ldots, 8$ which are nothing but the nodal coordinates of the transformed element. The shape functions are:

$$N_i = \frac{1}{8}(1 + \xi_i\,\xi)(1 + \eta_i\,\eta)(1 + \zeta_i\,\zeta) \tag{7.22}$$

The respective values of ξ_i, η_i, ζ_i, for $i = 1,\ldots, 8$ are given below in a vector form.

$$\begin{aligned} \xi_i &\Rightarrow (-1 \;\; 1 \;\; 1 \;\; -1 \;\; -1 \;\; 1 \;\; 1 \;\; -1) \\ \eta_i &\Rightarrow (-1 \;\; -1 \;\; 1 \;\; 1 \;\; -1 \;\; -1 \;\; 1 \;\; 1) \\ \zeta_i &\Rightarrow (-1 \;\; -1 \;\; -1 \;\; -1 \;\; 1 \;\; 1 \;\; 1 \;\; 1) \end{aligned} \tag{7.23}$$

One has to write down all the shape functions using these nodal value vectors and verify as to its appropriateness. This procedure also helps the programmer in determining the derivatives of these shape functions with respect to the ξ, η, ζ coordinates in an easy form.

Step 2. Internal strains

Since the internal strains at a point are derivatives of the displacements with respect to x, y, z axes, and displacements are expressed in (ξ, η, ζ) coordinates, one has to determine the transformation between (x, y, z) and (ξ, η, ζ) coordinates. Observing that the (x, y, z) coordinates can be expressed in terms of the nodal coordinates, therefore

$$\begin{aligned} x &= \sum_{i=1}^{8} N_i x_i \\ y &= \sum_{i=1}^{8} N_i y_i \\ z &= \sum_{i=1}^{8} N_i z_i \end{aligned} \tag{7.24}$$

The Jacobian of the transformation is expressed using the chain rule in partial differentiation and is given below as the matrix J

$$\begin{Bmatrix} \dfrac{\partial}{\partial \xi} \\ \dfrac{\partial}{\partial \eta} \\ \dfrac{\partial}{\partial \zeta} \end{Bmatrix} = \begin{bmatrix} \dfrac{\partial x}{\partial \xi} & \dfrac{\partial y}{\partial \xi} & \dfrac{\partial z}{\partial \xi} \\ \dfrac{\partial x}{\partial \eta} & \dfrac{\partial y}{\partial \eta} & \dfrac{\partial z}{\partial \eta} \\ \dfrac{\partial x}{\partial \zeta} & \dfrac{\partial y}{\partial \zeta} & \dfrac{\partial z}{\partial \zeta} \end{bmatrix} \begin{Bmatrix} \dfrac{\partial}{\partial x} \\ \dfrac{\partial}{\partial y} \\ \dfrac{\partial}{\partial z} \end{Bmatrix} = [J] \begin{Bmatrix} \dfrac{\partial}{\partial x} \\ \dfrac{\partial}{\partial y} \\ \dfrac{\partial}{\partial z} \end{Bmatrix} \tag{7.25}$$

and therefore,

$$\begin{Bmatrix} \dfrac{\partial}{\partial x} \\ \dfrac{\partial}{\partial y} \\ \dfrac{\partial}{\partial z} \end{Bmatrix} = [J]^{-1} \begin{Bmatrix} \dfrac{\partial}{\partial \xi} \\ \dfrac{\partial}{\partial \eta} \\ \dfrac{\partial}{\partial \zeta} \end{Bmatrix} \tag{7.26}$$

The element of the Jacobian matrix

$$J_{11} = \frac{\partial x}{\partial \xi} = \sum_1^8 N_{i,\xi}\, x_i, \quad J_{12} = \frac{\partial y}{\partial \xi} = \sum_1^8 N_{i,\xi}\, y_i, \quad J_{13} = \frac{\partial z}{\partial \xi} = \sum_1^8 N_{i,\xi}\, z_i,$$

and other coefficients can be expressed similarly. The derivatives of shape functions with respect to ξ, η, ζ can be easily expressed as

$$\begin{aligned} N_{i,\xi} &= \frac{1}{8}\xi_i\,(1 + \eta_i\,\eta)(1 + \zeta_i\,\zeta) \\ N_{i,\eta} &= \frac{1}{8}\eta_i\,(1 + \xi_i\,\xi)(1 + \zeta_i\,\zeta) \\ N_{i,\zeta} &= \frac{1}{8}\zeta_i\,(1 + \eta_i\,\eta)(1 + \xi_i\,\xi) \end{aligned} \tag{7.27}$$

Now the determinant and the inverse of the Jacobian has to be determined. The internal strain vector is expressed as

$$\{\varepsilon\} = \begin{Bmatrix} \dfrac{\partial u}{\partial x} \\ \dfrac{\partial v}{\partial y} \\ \dfrac{\partial w}{\partial z} \\ \dfrac{\partial u}{\partial y} + \dfrac{\partial v}{\partial x} \\ \dfrac{\partial v}{\partial z} + \dfrac{\partial w}{\partial y} \\ \dfrac{\partial w}{\partial x} + \dfrac{\partial u}{\partial z} \end{Bmatrix} = \begin{bmatrix} 1 & 0 & 0 & 0 & 0 & 0 & 0 & 0 & 0 \\ 0 & 0 & 0 & 0 & 1 & 0 & 0 & 0 & 0 \\ 0 & 0 & 0 & 0 & 0 & 0 & 0 & 0 & 1 \\ 0 & 1 & 0 & 1 & 0 & 0 & 0 & 0 & 0 \\ 0 & 0 & 0 & 0 & 0 & 1 & 0 & 1 & 0 \\ 0 & 0 & 1 & 0 & 0 & 0 & 1 & 0 & 0 \end{bmatrix} \begin{Bmatrix} \dfrac{\partial u}{\partial x} \\ \dfrac{\partial u}{\partial y} \\ \dfrac{\partial u}{\partial z} \\ \dfrac{\partial v}{\partial x} \\ \dfrac{\partial v}{\partial y} \\ \dfrac{\partial v}{\partial z} \\ \dfrac{\partial w}{\partial x} \\ \dfrac{\partial w}{\partial y} \\ \dfrac{\partial w}{\partial z} \end{Bmatrix} = [A] \begin{bmatrix} J^{-1} & & \\ & J^{-1} & \\ & & J^{-1} \end{bmatrix} \begin{Bmatrix} \dfrac{\partial u}{\partial \xi} \\ \dfrac{\partial u}{\partial \eta} \\ \dfrac{\partial u}{\partial \zeta} \\ \dfrac{\partial v}{\partial \xi} \\ \dfrac{\partial v}{\partial \eta} \\ \dfrac{\partial v}{\partial \zeta} \\ \dfrac{\partial w}{\partial \xi} \\ \dfrac{\partial w}{\partial \eta} \\ \dfrac{\partial w}{\partial \zeta} \end{Bmatrix}$$

$$= [A]\begin{bmatrix} J^{-1} & & \\ & J^{-1} & \\ & & J^{-1} \end{bmatrix} \begin{Bmatrix} \sum N_{i,\xi} u_i \\ \sum N_{i,\eta} u_i \\ \sum N_{i,\zeta} u_i \\ \sum N_{i,\xi} v_i \\ \sum N_{i,\eta} v_i \\ \sum N_{i,\zeta} v_i \\ \sum N_{i,\xi} w_i \\ \sum N_{i,\eta} w_i \\ \sum N_{i,\zeta} w_i \end{Bmatrix} = [A]\begin{bmatrix} J^{-1} & & \\ & J^{-1} & \\ & & J^{-1} \end{bmatrix} \{S\}\{u\} = [B]\{u\} \tag{7.28}$$

where the matrix A is the matrix of order 6×9 containing 0 and 1 in the above equation and matrix S contains derivatives of the shape functions with respect to ξ, η, ζ. The stiffness matrix can now be expressed and evaluated as

$$[k] = \iiint_{vol} [B]^t \, [D] \, [B] \, dvol = \sum_i \sum_j \sum_k w_i \, w_j \, w_k \left\{[B]^t [D] \, [B] \, |\, J \,|\right\}_{i,j,k} \tag{7.29}$$

where i,j,k shows the summation of the Gauss quadrature integration. A two point integration is sufficient in this calculation.

7.4 SUMMARY

For analysis of three-dimensional solids, a four noded tetrahedron finite element is presented. This element has constant strains and is used in nonlinear dynamic analysis of solids. Next is the more popular eight noded hexahedron element and this is transformed into an eight noded cube and using Jacobian of transformation and numerical integration, the element stiffness matrix is computed. This element is very popular among engineers to solve problems involving complex solid machine parts and in geomechanics.

REFERENCES

1. Grandin, H, "Fundamentals of The Finite Element Method", Macmillan Publishing Company, New York, 1986.
2. Krishnamoorthy, C.S, "Finite Element Analysis: Theory and Programming", Tata McGraw Hill Publishing Company, New Delhi, 1987.
3. Zienkiewicz, O.C. and Taylor, R.L, "The Finite Element Method", Vols. I & II, McGraw Hill Book Company, London, 1989.

8

CHAPTER

One-Dimensional Elements (Beam Elements): C_1-Continuity

8.1 INTRODUCTION

The one-dimensional bar elements with C_0-continuity are useful to analyze axially loaded longitudinal bars and trusses. To analyze laterally loaded beams, continuity of slopes of the elements at the common nodes is required. Elements with slope continuity at the interelement nodes are defined as those with at least C_1-continuity. This condition can also be adopted in the analysis of bar elements, by incorporating strain continuity between elements; a condition that is not necessary for analysis of axially loaded bars. The reason is that the energy expression for the bar element is a function of the first derivative, while the energy expression for the beam contains functions of the second derivatives. In order that the beam energy function for the assembly of several beam elements to be integrable, the global displacement function must have at least continuity of the first derivative at the common nodes. This element is used extensively for the analysis of two or three dimensional framed structures and arches. In order to clearly understand the development of the finite element analysis of the beam elements, it is appropriate to briefly review the equations in the classical Euler or sometimes called Euler-Bernoulli beam theory. For a comprehensive reading, any good text book on strength of materials[3,4,7,9] is recommended.

Even though many structural engineers might have studied analyses of continuous beams and frames from matrix structural analysis,[6,11,12] studying these through a finite element view point clarifies many issues that are hidden. Besides, later in this chapter the beam element concepts are extended to include the shear deformation using the Timoshenko beam theory.[3,4] Both Euler beam and Timoshenko beam elements are further extended to develop finite elements for analyses of plates using both Kirchoff and Mindlin theories of plates[5,6,8,13] in the next chapter. A thorough understanding of these of simple one-dimensional beam elements makes the extension to complex two-dimensional plate finite elements rather easy.

8.2 EULER BEAM EQUATIONS

A consistent sign convention for internal moments and shear forces as given in Chapter 1, will be convenient especially when these models are extended to analyze two- and three-dimensional frames and arches. Upon examining the derivation of the governing beam equations, it is interesting to find that the steps necessary to derive the continuous differential equations are exactly parallel to those used in the development of the finite element equations.

Step 1. Assumed Displacement Functions

In order to do this, first of all the assumptions involved in the classical Euler beam theory are examined. Here the beam is assumed to be:

1. linearly elastic, isotropic and homogeneous, such that Hooke's law is valid,
2. prismatic and symmetric with respect to the plane of bending,
3. undergoing small deflections, and the plane cross section before bending remain plane and normal to the deflected centroidal axis, and
4. free from normal stresses perpendicular to the axis of the beam.

Let $v(x)$ be the lateral displacement function of the axis of the beam where the u- displacement is zero as shown in Fig. 8.1.

Based on the last two assumptions, the displacements $\bar{u}$ and $\bar{v}$ at any general point (x,y) in the beam is therefore

$$\bar{u} = -y\frac{dv}{dx}$$
$$\bar{v} = v \tag{8.1}$$

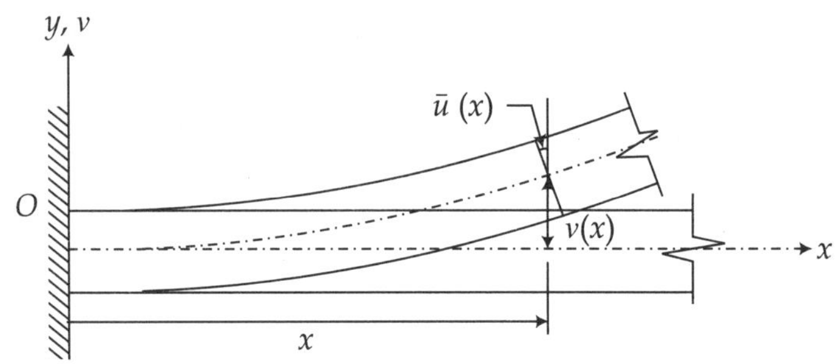

Fig. 8.1 Laterally Loaded Beam

Step 2. Internal Strain at any Point (x, y) in the Beam

Using linear strain displacement equations

$$\varepsilon_x = \frac{d\bar{u}}{dx} = -y\frac{d^2v}{dx^2}$$

$$\varepsilon_y = \frac{d\bar{v}}{dy} = 0,$$

$$\varepsilon_z = 0$$

$$\gamma_{xy} = \frac{\partial \bar{u}}{\partial y} + \frac{\partial \bar{v}}{\partial x} = 0 \tag{8.2}$$

and $\gamma_{yz} = \gamma_{xz} = 0$

Step 3. Corresponding Internal Stress

Using Hooke's law,

$$\sigma_x = E\,\varepsilon_x = -y\,E\,\frac{d^2 v}{dx^2} \tag{8.3}$$

where E is the Young's modulus of elasticity of the material of the beam.

Step 4. Equilibrium of Forces

There are four equations using equilibrium to be considered here. The first two deals with the internal axial force $p(x)$ and internal bending moment $m(x)$ as shown in Fig. 8.2(a). The third one deals with shear force $V(x)$ and applied lateral force $\tilde{q}$ and the fourth one is to develop the relationship of shear force V with the bending moment $m(x)$. All these equations are combined together to get the final governing equilibrium equation of the beam.

With the absence of the internal axial force,

$$p(x) = \int_A \sigma_x \, dA \tag{8.4}$$

$$= -E\,\frac{d^2 v}{dx^2}\int_A y\,dA = 0$$

For nonzero value of $\frac{d^2 v}{dx^2}$, the value of $\int_A y\,dA$ must be equal to zero. This indicates that the longitudinal x-axis passes through the centroid of the beam. The bending moment

$$m(x) = \int_A \sigma_x \, y\,dA = -E\frac{d^2 v}{dx^2}\int_A y^2 dA$$

$$= -EI\,\frac{d^2 v}{dx^2} \tag{8.5}$$

where $I = \int_A y^2 dA$ is the moment of inertia of the area of cross section of the beam with respect to the neutral axis. It is interesting to note that while the shear strain in the beam is assumed to be zero, there will be an internal shear force which exists in the beam to keep it in equilibrium. Considering equilibrium of vertical forces on an element of length dx, as shown in Fig. 8.2.

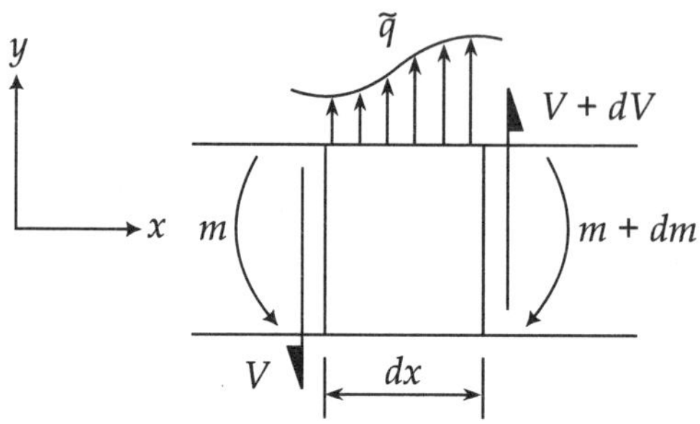

Fig. 8.2 Equilibrium of Forces in the Beam

$V + dV - V + \tilde{q}\ dx = 0$ which leads to

$$\frac{dV}{dx} = -\tilde{q} \tag{8.6}$$

Considering equilibrium of moments about O,

$$-m - dm + m + (V + dV)dx + \tilde{q}\ dx\ \frac{dx}{2} = 0, \quad \text{which leads to}$$

$$\frac{dm}{dx} = V \tag{8.7}$$

Substituting Eqs. (8.6) and (8.7) into Eq. (8.5),

$$\frac{d^2}{dx^2}\left(EI\frac{d^2v}{dx^2}\right) = \tilde{q} \tag{8.8}$$

which is the required equation in the domain of the beam. Now, the strain energy stored in the beam during bending can be written as

$$\mathbf{U} = \frac{1}{2}\int_L m\,d\theta$$

$$= \frac{1}{2}\int_L -EI\frac{d^2v}{dx^2}\,d\left(-\frac{dv}{dx}\right)$$

$$= \frac{1}{2}\int_L EI\left(\frac{d^2v}{dx^2}\right)^2 dx \tag{8.9}$$

where L denotes the length of the beam, and therefore the total potential energy function

$$\mathbf{\Phi} = \mathbf{U} + \mathbf{V}$$

$$= \frac{1}{2}\int_L EI\left(\frac{d^2v}{dx^2}\right)^2 dx - \int_L \tilde{q}v\,dx - \sum_{i=1}^{n_1}\tilde{F}_i\,v_i - \sum_{i=1}^{n_2}\tilde{M}_i\theta_i \tag{8.10}$$

where $\widetilde{M}_i$ and $\widetilde{F}_i$ are the prescribed external moments and forces at various discrete points along the beam and n_1 and n_2 are the number of prescribed forces and moments on the beam, respectively (See Fig. 8.3)

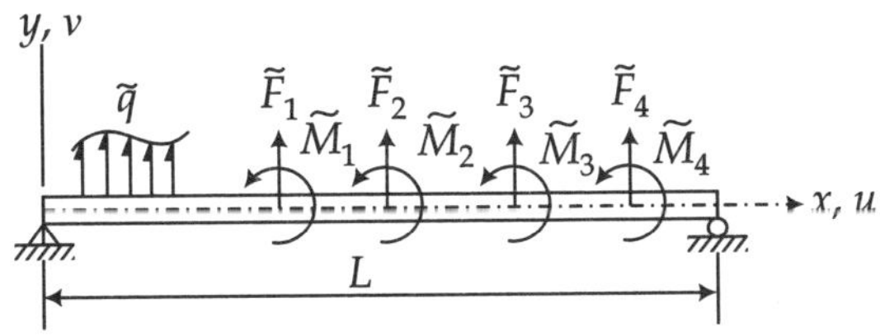

Fig. 8.3 Prescribed Forces and Displacements

8.3 FINITE ELEMENT ANALYSIS OF A BEAM ELEMENT

The stiffness matrix of the beam element of length l and having a uniform moment of inertia I, with two nodes as shown in Fig. 8.4 is developed by the four steps.

Step 1. Assume Displacement Function

To keep the continuity of the slopes at the common nodes, two parameters are used, such as the lateral displacement and slope at each node, making a total of four degrees of freedom for the element. Thus the element displacement vector has the following form $(v_1\ \theta_1\ v_2\ \theta_2)^t$. With four degrees of freedom, a cubic polynomial in x has to be used to define the displacement function $v(x)$ which has four adjustable parameters, and this is given below:

$$v(x) = \alpha_1 + \alpha_2 x + \alpha_3 x^2 + \alpha_4 x^3 \tag{8.11}$$

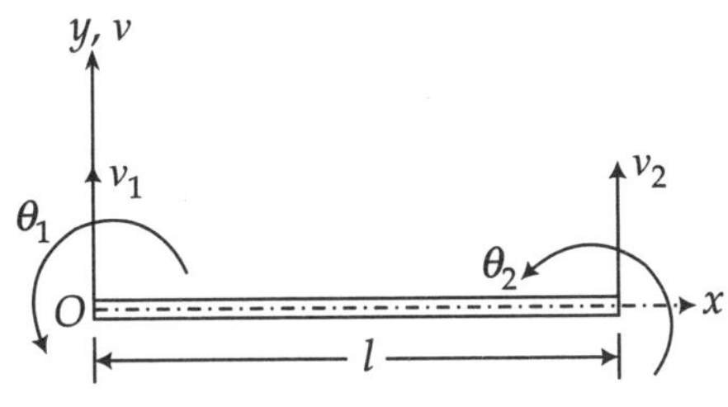

Fig. 8.4 A Beam Element with 4 Degrees of Freedom

Upon examining the above function, both rigid body motions of the element such as the lateral displacement and rotation of the element are expressed by α_1 and α_2 of the polynomial, and the constant strain condition is expressed by the term α_3. See Eq. (8.2). The corresponding shape functions for the element can be derived as follows. Taking O as the origin of the local coordinate system as shown in Fig. 8.4, the nodal parameters can be written as:

$$\begin{aligned} v_1 &= \alpha_1 \\ \theta_1 &= \alpha_2 \\ v_2 &= \alpha_1 + \alpha_2 l + \alpha_3 l^2 + \alpha_4 l^3 \\ \theta_2 &= \alpha_2 + 2\alpha_3 l + 3\alpha_4 l^2 \end{aligned} \tag{8.12}$$

Solving for α's

$$\alpha_1 = v_1$$

$$\alpha_2 = \theta_1$$

$$\alpha_3 = -\frac{3}{l^2}v_1 - \frac{2}{l}\theta_1 + \frac{3}{l^2}v_2 - \frac{1}{l}\theta_2 \tag{8.13}$$

$$\alpha_4 = \frac{2}{l^3}v_1 + \frac{1}{l^2}\theta_1 - \frac{2}{l^3}v_2 + \frac{1}{l^2}\theta_2$$

and substituting in Eq. (8.11),

$$v(x) = N_1 v_1 + N_2\theta_1 + N_3 v_2 + N_4\theta_2 \tag{8.14}$$

where

$$N_1 = 1 - \frac{3x^2}{l^2} + \frac{2x^3}{l^3}$$

$$N_2 = x\left(1 - \frac{2x}{l} + \frac{x^2}{l^2}\right)$$

$$N_3 = \frac{x^2}{l^2}\left(3 - 2\frac{x}{l}\right)$$

$$N_4 = \frac{x^2}{l}\left(\frac{x}{l} - 1\right) \tag{8.15}$$

The shape functions N_i, $i = 1,\ldots, 4$ are called Hermite polynomials[1,6,13] and they are plotted in Fig. 8.5. It is seen that each of these shape functions has a value equal to unity at the node corresponding to the particular degree of freedom, and zero corresponding to all remaining degrees of freedom.

Step 2. Internal Strain

The internal strain for the beam element is designated by the second derivative of the displacement, which is equal to

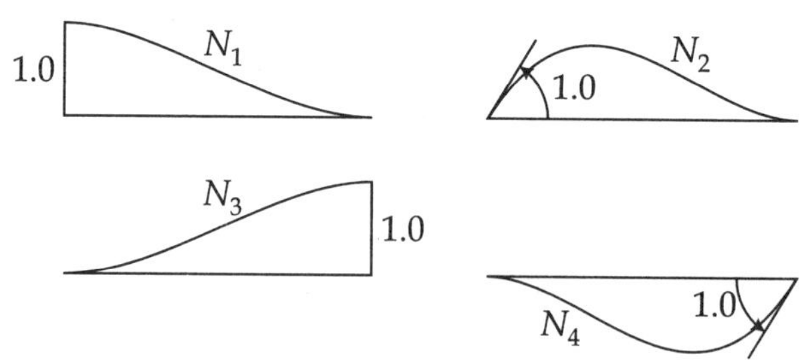

Fig. 8.5 Shape Functions for the Beam Element

$$\frac{d^2 v}{dx^2} = \sum_{i=1}^{4} \frac{d^2 N_i}{dx^2} u_i$$

$$= \frac{1}{l^2}\left[\left(-6+12\frac{x}{l}\right) \quad -4l+6x \quad 6-\frac{12x}{l} \quad 6x-2l\right]\{u\} \tag{8.16}$$

$$= [B]\{u\}$$

Step 3. Internal Stress

The corresponding internal stress is designated by the internal bending moment $m(x)$ which can be written as:

$$m(x) = -EI\frac{d^2v}{dx^2} = -EI[B]\{u\} \tag{8.17a}$$

which varies linearly within the element, and the shear force $V(x)$ is:

$$v(x) = -EI\frac{d^3v}{dx^3} = -\frac{EI}{l^3}(12v_1 + 6l\theta_1 - 12v_2 + 6l\theta_2) \tag{8.17b}$$

which yields a constant value of shear force within the element.

Step 4. Element Stiffness Matrix

Now the expressions for the element stiffness matrix and the equivalent load vector can be written as:

$$[k] = \int_0^l [B]^t (EI)[B]\,dx$$

$$= \frac{EI}{l^3}\begin{bmatrix} 12 & 6l & -12 & 6l \\ & 4l^2 & -6l & 2l^2 \\ & & 12 & -6l \\ & & & 4l^2 \end{bmatrix} \tag{8.18}$$

and

$$\tilde{f}_i = -\frac{\partial \mathbf{V}}{\partial u_i}, \quad \text{for} \quad i = 1, 4 \tag{8.19}$$

$$= + \int_l N_i\,\tilde{q}\,dx$$

Eq. (8.18) can be derived by minimizing the potential energy function expressed in Eq. (8.10).

8.3.1 Equivalent Nodal Forces

When forces are applied at the nodes, they are designated as global nodal forces corresponding to the global nodal numbering of the system. However, in beam problems it is normal to have forces and moments distributed within the beam element, in which case one has to determine the equivalent nodal forces due to these distributed loads. The minimum potential energy theorem or the principle of virtual work can be used to determine these equivalent or consistent nodal forces based on the energy principle. Using the equations given in Eq. (8.19) these equivalent nodal forces for different types of distributed forces, as illustrated in Fig. 8.6(a,b,c and d) are obtained as given below.

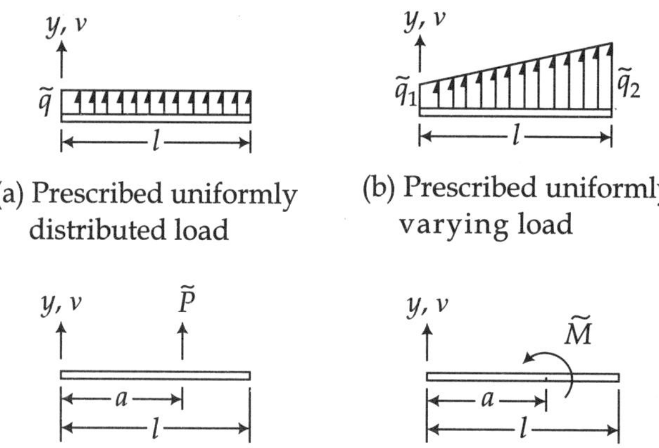

(a) Prescribed uniformly distributed load

(b) Prescribed uniformly varying load

(c) A concentrated load on the beam

(d) A concentrated Moment on the beam

Fig. 8.6 Applied Forces on a Beam Element

Case a. Uniformly Distributed Force $\tilde{q}$

Using the Eq. (8.19), the four consistent forces are obtained such as,

$$\tilde{f}_1 = +\int_0^l N_1 \tilde{q}\, dx = \frac{\tilde{q}l}{2} = \tilde{f}_2$$

$$\tilde{M}_1 = +\int_0^l N_2 \tilde{q}\, dx = \frac{\tilde{q}l^2}{12} = -\tilde{M}_2 \qquad (8.20)$$

Examining these equations, it can be seen that these equations match exactly with the equivalent forces used in the conventional beam analysis, where the equivalent forces are computed as the reversed fixed end forces and moments.

Case b. Uniformly Varying Force Across the Span

If the prescribed force varies uniformly from one end to the other, it can be interpolated using the linear Lagrange functions given in Eq. (3.8), $L_1 = \left(1 - \frac{x}{l}\right)$ and $L_2 = \frac{x}{l}$. Using Eq.(8.19), the equivalent forces are:

$$\tilde{f}_1 = +\int_0^l \left[\left(1 - \frac{x}{l}\right)\tilde{q}_1 + \frac{x}{l}\tilde{q}_2\right] N_1\, dx = \frac{l}{20}\,(7\tilde{q}_1 + 3\tilde{q}_2)$$

and it can be shown

$$\tilde{M}_1 = \frac{l^2}{60}(3\tilde{q}_1 + 2\tilde{q}_2)$$

$$\tilde{f}_2 = \frac{l}{20}(3\tilde{q}_1 + 7\tilde{q}_2), \quad \text{and}$$

$$\tilde{M}_2 = -\frac{l^2}{60}(2\tilde{q}_1 + 3\tilde{q}_2). \qquad (8.21)$$

Case c. A Concentrated Load P within the Span of the Beam

Here the location of the concentrated load $\tilde{P}$ as shown in Fig. 8.6(c). The work potential function due to this concentrated load is

$$\mathbf{V} = -\tilde{P}v|_{x=a} = -\tilde{P}\sum_{i=1}^{4} N_i|_{x=a}\, u_i$$

and therefore, the equivalent forces are:

$$\tilde{f}_1 = -\frac{\partial \mathbf{V}}{\partial v_1} = \tilde{P}\left(1 - \frac{3a^2}{l^2} + \frac{2a^3}{l^3}\right)$$

$$\tilde{M}_1 = -\frac{\partial \mathbf{V}}{\partial \theta_1} = \tilde{P}\left(a - \frac{2a^2}{l} + \frac{a^3}{l^2}\right)$$

$$\tilde{f}_2 = -\frac{\partial \mathbf{V}}{\partial v_2} = \tilde{P}\left(\frac{3a^2}{l^2} - \frac{2a^3}{l^3}\right)$$

and

$$\tilde{M}_2 = -\frac{\partial \mathbf{V}}{\partial \theta_2} = \tilde{P}\left(-\frac{a^2}{l} + \frac{a^3}{l^2}\right)$$

Case d. A Concentrated Moment $\tilde{M}$ Applied within the Span of the Beam

This concentrated moment $\tilde{M}$ is shown in Fig. 8.6(d). The work potential due to this moment can be written as

$$\mathbf{V} = -\tilde{M}\left.\frac{dv}{dx}\right|_{x=a} = -\tilde{M}\sum_{i=1}^{4}\left.\frac{dN_i}{dx}\right|_{x=a} u_i \tag{8.22}$$

and therefore the equivalent forces are:

$$\tilde{f}_1 = \tilde{M}\left(-\frac{6a}{l^2} + \frac{6a^2}{l^3}\right)$$

$$\tilde{M}_1 = \tilde{M}\left(1 - \frac{4a}{l} + \frac{3a^2}{l^2}\right)$$

$$\tilde{f}_2 = \tilde{M}\left(\frac{6a}{l^2} - \frac{6a^2}{l^3}\right)$$

and

$$\tilde{M}_2 = \tilde{M}\left(-\frac{2a}{l} + \frac{3a^2}{l^2}\right) \tag{8.23}$$

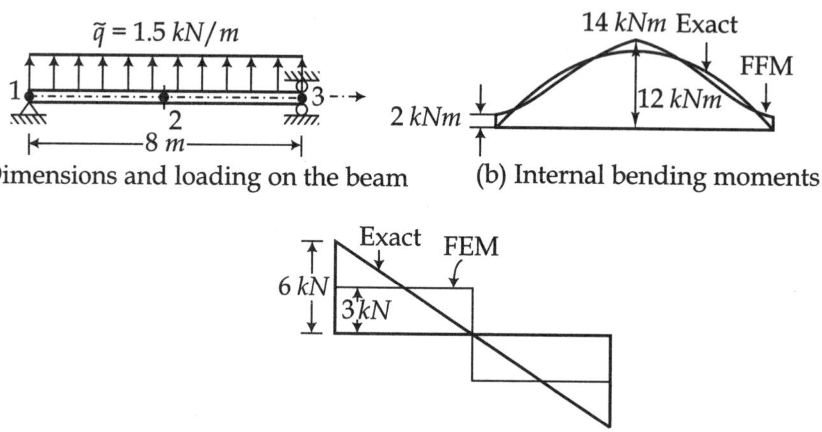

(a) Dimensions and loading on the beam

(b) Internal bending moments

(c) Internal shear forces

Fig. 8.10 An Example of a Simple Beam Problem

Similarly the equivalent forces for any given prescribed force configuration can be determined by differentiating the work potential function with respect to the nodal displacements.

Example Problem 8.1 A simply Supported Beam

A simple numerical example is shown here in order to develop confidence in the analysis procedure. This example also illustrates the various discrepancies between the finite element and the classical beam results. After getting the global displacements, for calculation of internal shear forces and bending moments within the beam, structural engineers use equilibrium of forces and moments within the beam, while finite element solutions yield different values for internal shear and moments.

The beam is simply supported at the ends and has a uniform cross section and carries a uniformly distributed load. The dimensions of the example beam along with other details are given in the Fig. 8.7(a).

Here the beam is divided into two finite elements, thus length of each element, l = 4.0 m., $EI = 10^4$ kNm^2, and $\tilde{q}$ = 1.5 kN/m. the element stiffness matrix for both elements are the same and they are calculated as:

$$[k^1] = [k^2] = \frac{EI}{64}\begin{bmatrix} 12 & 24 & -12 & 24 \\ 24 & 64 & -24 & 32 \\ -12 & -24 & 12 & -24 \\ 24 & 32 & -24 & 64 \end{bmatrix}$$

and the equivalent nodal forces for each element are the same and they are:

$$\{\tilde{f}^1\} = \{\tilde{f}^2\} = \begin{Bmatrix} 3 \\ 2 \\ 3 \\ -2 \end{Bmatrix}$$

The superscripts 1 and 2 on the matrix k and vector $\tilde{f}$ denote the element numbers. After assembling these element matrices, the global stiffness matrix becomes

$$[K]\{u\} = \frac{EI}{64}\begin{bmatrix} 12 & 24 & -12 & 24 & & \\ 24 & 64 & -24 & 32 & & \\ -12 & -24 & 24 & 0 & -12 & 24 \\ 24 & 32 & 0 & 128 & -24 & 32 \\ & & -12 & -24 & 12 & -24 \\ & & 24 & 32 & -24 & 64 \end{bmatrix}\begin{Bmatrix} \tilde{v}_1 \\ \theta_1 \\ \tilde{v}_2 \\ \theta_2 \\ \tilde{v}_3 \\ \theta_3 \end{Bmatrix} = \begin{Bmatrix} 3+\overline{R}_1 \\ 2+\tilde{M}_1 \\ 6 \\ 0 \\ 3+\overline{R}_3 \\ -2+\tilde{M}_3 \end{Bmatrix}$$

The boundary conditions are:

displacements $\tilde{v}_1 = \tilde{v}_3 = 0$ and applied external moments at the ends being hinged $\tilde{M}_1 = \tilde{M}_3 = 0$. Therefore the unknown reactions $\overline{R}_1$ and $\overline{R}_3$ at nodes 1 and 3 respectively are added in the above equation.

The modified global stiffness matrix equation thus becomes:

$$[K]\{u\} = \frac{EI}{64}\begin{bmatrix} 1 & 0 & 0 & 0 & & \\ 24 & 64 & -24 & 32 & & \\ -12 & -24 & 24 & 0 & -12 & 24 \\ 24 & 32 & 0 & 128 & -24 & 32 \\ & & 0 & 0 & 1 & 0 \\ & & 24 & 32 & -24 & 64 \end{bmatrix}\begin{Bmatrix} v_1 \\ \theta_1 \\ v_2 \\ \theta_2 \\ v_3 \\ \theta_3 \end{Bmatrix} = \begin{Bmatrix} 0 \\ 2 \\ 6 \\ 0 \\ 0 \\ -2 \end{Bmatrix}$$

and the solution of these equations yields:

for $EI = 10^4$ kN m^2,

$v_1 = 0$ $\qquad\qquad$ $\theta_1 = 0.0032$ rad

$v_2 = 0.008$ m $\qquad\qquad$ $\theta_2 = 0$ rad

$v_3 = 0$ $\qquad\qquad$ $\theta_3 = -0.0032$ rad

When these results are compared with the exact nodal displacements using the beam theory, the same results are obtained, i.e.,

v_2 at center = 0.008 m.

$\theta_1 = -\theta_3 = 0.0032$ rad.

Now a question arises, how was it possible to get the exact solution for displacement and slopes at the nodes, while only a cubic polynomial within the element was used as the displacement function in the analysis. In reality, the differential equation Eq. (8.8) yields a quartic displacement function in the entire domain of the beam. The reason for getting the some values at the nodes is that for this problem both methods yield the same potential energy in the beam.

Inernal Bending Moments

By substituting these nodal displacement values into Eqs. (8.16), (8.17a) and (8.17b), the internal bending moments and shear forces within the elements are obtained. As explained earlier, in this beam element, the internal bending moment varies linearly and the shear force is a constant within each element. For element 1, the values of the internal bending moments at the nodes are:

$$m_1 = 2.0 \text{ kNm. and } m_2 = 14.0 \text{ kNm.}$$

The corresponding exact values are 0 and 12.0 kNm. respectively. The value of the shear forcein element 1 is a constant and equal to 3.0 kN, while the actual values are 6.0 kN at node 1 and varying linearly to zero at node 2. The internal moments and the shear forces are graphically shown in Fig. 8.7(b) and (c) and these values are compared with the exact values using the elementary beam theory.

An interesting observation here is that since a cubic polynomial is assumed for the displacement function, the bending moment being the second derivative of the displacement, varies linearly, and the shear force being the third derivative remains as a constant within the element. The fourth derivative of the cubic polynomial is zero and therefore it does not match with any of the applied distributed force on the element.

If four elements are used for the finite element analysis, then the results improve as indicated, and it can be shown that as the number of elements are increased, the finite element solution converges to the exact solution.

Reactions at the Ends

The shear forces and moments at the ends of the elements were found to be different from the conventional reactions of the beam at the edges. The differences observed are due to the equivalent forces applied at the nodes of the elements. If the energy theorem is applied correctly including the reactions, then the correct values of the reactions at the ends of the beam can be obtained. For example, let us consider the entire beam subjected to reactive forces and moments at the ends such as $\bar{R}_1, \bar{M}_1, \bar{R}_3$ and $\bar{M}_3$ at the nodes 1 and 3 respectively. In other words, these forces keep the entire beam in equilibrium. For the given problem, the end moments are prescribed, each equal to zero because of the simply supported boundary conditions at the ends. Taking the global stiffness matrix equation before applying the prescribed displacement boundary conditions into the equations,

$$[K]\{u\} = \frac{EI}{64}\begin{bmatrix} 12 & 24 & -12 & 24 & & \\ 24 & 64 & -24 & 32 & & \\ -12 & -24 & 24 & 0 & -12 & 24 \\ 24 & 32 & 0 & 128 & -24 & 32 \\ & & -12 & -24 & 12 & -24 \\ & & 24 & 32 & -24 & 64 \end{bmatrix}\begin{Bmatrix} v_1 \\ \theta_1 \\ v_2 \\ \theta_2 \\ v_3 \\ \theta_3 \end{Bmatrix} = \begin{Bmatrix} 3+\bar{R}_1 \\ 2+\bar{M}_1 \\ 6 \\ 0 \\ 3+\bar{R}_3 \\ -2+\bar{M}_3 \end{Bmatrix}$$

Substituting the values of the calculated global displacement vector into this equation,

$\overline{R}_1 = -6.0$ kN

$\overline{M}_1 = 0$

$\overline{R}_3 = -6.0$ kN and

$\overline{M}_3 = 0$ respectively.

Truly, this technique has been used in structural analysis of beams for obtaining more correct end forces in the beam element from the element nodal displacement vectors. Using similar notations, and examining the element stiffness matrix equation for the first element,

$$\frac{EI}{64}\begin{bmatrix} 12 & 24 & -12 & 24 \\ 24 & 64 & -24 & 32 \\ -12 & -24 & 12 & -24 \\ 24 & 32 & -24 & 64 \end{bmatrix}\begin{Bmatrix} v_1 \\ \theta_1 \\ v_2 \\ \theta_2 \end{Bmatrix} = \begin{Bmatrix} 3+\overline{R}_1 \\ 2+\overline{M}_1 \\ 3+\overline{R}_2 \\ -2+\overline{M}_2 \end{Bmatrix} = \begin{Bmatrix} -3 \\ 2 \\ 3 \\ -14 \end{Bmatrix}$$

The force vector on the right hand side is obtained after multiplying the first element stiffness matrix with the element nodal displacements. Thus the equilibrating forces on the beam element one are:

$\overline{R}_1 = -6$ kN

$\overline{M}_1 = 0$

$\overline{R}_2 = 0$ and

$\overline{M}_2 = 12.0$ kNm respectively.

Incidentally, these values are the exact values of the internal forces and moments in the element 1 for this problem. Similarly, the internal forces and moments in element 2 can also be calculated. The discrepancy between the moments and shear forces calculated using the finite element method and the exact values becomes smaller as the number of elements are increased in the overall discretization.

8.4 ANALYSIS OF NON UNIFORM BEAM ELEMENTS

In structural engineering, often non uniform beams have to be analyzed, particularly beams having variable moment of inertia. If the moment of inertia of the beam is expressed as a function of x, then the element stiffness matrix and the associated nodal forces can be developed using the minimum potential energy theorem. For example, let the beam element has a moment of inertia varying linearly from one end to the other as shown in Fig. 8.8.

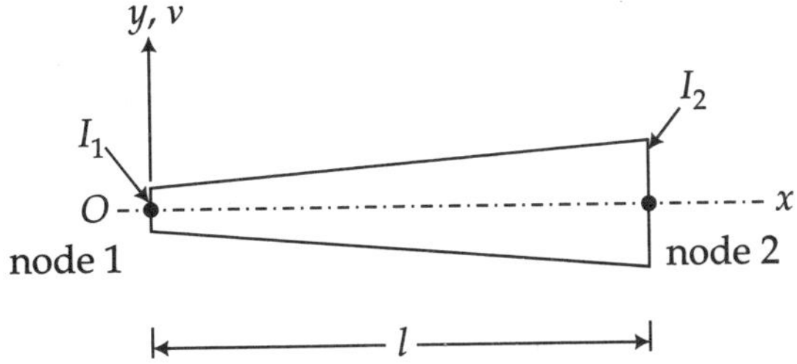

Fig. 8.8 A Beam Element with Variable Moment of Inertia

In such a case, the moment of inertia $I(x)$ can be linearly interpolated using the linear Lagrange interpolation functions L_1 and L_2 given in Eq. (3.8). Thus the potential energy function becomes

$$\Phi = \mathbf{U} + \mathbf{V} = \frac{1}{2}\int_0^l EI_x \left(\frac{d^2 v}{dx^2}\right)^2 - \int_0^l \tilde{q}\, v\, dx$$

and
$$I_x = \left(1 - \frac{x}{l}\right) I_1 + \frac{x}{l} I_2 \tag{8.24}$$

where I_1 and I_2 are the values of moment of inertia at the nodes 1 and 2 respectively, therefore the element stiffness matrix becomes

$$[k] = E\int_0^l \left[\left(1 - \frac{x}{l}\right) I_1 + \frac{x}{l} I_2\right] [B]^t [B] dx \tag{8.25}$$

It is interesting to note that the consistent nodal forces in the finite element formulation is independent of the variation of the moment of inertia of the beam element.

8.5 BEAM ON ELASTIC FOUNDATION

Beam on elastic foundation[2,9] concept was initiated by Winkler (1867) where he assumed that a beam is supported by continuous closely spaced linear springs that represents the foundation (See Fig. 8.9). The spring constant k_s is derived from a term called the 'modulus of subgrade reaction' by geotechnical engineers who employ this concept for modeling many of their soil structure interaction problems. Here, the k_s has the unit of pressure, not to be confused with unit of pressure/displacement as expressed in many text books on geotechnical engineering to represent modulus of subgrade reaction as a property of the soil. Special care has to be taken to convert this property into proper units before using it in Eq. (8.26). The governing equation for beams on elastic foundation is

$$\frac{d^2}{dx^2}\left(EI_x \frac{d^2 v}{dx^2}\right) + k_s v = \tilde{q} \tag{8.26}$$

with proper boundary conditions on the ends of the beam, as given in Chapter 1. Here the y-axis is taken as downward positive in order to keep a practical soil support at the bottom of the beam, and consistent sign conventions are used for the beam parameters.

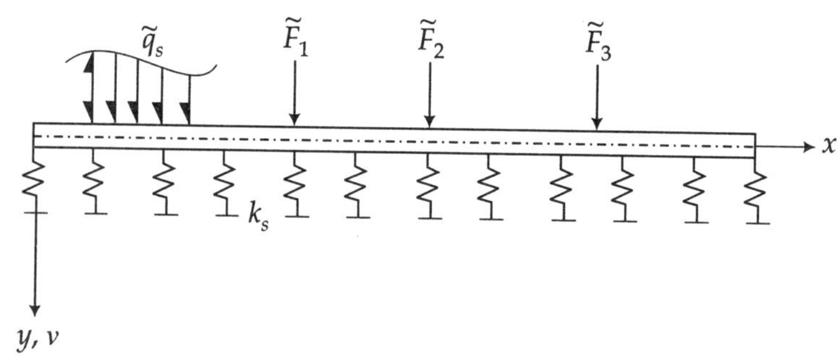

Fig. 8.9 A Beam on Elastic Foundation

8.5.1 Finite Element Model for the Beam on Elastic Foundation

The total potential energy function for the beam element of length l and the corresponding elastic foundation is

$$\Phi = \frac{1}{2}\int_0^l EI_x \left(\frac{d^2v}{dx^2}\right)^2 dx + \frac{1}{2}\int_0^l k_s\, v^2 dx + V \tag{8.27}$$

Assuming 4 d.o.f. the stiffness matrix of the beam element is the same as given in Eq. (8.18), with uniform stiffness EI_x for the element. Substituting Eq. (8.14) for $v(x)$ in the above equation for the energy in the elastic foundation and differentiating with respect to the nodal parameters, the stiffness matrix for the elastic foundation is obtained as,

$$[k_{soil}] = k_s \int_0^l [N]^t\,[N]\,dx$$

$$= \frac{k_s\, l}{420}\begin{bmatrix} 156 & 22l & 54 & -13l \\ & 4l^2 & 13l & -3l^2 \\ & & 156 & -22l \\ & & & 4l^2 \end{bmatrix} \tag{8.28}$$

thus for each element the stiffness equation will be

$$([k_{beam}] + [k_{soil}])\{u_e\} = \{f_e\} \tag{8.29}$$

Example Problem 8.2 Determine the deflection in a uniform beam resting on an elastic foundation as shown in the Fig. 8.10.

Problem data

EI = 10000 kN m^2, total length of the beam = 8.0 m. and the soil spring constant is k_s = 1200 kN/m^2. The force at the center of the beam is $\tilde{P}$ = 120 kN.

Solution 1: With a total of two elements, by symmetry only half of the beam has to be modeled, i.e., one element.

$$l = 4.0 \text{ m.}$$

$$[k_{beam}] = \frac{EI}{64}\begin{bmatrix} 12 & 24 & -12 & 24 \\ 24 & 64 & -24 & 32 \\ -12 & -24 & 12 & -24 \\ 24 & 32 & -24 & 64 \end{bmatrix}$$

$$[k_{soil}] = \frac{k_s\,(4)}{420}\begin{bmatrix} 156 & 88 & 54 & -52 \\ 88 & 64 & 52 & -48 \\ 54 & 52 & 156 & -88 \\ -52 & -48 & -88 & 64 \end{bmatrix} \quad \text{and } \{f\} = \begin{Bmatrix} 60 \\ 0 \\ 0 \\ 0 \end{Bmatrix}$$

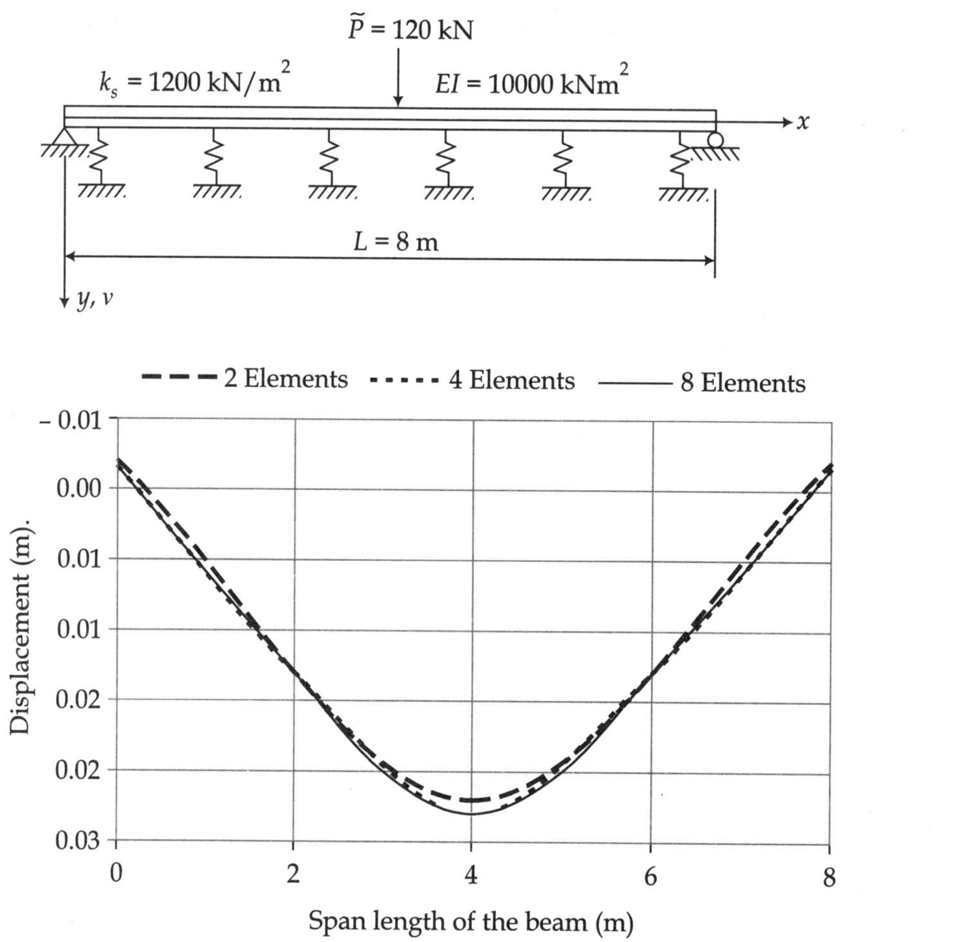

Fig. 8.10 Beam on Elastic Foundation Problem

Adding $[k_{beam}]$ and $[k_{soil}]$, the global stiffness matrix is obtained and the nodal displacements are solved with respect to the force vector $\{f\}$ and the result is:

$$\{u\} = \begin{Bmatrix} 0.023 \\ 0 \\ -.002 \\ -.00738 \end{Bmatrix} \text{ m.}$$

The displacement pattern is plotted in Fig. 8.10.

Solution 2: The same problem is solved with a total of 4 elements and taking advantage of symmetry only two elements are used. The result for this discretization is:

$$\{u\}^t = \{0.023 \quad 0 \quad .013 \quad -.00710 \quad -.00157 \quad -.007453\}^t \text{ m.}$$

Solution 3: The problem is solved again with a total of 8 elements and with symmetry only 4 elements are used in the calculations. The result for this discretization is:

$$\{u\}^t = \{.023 \quad 0 \quad .02 \quad -.005286 \quad .013 \quad -.007114 \quad .005908 \quad -.007456 \quad -.001554 \quad -.007462\}^t \text{ m}$$

All the three solutions are used to plot the displacement pattern in Fig. 8.10. It is found that for the above problem, a total of 8 elements are sufficient to get a satisfactory convergent solution.

8.6 ANALYSIS OF TWO DIMENSIONAL ARCHES AND RIGID FRAMES

A simple two dimensional arch as illustrated in Fig. 8.11 can be assumed to be composed of arch finite elements, which are combinations of bar and beam elements.[11,12]

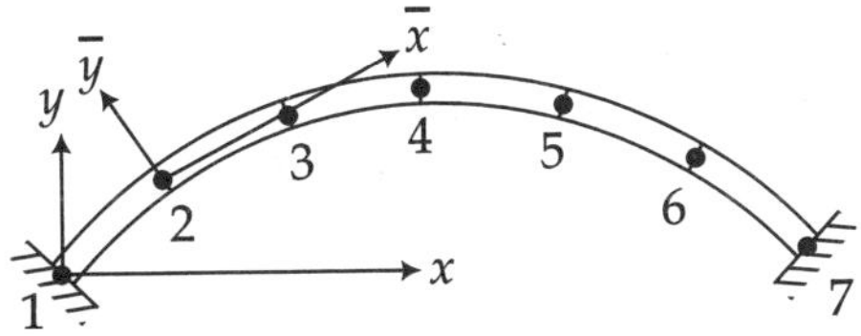

Fig. 8.11 Discretization of an Arch

Even though it is possible to develop curved bar and beam elements, as a first approximation, straight bar and beam elements are combined to model the arch finite element. In the local coordinate system, the combined arch element, has three d.o.f. per node as shown in Fig. 8.12, the displacement vector can be written as $\{\bar{u}_1\ \bar{v}_1\ \bar{\theta}_1\ \bar{u}_2\ v_2\ \bar{\theta}_2\}^t$.

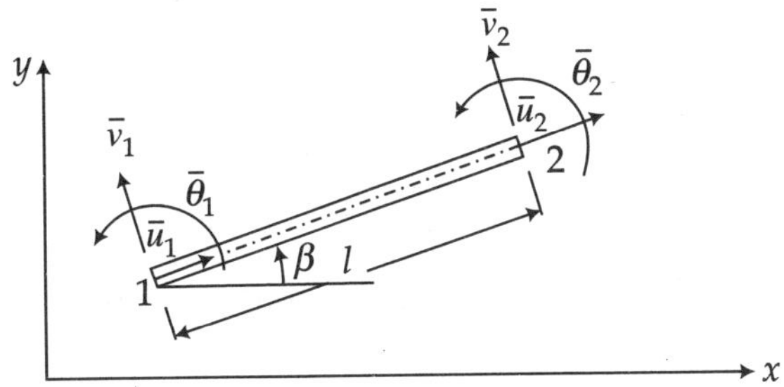

Fig. 8.12 A Combined Bar and Beam Element

and the combined stiffness matrix can be written as

$$[k]_{local} = \begin{bmatrix} \frac{EA}{l} & 0 & 0 & -\frac{EA}{l} & 0 & 0 \\ 0 & \frac{12EI}{l^3} & \frac{6EI}{l^2} & 0 & -\frac{12EI}{l^3} & \frac{6EI}{l^2} \\ 0 & \frac{6EI}{l^2} & \frac{4EI}{l} & 0 & -\frac{6EI}{l^2} & \frac{2EI}{l} \\ -\frac{EA}{l} & 0 & 0 & \frac{EA}{l} & 0 & 0 \\ 0 & -\frac{12EI}{l^3} & -\frac{6EI}{l^2} & 0 & \frac{12EI}{l^3} & -\frac{6EI}{l^2} \\ 0 & \frac{6EI}{l^2} & \frac{2EI}{l} & 0 & -\frac{6EI}{l^2} & \frac{4EI}{l} \end{bmatrix} \tag{8.30}$$

8.6.1 Transformation of Equations to Global Coordinates

Here the element nodal displacement vector in local coordinates has to be written in terms of the corresponding global displacement vector using the direction cosines of the two coordinate axes. This relationship has been shown previously for the case of the analysis of trusses, and therefore only the final equation is shown here, i.e., for one node,

$$\begin{Bmatrix} \bar{u} \\ \bar{v} \\ \bar{\theta} \end{Bmatrix}_{local} = \begin{bmatrix} c & s & 0 \\ -s & c & 0 \\ 0 & 0 & 1 \end{bmatrix} \begin{Bmatrix} u \\ v \\ \theta \end{Bmatrix}_{global}$$

$$\{\bar{u}_i\}_{local} = [L]\{u_i\}_{global}\text{, for node } i. \tag{8.31}$$

where $c = \cos\beta$, $s = \sin\beta$, and β is the angle between the two coordinate axes as shown in the Fig. 8.12 and matrix [L] is 3 by 3. The subscripts local and global pertain to components with respect to the local and global coordinate systems. Therefore for the element displacement vector,

$$\{\bar{u}\}_{local} = \begin{Bmatrix} \bar{u}_1 \\ \bar{u}_2 \end{Bmatrix}_{local} = \begin{bmatrix} L & 0 \\ 0 & L \end{bmatrix} \begin{Bmatrix} u_1 \\ u_2 \end{Bmatrix}_{global}$$

$$= [T]\{u\}_{global} \tag{8.32}$$

The {**u**} displacement vector has six components. Finally, the global element stiffness matrix for the arch element is given by

$$[k]_{global} = [T]^t [k]_{local} [T] \tag{8.33}$$

The methodology for the analysis of the complete arch using these straight elements is same as discussed for the other types of finite element analysis. Extending this idea, similar to the analysis of trusses, one can utilize the same computer code that is developed for the analysis of arches, to analyze rigid frames where each finite element is also subjected to axial compression in addition to bending.

Example Problem 8.3 A circular arch fixed at both ends with a span of 8.0 m. and a rise of 2.0 m. as shown in Fig. 8.13 is taken as a simple example problem. This is solved using two elements, four elements and eight elements for the entire arch or taking advantage of symmetry only half the number of elements are used in the analysis. $E = 1.5 \times 10^7$ kN/m^2, $I = 7.2 \times 10^{-3}$ m^4, area $A = 0.24$ m^2. Applied force $\tilde{P}$ in the middle is – 60 kN.

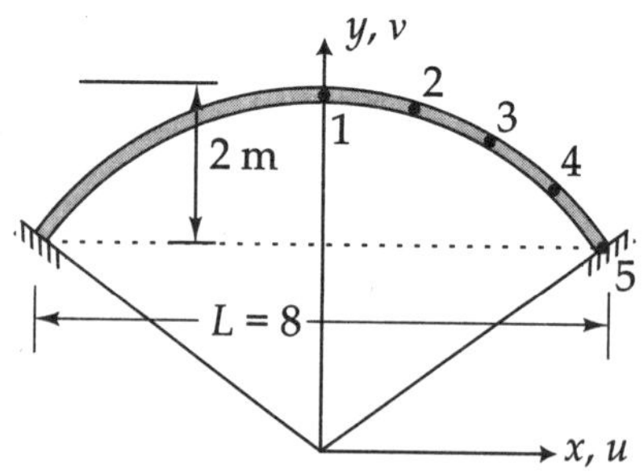

Fig 8.13 A Simple Arch Problem (8 element discretization)

Solution 1 Vertical Displacement at node 1 are computed for the three different discretiztions, taking advantage of symmetry.

v_1 for one element (right side) = – 0.3476 mm

v_1 for two elements = – 0.5865 mm

v_1 for four elements = – 0.6019 mm.

The internal axial forces, shear forces and moments as calculated in the respective local coordinates are also plotted in Fig 8.13.

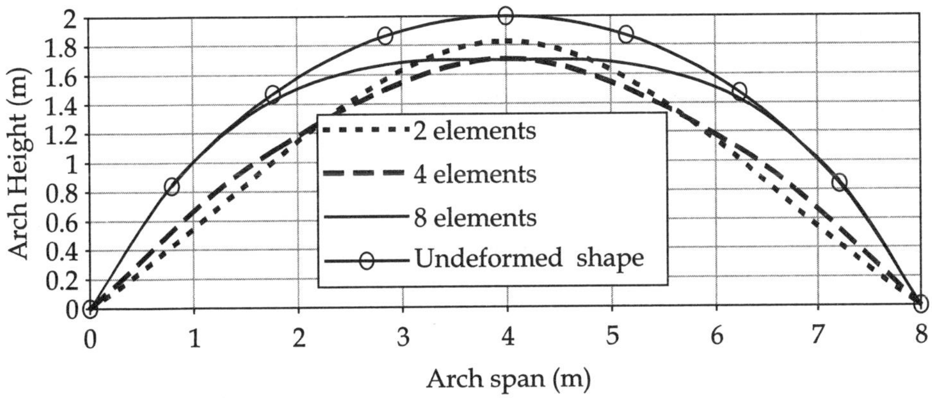

Fig. 8.14 Arch: Deformed and Undeformed Shapes (Vertical Deflection is Multiplied by 1000)

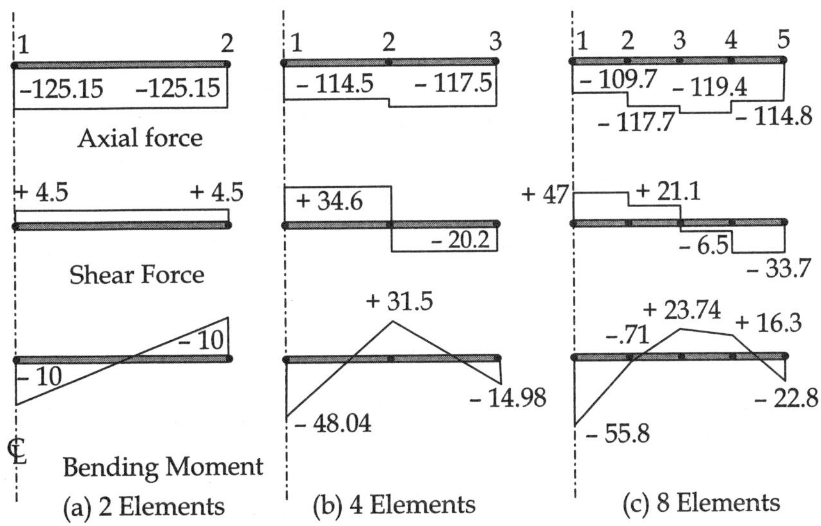

Fig. 8.15 Solution for an Arch for Different Elements (for half of the arch)

8.7 THREE DIMENSIONAL FRAMES

By creating six degrees of freedom per node of an arch element by adding a bending in the perpendicular direction and axial torsion, a twelve degree of freedom element [6,11,12] can be constructed. The degrees of freedom are: ($u \quad v \quad w \quad \theta_x \quad \theta_y \quad \theta_z$). The steps are very similar to the development of three dimensional truss as explained in Chapter 3. A finite element program using 6 d.o.f. per node is extensively used by structural engineers for three dimensional frame analysis and design.

8.8 FINITE ELEMENT ANALYSIS OF BEAMS INCLUDING SHEAR DEFORMATION

Classical Euler-Bernoulli beam theory neglects the deformation of a beam due to shear, even though there is a formula to compute shear stresses in the beam due to an internal shear force. This confuses most beginners in solid mechanics. But using theory of elasticity[10], Timoshenko has shown that effect of shear deformations in a beam with a rectangular cross section, is negligible, when the length of the beam is larger than 10 times its depth, even though maximum shear stress in the beam can be substantially large enough to be paid attention to. As the beam becomes shorter, shear deformations become more pronounced and Timoshenko developed a new beam theory[3,4] which includes shear deformation in the beam. Since, structural beams fall into the category of long beams, structural engineers do not pay attention to the Timoshenko beam theory. The reason for neglect of the new theory is apparent; most beams are long and solving a real beam problem using this new theory is rather cumbersome.

One may wonder what led to the sudden interest in the Timoshenko beam theory? The interest in this theory came in a round about way. Just as Kirchhoff extended Euler beam theory to theory of plates, Mindlin extended Timoshenko theory to formulate a new theory of plates including shear deformation in the plates.[5,6,8,13] Mindlin plate theory did not get attention among engineers as it was rather too difficult to solve his plate equations. During the quest of developing a quadrilateral finite element for the analysis of plates, researchers[5] found that Mindlin theory can be used instead of the classical Kirchhoff theory of plates. The reason for this choice is to help the formulation of the finite element analysis of plates and this is explained in Chapter 9. During this attempt, researchers[5,8,13] wanted to experiment with the Timoshenko beam theory and made strategies for making successful finite element stiffness matrix equations of beams and plates.

8.8.1 Assumptions in the Timoshenko Beam Theory

Realizing that the transverse shear stress in a beam with rectangular cross section varies quadratically, the corresponding longitudinal displacement has to have a cubic variation. This is too complex. Timoshenko adopted all assumptions used in the Euler beam theory, including 'the plane sections before bending remain plane after bending', but assumed as not normal to the longitudinal axis.[3,4,7] This condition is due to the effect of the shear deformation as a first approximation. Thus, instead of Eq.(8.1) in the Euler beam theory, Timoshenko assumed that the corresponding longitudinal displacement at a point (x, y) within the beam

$$u = -y\theta_z \tag{8.34}$$

where θ_z is the rotation of the normal plane in the direction of the z-axis. The shear strain at any point γ_{xy} in the beam is:

$$\gamma_{xy} = \frac{du}{dy} + \frac{dv}{dx} = -\theta_z + \frac{dv}{dx},$$

and therefore
$$\theta_z = \frac{dv}{dx} - \gamma_{xy} \tag{8.35}$$

γ_{xy} is an average shear deformation of the cross section of the beam, which causes some additional rotation of the plane due to the shear deformation. See Fig. 8.16. Note that rotation is positive in the anticlockwise direction.

Since Timoshenko assumed the deformed section to remain plane, a correction factor was necessary to match the shear strain energy in the beam with an approximate deformation with the actual shear energy in the beam. The value of α depends on the cross section of the beam. For a beam with a rectangular cross section[7], the value of α is approximately equal to 5/6. If the shear deformation γ_{xy} is ignored, then

$$\theta_z \equiv \frac{dv}{dx},$$

then Timoshenko theory matches with Euler beam theory. From here on in this formulation, θ_z, and γ_{xy} are identified as θ and γ respectively.

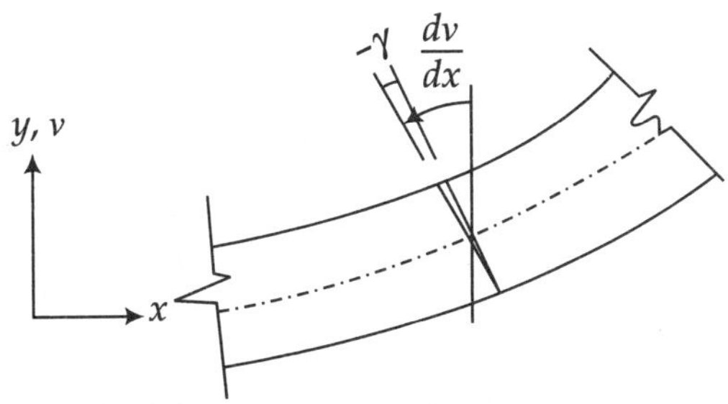

Fig. 8.16 Deformation of a Beam Including Shear Strain

8.8.2 Total Potential Energy Function

In contrast with the Euler beam, there is a shear strain energy function $\mathbf{U}_s$ in addition to the bending strain energy function $\mathbf{U}_b$. Thus, the total potential energy function

$$\Phi = \mathbf{U}_b + \mathbf{U}_s + \mathbf{V} \tag{8.36}$$

$$= \frac{1}{2}\int_0^L EI\left(\frac{d\theta}{dx}\right)^2 dx + \frac{1}{2}\int_0^L \alpha GA\,(\gamma)^2\,dx + \mathbf{V}$$

where G and A are the shear modulus and area of cross section of the beam. Here α is an adjustable parameter to allow for the difference in the distribution of shear strain in the crossection of the beam as defined earlier. **V** is the work potential.

8.8.3 Formulation of the Finite Element Stiffness Matrix

The basic technique used for the formulation of the stiffness matrix is the same as used before. Two degrees of freedom (v and θ) per node are used here also. The element shown here has 4 degrees of freedom as in the previous case.

Step 1. Assumed Displacement Functions

Since the total rotation θ is no longer a pure function of $v(x)$, v and θ are independently interpolated using linear interpolation functions. Introducing $\theta(x)$ as a function of x within the element,

$$v(x) = \left(1 - \frac{x}{l}\right) v_1 + \frac{x}{l} v_2$$

and,

$$\theta(x) = \left(1 - \frac{x}{l}\right) \theta_1 + \frac{x}{l} \theta_2 \tag{8.37}$$

Step 2. Internal Strains

The internal strains are the derivative of θ (similar to the second derivative in the classical Euler theory) and the shear strain.

Symbolically,

$$\{\varepsilon\} = \begin{Bmatrix} \dfrac{d\theta}{dx} \\ \gamma \end{Bmatrix} = \begin{bmatrix} 0 & -\dfrac{1}{l} & 0 & \dfrac{1}{l} \\ -\dfrac{1}{l} & -\left(1 - \dfrac{x}{l}\right) & \dfrac{1}{l} & -\dfrac{x}{l} \end{bmatrix} \begin{Bmatrix} v_1 \\ \theta_1 \\ v_2 \\ \theta_2 \end{Bmatrix} = [B]\{u\} \tag{8.38}$$

Step 3. Internal Stresses

The internal stresses in a beam can be written symbolically represented by the internal bending moment and the shear force and the stress-strain relations can be written as:

$$\{\sigma\} = \begin{Bmatrix} m \\ V \end{Bmatrix} = \begin{bmatrix} EI & 0 \\ 0 & \alpha GA \end{bmatrix} \begin{Bmatrix} \dfrac{d\theta}{dx} \\ \gamma \end{Bmatrix} = [D]\{\varepsilon\} \tag{8.39}$$

Step 4. Element Stiffness Matrix and the Element Force Vector

Following the standard equation for the element stiffness matrix,

$$[k] = \int_0^l [B]^t [D][B]\, dx \tag{8.40}$$

and the element force vector is

$$f_i = -\frac{\partial \mathbf{V}}{du_i} \text{ for } i = 1,..,4.$$

The Eq. (8.40) for the element stiffness matrix is integrated and given as a sum of contributions from bending and shear separately as given below.

$$[k] = [k_b] + [k_s] \tag{8.41}$$

where
$$[k_b] = \frac{EI}{l}\begin{bmatrix} 0 & 0 & 0 & 0 \\ 0 & 1 & 0 & -1 \\ 0 & 0 & 0 & 0 \\ 0 & -1 & 0 & 1 \end{bmatrix}$$

and
$$[k_s] = \frac{\alpha GA}{l}\begin{bmatrix} 1 & \frac{l}{2} & -1 & \frac{l}{2} \\ \frac{l}{2} & \frac{l^2}{3} & -\frac{l}{2} & \frac{l^2}{6} \\ -1 & -\frac{l}{2} & 1 & -\frac{l}{2} \\ \frac{l}{2} & \frac{l^2}{6} & -\frac{l}{2} & \frac{l^2}{3} \end{bmatrix}$$

Here $[k_b]$, $[k_S]$ are the stiffness matrices due to bending and shear respectively. The beam problem given in Fig. 8.7 is solved using the above matrices for the beam elements. The displacements are shown in the Fig. 8.17. It can be observed that the solution of the problem with two elements has considerable error as compared to the solution of the beam with Euler beam elements which gives almost exact displacements for two elements in the domain. Thus, the new element is very stiff. The reason does not pertain to the quality of the Timoshenko beam theory, rather, the lack of accuracy is in the finite element approximation of the assumed displacement functions. Since from Eq. (8.35),

$$\gamma = \frac{dv}{dx} - \theta \tag{8.35}$$

and for long beams, the shear deformation is small,

$$\theta \cong \frac{dv}{dx}$$

Since a linear function for $v(x)$ is assumed, then

$$\frac{dv}{dx} = \frac{v_2 - v_1}{l} = \text{a constant},$$

contrary to the linear assumption of slope. Thus, when the beam is long, instead of shear strain γ to be neglibible, Eq. (8.35) yields some unrealistic value. Thus, the solution using this element is poor as seen in Fig. 8.17 and this is due to the dominance of the shear strain energy in this formulation. This phenomena is called 'shear locking' in finite element literature. The solution improves slowly with 2, 4 and 8 elements as shown in the Fig. 8.17.

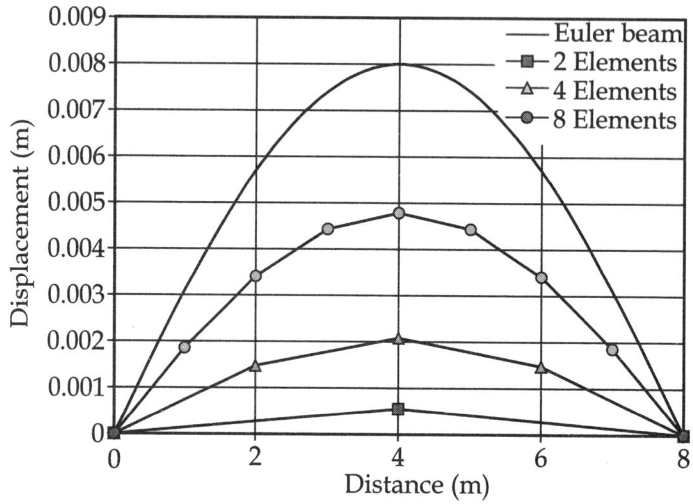

Fig. 8.17 Results using Beam Elements with Complete Integration

8.8.4 Reduced Integration Schemes

To reduce the effect of shear locking, researchers[5] used a reduced numerical integration technique. In the evaluation of the stiffness matrix, that part of the stiffness matrix pertaining to bending energy is integrated exactly with a one-point Gauss integration, while the part for the shear energy requires a two-point integration scheme. By reducing the number of points of integration from two to one, only four terms containing are altered and reduced from a factor 1/3 to 1/4 and they are k_{22}, k_{24}, k_{42} as shown in Eq. (8.42). The final element stiffness matrix can be written as a combination of terms with bending stiffness k_b and the new shear stiffness k_s.

$$[k_s] = \frac{\alpha GA}{l}\begin{bmatrix} 1 & \frac{l}{2} & -1 & \frac{l}{2} \\ \frac{l}{2} & \frac{l^2}{4} & -\frac{l}{2} & \frac{l^2}{4} \\ -1 & -\frac{l}{2} & 1 & -\frac{l}{2} \\ \frac{l}{2} & \frac{l^2}{4} & -\frac{l}{2} & \frac{l^2}{4} \end{bmatrix} \tag{8.42}$$

The final stiffness matrix becomes more flexible and yields better solution as shown in Fig. 8.18. Using the reduced integration method, the results are much better, one still needs to employ more number of elements to get a good solution as compared to the Euler beam element which yields good results even with small number of elements.

8.8.5 Modified Equations by Reddy[8]

For $\theta(x)$ to be linear within the element, $v(x)$ must be at least interpolated by a quadratic function. This condition requires at least three displacements, two at the ends and a third one in the middle of the element. With two rotation parameters at each ends, the stiffness matrix becomes 5 by 5. Reddy[8] has shown that by using 5 nodal parameters and with reduced

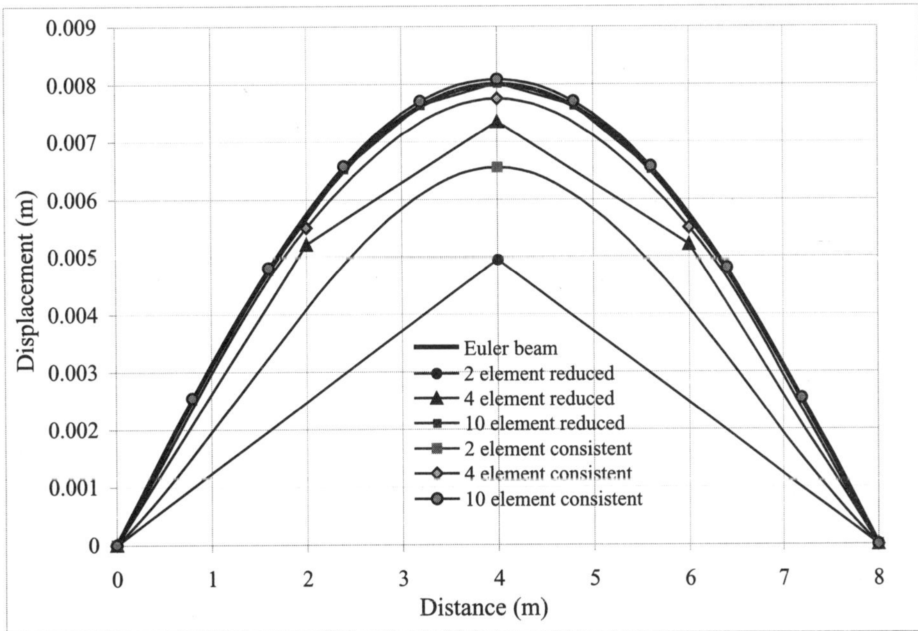

Fig. 8.18 Results using Beam Elements with Reduced Integration and Consistent Interpolation

integration, one can improve the solution even with fewer number of elements. Reddy calls this element as 'Consistent Interpolation Element'. This element has three displacements and two slopes, and to make the degrees of freedom equal to two per node, Reddy used a static condensation technique to remove the middle displacement. It is interesting to note that the consistent integration element with static condensation and reduced integration element result in the same stiffness matrix given by $[k] = [k_b] + [k_s]$ where matrix k_s is modified by Eq. (8.42) with reduced integration. But he obtained a new force vector which is the same as that in the Euler beam element in the case of a uniformly distributed force on the element. The revised force vector makes the results to converge faster to exact solution as shown in Fig. 8.18. This problem can also be solved using theory of elasticity as a plane stress problem and the result is shown in the Table 8.1.

Table 8.1 Results of the Dispalcements and Slopes for the Beam

	2 element			
	$x = 4$		$x = 8$	
	v_1	θ_1	v_2	θ_2
Reduced	0.00495	0	0	-0.0024
Consistent	0.00655	0	0	-0.0032
Euler Beam	0.008	0	0	-0.0032
Plane Stress	0.008392			

	4 element					
	$x = 4$		$x = 6$		$x = 8$	
	v_1	θ_1	v_2	θ_2	v_3	θ_3
Reduced	0.00735	0	0.00521	-0.0021	0	-0.003
Consistent	0.00775	0	0.00551	-0.0022	0	-0.0032
Euler Beam	0.008	0	0.0057	-0.0022	0	-0.0032
Plane Stress	0.008392					

	10 element					
	$x = 4$		$x = 5.6$		$x = 8$	
	v_1	θ_1	v_3	θ_3	v_6	θ_6
Reduced	0.008019	0	0.006521	-0.00180	0	-0.003168
Consistent	0.008083	0	0.006575	-0.00182	0	-0.003200
Euler Beam	0.008	0	0.006505	-0.00182	0	-0.0032
Plane Stress	0.008392					

This section is provided only as an introduction to application of finite element analysis to solve the Timoshenko theory of beams even though from a practical point of view, none will use this theory to analyse beams except for very rare cases where the beam is very thick.

8.9 SUMMARY

One-dimensional elements with C_1-continuity is developed for the analysis of beams. The elementary theory of beams, known as Euler beam theory is presented to develop the beam element stiffness matrix and the equivalent force vector due to applied prescribed forces. In a simple example beam problem, the differences between the techniques in finite element analysis and classical structural analysis are shown in detail for clarity in understanding of these techniques. Particular emphasis is made to show the computation of internal bending moments and shear forces within each element in addition to the reaction moments and forces at the ends of the element.

Finite element analysis of beams on elastic foundation is also presented. Combining the bar and beam elements, methods for analyses of arches and frames are shown. Timoshenko beam theory including the shear strain energy is introduced and the corresponding finite beam element stiffness matrix is derived.

Just as the theory for bar elements is extended to the analysis of two- and three-dimensional solids, the theory for beam elements is extended to analyze plates and shells in the next chapter.

PROBLEMS

8.1 Derive equivalent loads at the nodes due to an applied bending moment M in the middle. The length of the beam element is l.

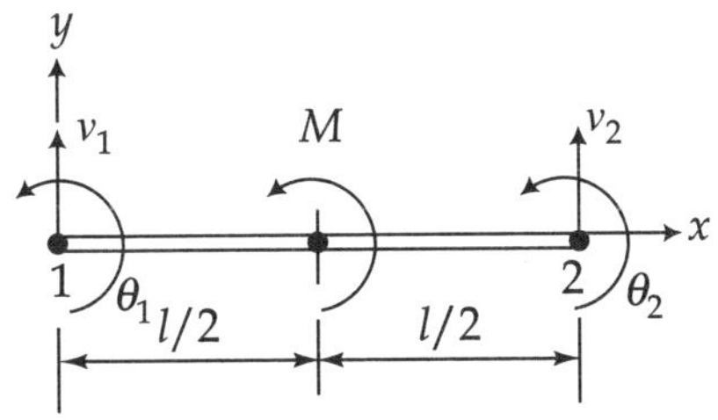

Fig. P8.1

8.2 Determine the equivalent loads at the nodes for a concentrated load $\tilde{P}$ as shown.

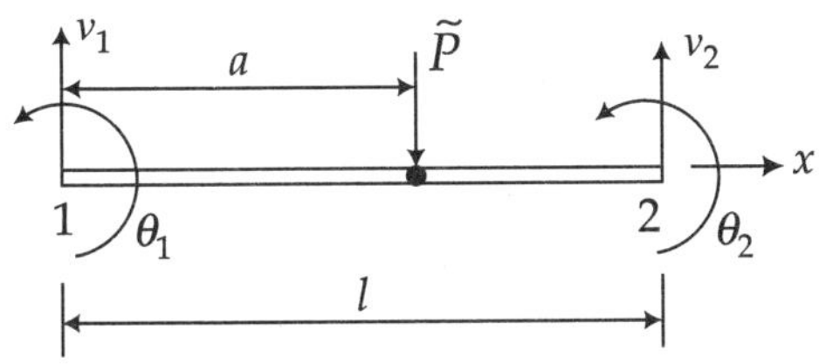

Fig. P8.2

8.3 A beam element with three d.o.f. can be developed using quadratic interpolation functions. Assume that EI is uniform.

(a) Develop element stiffness matrix

(b) Calculate the equivalent loads for

(i) a uniformly distributed load $\tilde{q}$

(ii) a concentrated load $\tilde{P}$ at a, distance from node 1.

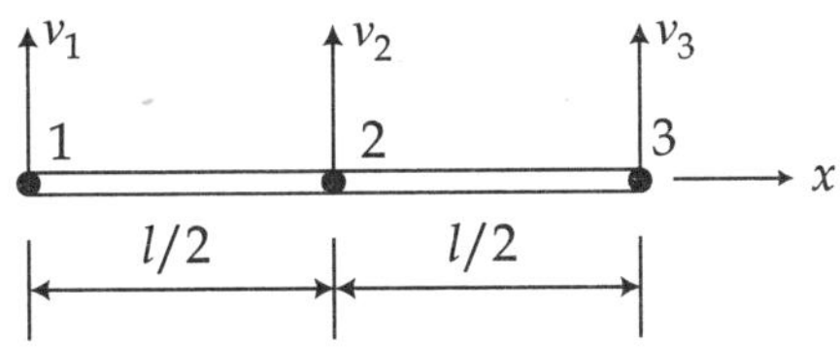

Fig. P8.3

8.4 The displacement function of a beam elment of length l = 3a is

$$v = \alpha_1 + \alpha_2 x + \alpha_3 x^2 + \alpha_4 x^3$$

The element has 4 nodes with one d.o.f. per node. The beam has uniform stiffness EI.

(i) Derive the stiffness matrix and the consistent loads for a uniformly distributed load $\tilde{q}$ /unit length.

(ii) Solve a simply supported beam of length L carrying a uniformly distributed load $\tilde{q}$ /unit length using

(a) one lement

(b) two element

(c) three elements.

(iii) Compare the results with exact solution.

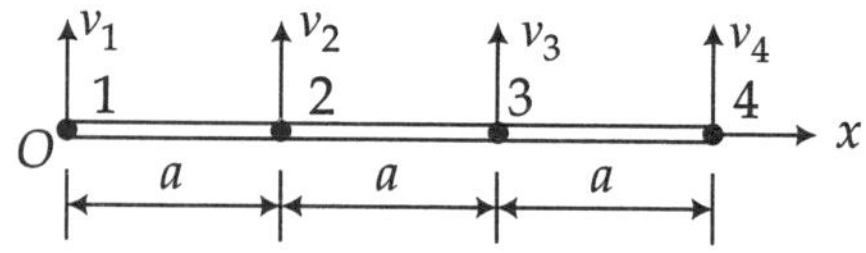

Fig. P8.4

8.5 Solve the following beam problem by using 2, 4 and 6 beam elements. Plot the displacement and stresses along the beam to observe the convergence when the number of elements is increasing. Beam has uniform *EI*.

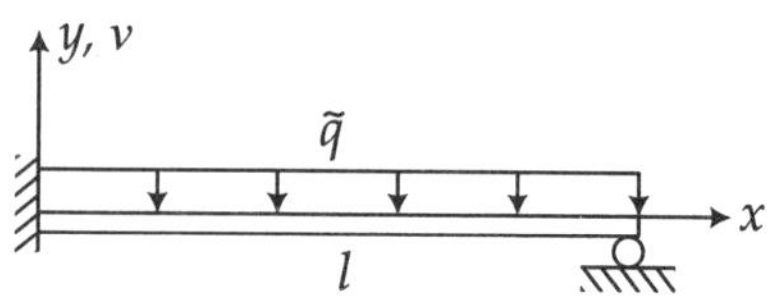

Fig. P8.5

8.6 The beam given below with $E = 225$ GPa, $b = 20$ cm, $h = 45$ cm, $l = 2$ m, $\tilde{q}_0 = 40$ kN/m, is supported at the end with a spring $k = 120$ kN/m.

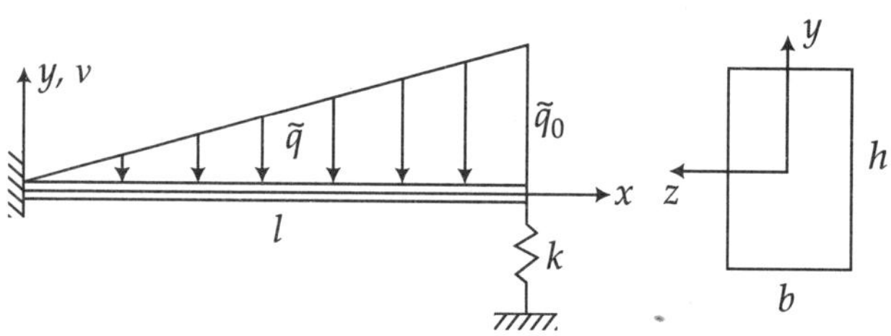

Fig. P8.6

By considering two beam elements only

(a) Find the equivalent loads

(b) Develop global stiffness matrix, K

(c) Solve for displacements

(d) Solve for internal strains and stresses

8.7 A tapered composite beam consists of two different parts, one part with steel and the other with aluminum. The modulus of elasticity of steel, E_s is 200 GPa and the modulus of elasticity of Aluminum, E_a is 100 GPa. Heights are h_1 = 40 cm, h_2 = 20 cm and h_3 = 10 cm. Width b of the composite beam is constant and equal to 20 cm.

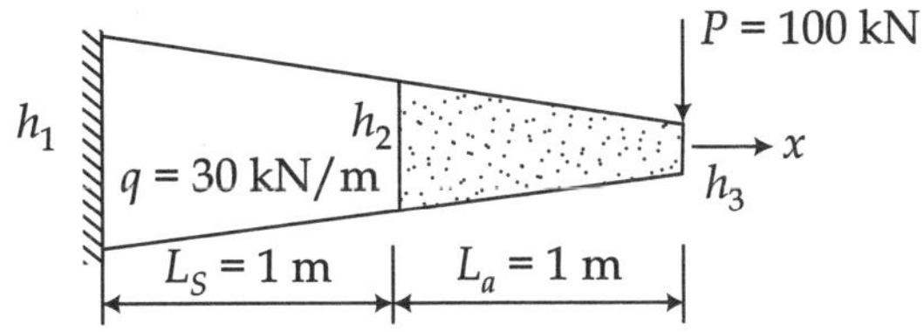

Fig. P8.7

Find the displacements, reactions, strains and stresses by using beam finite elements. What are the stresses in the beams at the junction ?

8.8 A uniform beam is supported by uniformly distributed springs having a spring constant k_s force/unit displacement/unit length. Solve this problem using two and four elements separately. EI is uniform. Calculate the displacements, moments and shears at nodes and plot them.

$$E = 100 \text{ GPa}, b = 40 \text{ cm}, h = 70 \text{ cm}, l = 5 \text{ m}, \tilde{q} = 40 \text{ kPa and } k_s = 500 \text{ kN/m}^2.$$

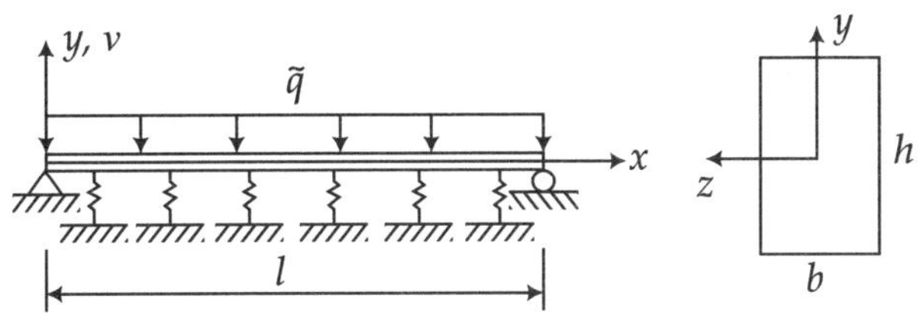

Fig. P8.8

REFERENCES

1. Bathe, K.J, "Finite Element Procedures", Prentice Hall, Englewood Cliffs, New Jersey, 1996.
2. Boresi, A. P, Sidebottom, O.M, Seely, Fred. B. and Smith, J.O., "Advanced Mechanics of Materials", John Wiley & Sons, New York, 1978.
3. Dym, C.L. and Shames, I.H, "Solid Mechanics: A Variational Approach", McGraw Hill, New York, 1973.
4. Fung, Y.C, "Foundations of Solid Mechanics", Prentice Hall, Englewood Cliffs, New Jersey, 1965.
5. Hughes, T.J.R, "The Finite Element Method", Prentice Hall, Englewood Cliffs, New Jersey, 1987.
6. Krishnamoorthy, C.S, "Finite Element Analysis, Theory and Programming", Tata McGraw-Hill Publishing Company, New Delhi, 1987.
7. Popov, Egor. P, "Engineering Mechanics of Solids", Prentice Hall, Englewood Cliffs, New Jersey, 1996.

8. Reddy, J.N, "An Introduction to the Finite Element Method", 2nd Ed. McGraw Hill Book Company, New York, 1993.
9. Timoshenko, S, "Strength of Materials", Van Nostrand Co.,New York, 1947.
10. Timoshenko, S.P. and Goodier, J.N, "Theory of Elasticity", McGraw Hill Book Company, New York, 1970.
11. Weaver, W. Jr., and Gere, J.M, "Matrix Analysis of Framed Structures", Van Nostrand Reinhold, New York, 1980.
12. Yang, T.Y, "Finite Element Structural Analysis", Prentice Hall, Englewood Cliffs, New Jersey, 1986.
13. Zienkiewicz, O.C. and Taylor, R.L, "The Finite Element Method", Vols. I & II, McGraw Hill Book Company, London, 1989.

9

CHAPTER

Application to Plate Bending Problems: C_1-Continuity

9.1 INTRODUCTION

C_1-continuity in finite plate elements means that a finite plate element should have continuity of slopes along and normal to its boundaries in addition to the continuity of displacement along its boundaries. The problem arises when finite elements are to be developed with the conventional degrees of freedom such as displacements and slopes at the corner nodes. Considerable research[3,6,7] has been done for the development of finite elements for the analysis of bending of plates with C_1-continuity with the above criteria. This is a very difficult task to achieve. However, researchers have found that even plate elements which do not have complete C_1-continuity along the boundaries of the element are also useful in overall spectrum of finite elements. During the early days of finite element research, the aim was to develop an element that has the least number of degrees of freedom and at the same time would accommodate any geometrical shape of the plate. In other words, most of them were interested to develop either triangular or quadrilateral plate elements. But, along with triangular finite elements, simple rectangular elements[7] were also developed and interestingly many structural plate problems can be solved with rectangular plate elements.

Since 1964, the continuity of slope requirements in plate elements became the most complicated issue. Perhaps, the primary reason for this, is that engineers were very familiar with the classical Kirchhoff theory of bending of thin plates and using the many available closed-form solutions. Therefore researchers concentrated on the use of Kirchhoff theory for the development of finite plate bending elements. A wide variety of both triangular and rectangular finite elements[7] have been developed with many degrees of freedom using Kirchhoff theory, and a comprehensive study of what has been done in the past is itself a major task. In the pursuit of development of quadrilateral elements, researchers[3] turned their attention towards the use of the Mindlin theory[1,3,6] of plates, where the vertical shear deformation in the plate is included in the overall analysis. Now there is a general tendency toward using the Mindlin theory for the finite element analysis of plates and shells adopting a more rigorous quadrilateral element. To differentiate these two types of approaches, one has to peep into the two theories of plates. Here, the Kirchhoff theory is first discussed, along with some simple triangular and rectangular finite element formulations.

9.2 KIRCHHOFF THEORY OF BENDING OF THIN PLATES

A brief account of the theory is presented in this section. In the discussion of all these theories, one should not forget to examine the assumptions used in the development of the theory. This theory is actually an extension of the classical Euler theory for the bending of beams and hence the assumptions are easily understood for those who have learned the Euler beam theory. The assumptions are:

1. the plate is thin, uniform and linearly elastic,
2. the plate undergoes small lateral deflections only,
3. middle surface of the plate is free from strains,
4. normal to the surface remain normal to the deformed middle surface, i.e., no transverse shear strains, and
5. the stress normal to the plate is ignored.

The coordinate system used for the development of the plate equations is shown in Fig. 9.1. The theory is developed in four steps as given below.

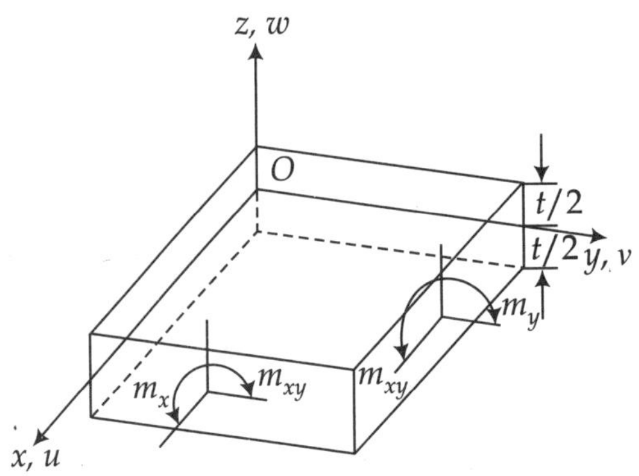

Fig. 9.1 Coordinate Axes and Positive Bending Moments on an Infinitesimal Plate Element

Step 1. Assume Displacement Functions

Using the above assumptions, the displacements at any general point (x, y, z) within the plate are:

$$u = -z\frac{dw}{dx}$$

$$v = -z\frac{dw}{dy}$$

$$w = w(x, y) \quad \text{of the middle surface.} \tag{9.1}$$

Step 2. Internal Strains

$$\varepsilon_x = \frac{\partial u}{\partial x} = -z\frac{\partial^2 w}{\partial x^2}$$

$$\varepsilon_y = \frac{\partial v}{\partial y} = -z\frac{\partial^2 w}{\partial y^2}$$

$$\gamma_{xy} = \frac{\partial u}{\partial y} + \frac{\partial v}{\partial x} = -2z\frac{\partial^2 w}{\partial x\, \partial y}, \quad \text{and} \tag{9.2}$$

$$\gamma_{xz} = \gamma_{yz} = 0,$$

$$\sigma_z = 0$$

Step 3. Internal Stresses

The internal stresses are written in terms of the strains using the plane stress equations given in Eq. (2.26)

$$\begin{Bmatrix} \sigma_x \\ \sigma_y \\ \tau_{xy} \end{Bmatrix} = \begin{bmatrix} 1 & \mu & 0 \\ \mu & 1 & 0 \\ 0 & 0 & \dfrac{1-\mu}{2} \end{bmatrix} \begin{Bmatrix} \varepsilon_x \\ \varepsilon_y \\ \gamma_{xy} \end{Bmatrix}$$

and the internal bending moments m_x, m_y and m_{xy} per unit length along x and y axes are respectively defined as:

$$\begin{Bmatrix} m_x \\ m_y \\ m_{xy} \end{Bmatrix} = \begin{Bmatrix} \int_t \sigma_x z dz \\ \int_t \sigma_y z dz \\ \int_t \tau_{xy} z dz \end{Bmatrix} = -D \begin{bmatrix} 1 & \mu & 0 \\ \mu & 1 & 0 \\ 0 & 0 & \dfrac{1-\mu}{2} \end{bmatrix} \begin{Bmatrix} \dfrac{\partial^2 w}{\partial x^2} \\ \dfrac{\partial^2 w}{\partial y^2} \\ 2\dfrac{\partial^2 w}{\partial x\, \partial y} \end{Bmatrix} \tag{9.3}$$

denoted by

$$\{m\} = [D]\{\kappa\}$$

Here $D = \dfrac{Et^3}{12(1-\mu^2)}$, usually referred as the flexural rigidity of the plate, and the matrix $[D]$ = is the matrix representing the stress-strain relations and $\{\kappa\}$ is the vector of internal strains of the plate, the components of which are respectively the curvatures of the plate in the x,y plane, if the plate is undergoing small deflections. The respective positive internal moments are shown in Fig. 9.1.

Step 4. Minimization of the Total Potential Energy Function

The total energy function

$$\Phi = \mathrm{U} + \mathrm{V}$$

$$= \frac{1}{2}\iint_A \{m\}^t\{\kappa\}\,dx\,dy - \iint_A \tilde{q}w\,dx\,dy - \text{(terms due to external prescribed forces)}$$

$$= \frac{1}{2}\iint_A \{\kappa\}^t[D]\{\kappa\}\,dx\,dy - \iint_A \tilde{q}w\,dx\,dy - \text{etc.,} \tag{9.4}$$

where A denotes the domain of the plate. It may be noted that by minimizing Φ, one can obtain the classical plate equation $D\nabla^4 w = \tilde{q}$ and the respective boundary conditions, just as the beam equilibrium equations and boundary conditions are obtained in Eqs. (1.19) and (1.20). In the finite element formulations, the equilibrium equations of the plate are not directly used, instead the total energy function is employed to arrive at the system equations.

9.3 FINITE ELEMENTS USING KIRCHHOFF PLATE THEORY

Only two types of finite elements such as triangular and rectangular elements are discussed here. Many different types of triangular finite elements are reported in the literature.[6, 9] Here only one type is presented. The positive directions of the displacements and rotations at an arbitrary node i are shown in Fig. 9.2.

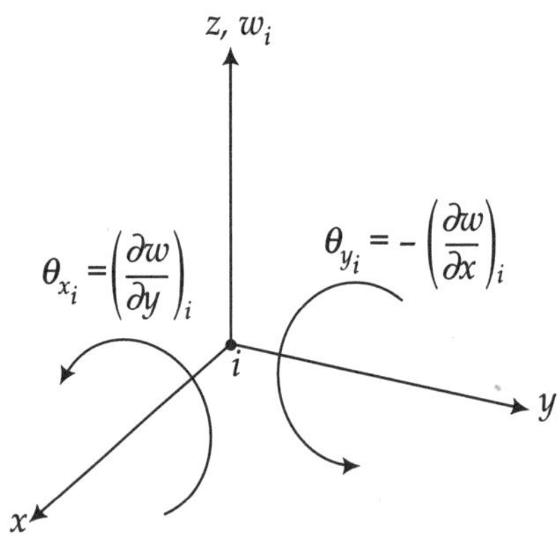

Fig. 9.2 Positive Nodal Displacements and Rotations

9.3.1 Triangular Plate Bending Finite Element

The simplest of these is a triangular element having three nodes with three degrees of freedom per node as shown in Fig. 9.3, resulting in a 9 d.o.f. element.

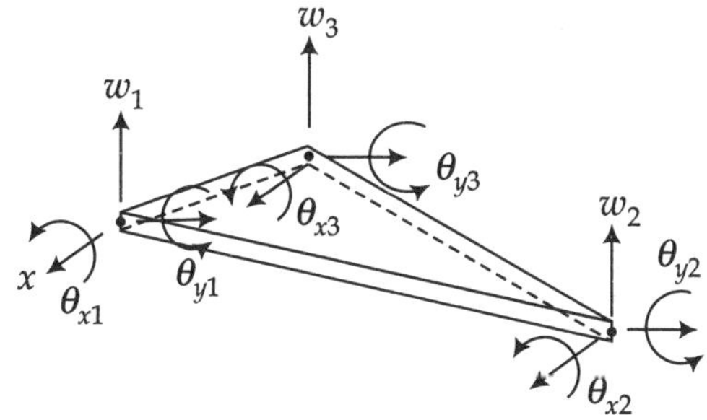

Fig. 9.3 Nodal Displacement Parameters on a Triangular Element

Step 1. Assume Displacement Function

Selection of the displacement function is the most difficult problem. Examining the Pascal triangle, a complete cubic polynomial results in 10 parameters. The degrees of freedom are the displacement and the slopes in x and y directions. In the early development of these functions, the term xy was omitted, to make a 9 term function, unfortunately, this omission, removes one of the constant strains (twist of the plate), which makes the finite element very stiff. Later, this term was included, but the terms x^2y and xy^2 were added together as one term, thus the displacement function will be

$$w(x, y) = [1 \; x \; y \; x^2 \; xy \; y^2 \; x^3 \; (x^2y + xy^2) \; y^3]\{\alpha\} \tag{9.5}$$

where $\{\alpha\}$ is a vector containing the nine α - parameters.

If one examines the above equation, for the direction, say 's', along one edge, w is cubic in s and is unique since it is defined by 4 variables, i.e., 2 dispalcements and 2 slopes at the ends. Therefore compatibility of displacements and slopes along that direction is guaranteed. However when the slope normal to that edge is developed, it is found that also to be cubic in 's', which requires four parameters to be unique, but only two are available. Hence the compatability of slope normal to the edge is not guaranteed, and hence this element is called *"a* nonconformable element".

Since it is rather too complex to develop the shape functions, an alternate method is employed for their derivation. For the node i, the displacement vector, is illustrated in Fig. 9.2.,

$$\{u_1\} = \begin{Bmatrix} w_i \\ \theta_{xi} \\ \theta_{yi} \end{Bmatrix} = \begin{Bmatrix} w_i \\ \left(\dfrac{\partial w}{\partial y}\right)_i \\ -\left(\dfrac{\partial w}{\partial x}\right)_i \end{Bmatrix}$$

$$= \begin{bmatrix} 1 & x & y & x^2 & xy & y^2 & x^3 & x^2y + xy^2 & y^3 \\ 0 & 0 & 1 & 0 & x & 2y & 0 & x^2 + 2xy & 3y^2 \\ 0 & -1 & 0 & -2x & -y & 0 & -3x^2 & -2xy - y^2 & 0 \end{bmatrix}_i \{\alpha\} \tag{9.6}$$

$$= [P_i]\{\alpha\}$$

and therefore, the nodal displacement vector for the element

$$\{u\} = \begin{bmatrix} P_1 \\ P_2 \\ P_3 \end{bmatrix} \{\alpha\} = [P]\{\alpha\} \tag{9.7}$$

and hence, $\{\alpha\} = [P]^{-1}\{u\}$

Step 2. Internal Strains

The internal strain vector is $\{\kappa\}$ is obtained by differentiating the Eq. (9.5) twice, and thus

$$\{\kappa\} = \begin{Bmatrix} \dfrac{\partial^2 w}{\partial x^2} \\ \dfrac{\partial^2 w}{\partial y^2} \\ 2\dfrac{\partial^2 w}{\partial x\,\partial y} \end{Bmatrix} = \begin{bmatrix} 0 & 0 & 0 & 2 & 0 & 0 & 6x & 2y & 0 \\ 0 & 0 & 0 & 0 & 0 & 2 & 0 & 2x & 6y \\ 0 & 0 & 0 & 0 & 2 & 0 & 0 & 8(x+y) & 0 \end{bmatrix} [P]^{-1}\{u\}$$

$$= [G][P]^{-1}\{u\} = [B]\{u\} \tag{9.8}$$

Step 3. Internal Moments

The equation for the internal moments is the same as given in Eq. (9.3).

Step 4. Element Stiffness Matrix

Thus, the equation for the element stiffness matrix is expressed as

$$[k] = \iint_A [B]^t [D][B] dx\,dy$$

$$= \left[P^{-1}\right] \int_A [G]^t [D][G] dx\,dy\,[P]^{-1} \tag{9.9}$$

Satisfactory solutions are obtained for plate analysis using this element for most cases except when any one of the triangular elements in the discretization is a right angled triangle, in such instances the corresponding *P* matrix becomes singular. For this reason, many analysts do not recommend the use of this element.

9.3.2 Rectangular Plate Bending Element

This element originally developed by Cheung[7] was one of the earlier finite element derived for the analysis of bending of thin plates. The element has 4 nodes and three degrees of freedom at each node thus making a 12 d.o.f. element, as shown in Fig. 9.4.

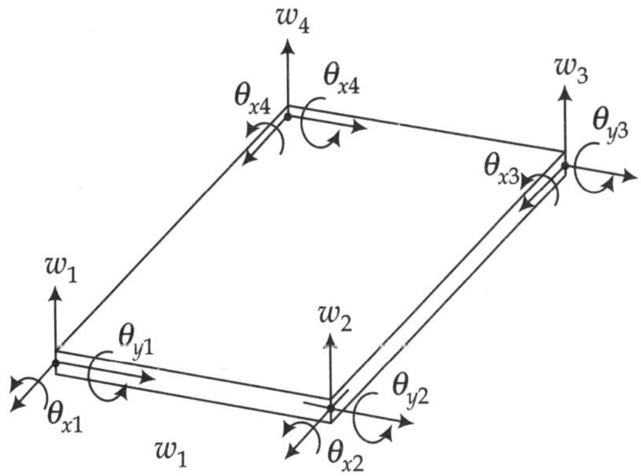

Fig. 9.4 A Rectangular Plate Element with 12 d.o.f.

Step 1. Assume Displacement Function

A twelve parameter polynomial is selected from the Pascal Triangle as given below.

$$w(x, y) = [1\ x\ y\ x^2\ xy\ y^2\ x^3\ x^2y\ xy^2\ y^3\ x^3y\ xy^3]\{\alpha\} \tag{9.10}$$

where $\{\alpha\}$ is a vector of 12 α-parameters. By inspection, it can verified that the rigid body modes and the constant strains conditions are easily satisfied in the above expression. But the expression does not satisfy the inter element compatibility completely. It can shown that the displacements and the slopes along the sides are completely compatible, while the slopes normal to the sides of the element are not necessarily compatible as in the case of the triangular element. For this reason this element is called a nonconforming element. The formulation of the element stiffness matrix can be explicitly developed and they are published in the literature[8]. This element is performing well for analyzing plate problems having a rectangular geometry. From the equation relating nodal displacements in terms of the α-parameters using the local coordinates, the α-parameters are derived in terms of the nodal coordinates and nodal displacements.

$$\{u\} = [P]\{\alpha\} \text{ and therefore } \{\alpha\} = [P]^{-1}\{u\} \tag{9.11}$$

Step 2. Internal Strains

The internal strains are in terms of the second derivatives as discussed before, and thus,

$$\{\kappa\} = \begin{Bmatrix} \dfrac{\partial^2 w}{\partial x^2} \\ \dfrac{\partial^2 w}{\partial y^2} \\ 2\dfrac{\partial^2 w}{\partial x\, \partial y} \end{Bmatrix}$$

$$= \begin{bmatrix} 0 & 0 & 0 & 2 & 0 & 0 & 6x & 2y & 0 & 0 & 6xy & 0 \\ 0 & 0 & 0 & 0 & 0 & 2 & 0 & 0 & 2x & 6y & 0 & 6xy \\ 0 & 0 & 0 & 0 & 2 & 0 & 0 & 4x & 4y & 0 & 6x^2 & 6y^2 \end{bmatrix} \{\alpha\}$$

$$= [G]\{\alpha\} = [G][P]^{-1}\{u\} \tag{9.12}$$

Step 3. Internal Moments

The internal moments are given in terms of the second derivatives as given in Eq. (9.3).

Step 4. Element Stiffness Matrix and the Load Vector

The element stiffness matrix is given as

$$[k] = [P^{-1}]^t \iint_A [G]^t\,[D]\,[G]dx\,dy\,[P]^{-1} \tag{9.13}$$

and has the same form as given in Eq. (9.10), and the corresponding load vector is

$$\tilde{f}_i = -\frac{\partial \mathbf{V}}{\partial u_i} \tag{9.14}$$

If $\tilde{q}$ is the uniformly distributed load on the plate, the equivalent load vector is

$$\{\tilde{f}\} = \tilde{q}ab\left\{1 \;\; -\frac{b}{3} \;\; \frac{a}{3} \;\; 1 \;\; \frac{b}{3} \;\; \frac{a}{3} \;\; 1 \;\; -\frac{b}{3} \;\; -\frac{a}{3} \;\; 1 \;\; \frac{b}{3} \;\; -\frac{a}{3}\right\}^t \tag{9.15}$$

where the size of the plate element is 2 *a x* 2*b*. There are a wide variety of plate bending elements in the literature, a good reference list for these elements are given in the text book by Zienkiewicz.[7]

Example Problem 9.1 An example problem is solved for a rectangular plate 2 m. X 3 m. in size as shown in Fig. 9.5 with uniform thickness equal to 0.02 m., uniformly loaded with $\tilde{q}$ = 0.3kPa., E = 10 GPa and Poisson's ratio μ = 0.25. By virtue of symmetry, only a one quarter of the plate is used in the discretization. The exact value of displacement at the center using Kirchhoff theory is 0.00433 m. and the value of the stress σ_x at the center is 1461.6 kPa.

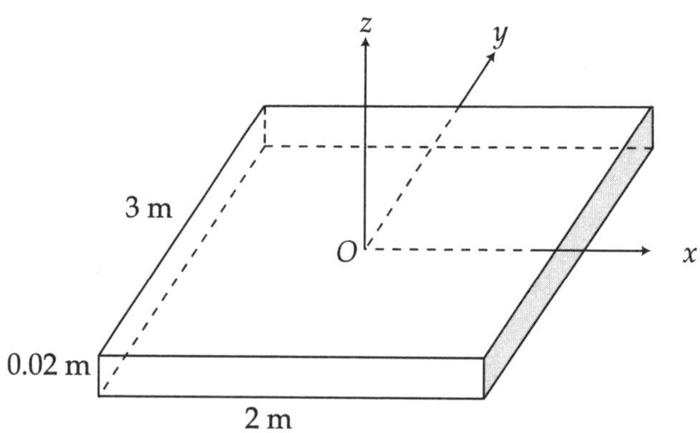

Fig. 9.5 Rectangular Plate with the size of 2m × 3m

The displacement and stresses are shown in Figs. 9.6 and 9.7 respectively.

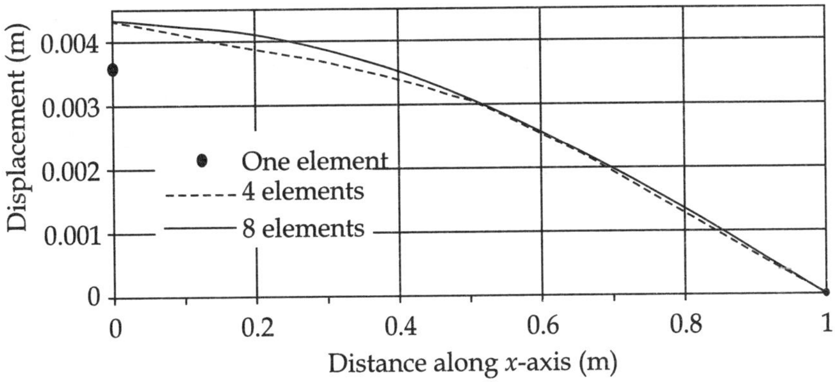

Fig. 9.6 Displacement Along the x-axis

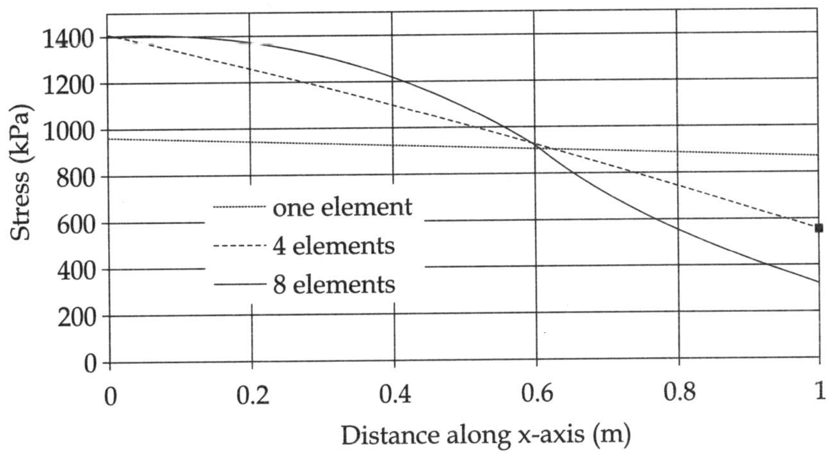

Fig. 9.7 σ_x Along the x-axis

Table 9.1 Comparison of Results

Number of elements	*Maximum displacement (m)*	σ_x*(max) (kPa)*
1	0.0034478	963.11
4	0.0042724	1407.0
16	0.0043298	1429.8
Exact value	0.00433	1461.8

9.4 MINDLIN THEORY OF BENDING OF PLATES

This theory is very similar to the Kirchhoff theory except that it includes the transverse shear deformations within the plate. For this reason, this theory is very suitable for analysis of thick plates. Contrary to the Kirchhoff theory, Mindlin assumed that the normal to the middle surface before bending remains straight but does not remain normal to the deflected middle surface, in other words, the angle of slope is due to an average shear strain (constant) at every cross section. The same concept is used in the Timoshenko beam theory. The following derivation was made by Vallabhan[7] and the theory can be formulated in the four steps given below:

Step 1. Assume Displacement Functions

Since the normal before bending does not remain normal after bending, the slopes of the middle surface cannot be used to define the u and v displacements. Thus, two new parameters θ_x and θ_y such as rotations of the cross section are introduced into the theory, with directions as defined in Fig. 9.2. Using these rotations, the displacements u, v and w at any arbitrary point at x, y and z within the plate are written as:

$$u = + z\theta_y$$
$$v = - z\theta_x$$
$$w = w(x,y) \text{ of the middle surface} \tag{9.16}$$

Step 2. Internal Strains

The internal strains are written using the strain-displacement equations given by Eq. (2.2). All components of strain except ε_z are non zero.

$$\varepsilon_x = + z\theta_{y,x} \tag{9.17}$$
$$\varepsilon_y = - z\theta_{x,y}$$
$$\varepsilon_z = 0$$
$$\gamma_{xy} = z(\theta_{y,y} - \theta_{x,x})$$
$$\gamma_{yz} = w_{,y} - \theta_x$$
$$\gamma_{xz} = w_{,x} + \theta_y$$

Subscript after a comma denotes derivative of the function with respect to the subscript.

Step 3. Internal Forces

Following the Eq. (9.3), the internal forces considered are the internal bending moments and the transverse shear forces and as in the Kirchoff theory, the in-plane force is equal to zero. The relationships for internal moments and transverse shear forces are given in two different equations as given below.

$$\{m\} = \begin{Bmatrix} m_x \\ m_y \\ m_{xy} \end{Bmatrix} = - D \begin{bmatrix} 1 & \mu & 0 \\ \mu & 1 & 0 \\ 0 & 0 & \dfrac{1-\mu}{2} \end{bmatrix} \begin{Bmatrix} -\theta_{y,x} \\ \theta_{x,y} \\ \theta_{x,x} - \theta_{y,y} \end{Bmatrix} = [D^*]\{\kappa\} \tag{9.18}$$

$$\{Q\} = \begin{Bmatrix} Q_x \\ Q_y \end{Bmatrix} = \begin{Bmatrix} \int \tau_{xz}\, dz \\ \int \tau_{yz}\, dz \end{Bmatrix} = \alpha G t \begin{bmatrix} 1 & 0 \\ 0 & 1 \end{bmatrix} \begin{Bmatrix} w_{,x} + \theta_y \\ w_{,y} - \theta_x \end{Bmatrix} = [\overline{D}]\{\gamma\} \tag{9.19}$$

The D is the flexural rigidity of the plate as defined in Eq. (9.3), the matrices D^* and $\overline{D}$ represent the respective stress-strain relations, κ and γ, the equivalent curvature vector and shear strain vector, all these are written in the same form as given in Eq. (9.3) to make a comparison of the equations. Here α is an adjustable parameter to account for the assumption of constant average shear strain in the z-direction of the plate. If the plate has orthotropic properties, the matrices D^* and $\overline{D}$ can be substituted appropriately to incorporate those properties.

Step 4. Minimization of the Total Potential Energy Function

The total energy function can be symbolically written as

$$\Phi = \frac{1}{2}\iint_A \{\kappa\}^t\left[D^*\right]\{\kappa\}dx\,dy + \frac{1}{2}\iint_A \{\gamma\}^t\left[\overline{D}\right]\{\gamma\}dx\,dy$$

$$- \iint_A \tilde{q}w\,dx\,dy - \int_S \tilde{M}_s\theta_s dS - \int_S \tilde{Q}_s w_s dS \tag{9.20}$$

which has to be minimized to obtain the necessary equilibrium equations. Here $\tilde{M}_s$, $\tilde{Q}_s$ represent prescribed moments and forces along the boundary of the element.

9.4.1 Finite Element Using Mindlin Plate Theory

The first element in this category is the quadrilateral element. After the development of a simple quadrilateral element, a family of quadrilateral elements were subsequently developed[3] with higher degrees of freedom and curved boundary edges. Only the simple 4 noded quadrilateral element with 3 d.o.f per node is discussed here. The degrees of freedom is the same for the rectangular element as shown in Fig. 9.4.

Step 1. Assume Displacement Functions

Following the concept in Timoshenko beam theory, in the Mindlin type of plate element, the rotations of the plate are independent of its lateral deflection. Considering a quadrilateral element with 4 nodes, the internal displacements and rotations with sign conventions as given in Fig. 9.2, can be interpolated using the shape functions that are developed for the quadrilateral plane stress elements given in Eq. (5.5), i.e.,

$$w(x, y) = \sum_{i=1}^{4} N_i w_i$$

$$\theta_x(x, y) = \sum_{i=1}^{4} N_i \theta_{xi}$$

$$\theta_y(x, y) = \sum_{i=1}^{4} N_i \theta_{yi} \tag{9.21}$$

In this way, complete compatibility of displacements and rotations along the inter element boundaries is satisfied in this element, but there is no direct relationship between w and θ as given in Kirchhoff theory.

Step 2. Internal Strains

The internal strains are symbolically represented by curvatures κ and transverse shear strains γ as given in Eqs. (9.17), (9.18) and (9.19) substituting into the assumed displacement functions, the B matrix can be developed as follows.

$$\begin{Bmatrix} \kappa \\ \gamma \end{Bmatrix} = \begin{Bmatrix} -\theta_{y,x} \\ \theta_{x,y} \\ \theta_{x,x} - \theta_{y,y} \\ w_{,x} + \theta_y \\ w_{,y} - \theta_x \end{Bmatrix} = [B_1 \;\; B_2 \;\; B_3 \;\; B_4]\begin{Bmatrix} u_1 \\ u_2 \\ u_3 \\ u_4 \end{Bmatrix} = [B]\{u\} \tag{9.22}$$

where $\{u_i\} = \begin{Bmatrix} w_i \\ \theta_{xi} \\ \theta_{yi} \end{Bmatrix}$ for $i = 1, 2, 3, 4$.

Derivation of Matrix $[B]$ is made in two parts, one for the curvatures and the other for the shear strain. Following the approach used in the derivation of the 4-noded quadrilateral plane stress element, a chain of matrices are developed from the strain vector to the nodal displacement vector as given below:

Part 1. For the Curvatures

$$\{\kappa\} = \begin{Bmatrix} -\theta_{y,x} \\ \theta_{x,y} \\ \theta_{x,x} - \theta_{y,y} \end{Bmatrix} = \begin{bmatrix} 0 & 0 & -1 & 0 \\ 0 & 1 & 0 & 0 \\ 1 & 0 & 0 & -1 \end{bmatrix}\begin{Bmatrix} \theta_{x,x} \\ \theta_{x,y} \\ \theta_{y,x} \\ \theta_{y,y} \end{Bmatrix} \tag{9.23}$$

The right hand side vector of Eq. (9.23) is expressed as derivatives of ξ and η using the inverse of the Jacobian matrix J (2 x 2) given in Eq. (5.37).

$$\begin{Bmatrix} \theta_{x,x} \\ \theta_{x,y} \\ \theta_{y,x} \\ \theta_{y,y} \end{Bmatrix} = \begin{bmatrix} J^{-1} & \\ & J^{-1} \end{bmatrix}\begin{Bmatrix} \theta_{x,\xi} \\ \theta_{x,\eta} \\ \theta_{y,\xi} \\ \theta_{y,\eta} \end{Bmatrix} \tag{9.24}$$

Using Eq. (9.21),

$$\theta_{x,\xi} = \sum_i N_{i,\xi}\theta_{xi} \text{ and } \theta_{x,\eta} = \sum_i N_{i,\eta}\theta_{xi} \text{ etc. for } i = 1,..\, 4 \tag{9.25}$$

Thus using Eq. (9.21) through (9.25) the relationship between the curvature vector and the nodal displacement vector is established,

$$\{\kappa\} = [B^*]\{u\} \tag{9.26}$$

Here the matrix B^* is a product of the matrices developed through Eq. (9.22) through (9.25).

Part 2. For the Shear Strains:

$$\{\gamma\} = \begin{Bmatrix} w_{,x} + \theta_y \\ w_{,y} - \theta_x \end{Bmatrix} = \begin{bmatrix} 1 & 0 & 0 & 1 \\ 0 & 1 & -1 & 0 \end{bmatrix} \begin{Bmatrix} w_{,x} \\ w_{,y} \\ \theta_x \\ \theta_y \end{Bmatrix} \tag{9.27}$$

and the right hand side vector is expressed in terms of ξ, η using the inverse of the Jacobian transformation matrix J and an identity matrix I of order 2, i.e.,

$$\begin{Bmatrix} w_{,x} \\ w_{,y} \\ \theta_x \\ \theta_y \end{Bmatrix} = \begin{bmatrix} J^{-1} & \\ & I \end{bmatrix} \begin{Bmatrix} w_{,\xi} \\ w_{,\eta} \\ \theta_x \\ \theta_y \end{Bmatrix} \tag{9.28}$$

Noting that using Eq. (9.21), w and θ are functions of ξ and η and the nodal displacements, taking derivatives,

$$w_{,\xi} = \sum_i N_{i,\xi} w_i \quad \text{and } w_{,\eta} = \sum_i N_{i,\eta} w_i \tag{9.29}$$

$$\theta_x = \sum_i N_i \theta_{xi} \quad \text{and } \theta_y = \sum_i N_i \theta_{yi} \text{ etc.} \quad \text{for } i = 1,2,3,4$$

the right hand side of Eq. (9.29) can be written in terms of the element nodal displacement vector. Thus the shear strain vector can be expressed in terms of the nodal displacement vector as:

$$\{\gamma\} = [\bar{B}]\{u\} \tag{9.30}$$

Step 3. Internal Forces

The relationship between the internal moments and transverse shear forces with the internal strains are given in Eqs. (9.18) and (9.19).

Step 4. Element Stiffness Matrix for the Plate Bending Element

At this moment the equation for the element stiffness matrix is written by combining the stiffness matrices due to curvature and shear strains.

$$\begin{aligned} [k] &= \iint_A [B^*]^t [D^*][B^*] \, dx \, dy + \iint_A [\bar{B}]^t [\bar{D}][\bar{B}] \, dx \, dy \\ &= \sum_i \sum_j W_i W_j ([B^*]^t [D^*][B^*] \| J \|)_{i,j} \\ &+ \sum_k \sum_l W_k W_l \left([\bar{B}]^t [\bar{D}][\bar{B}] \| J \| \right)_{k,l} \end{aligned} \tag{9.31}$$

$$= [k_b] + [k_s]$$

The indices i, j, k and l denote the respective numbering of the pivotal points and quantities W_i, etc., denote the weighting coefficients for the numerical integration.

Since the emphasis in analysis of plates was to discover a quadrilateral element for analysis of thin plates so as to fit into any geometry and theoretically it was accomplished. However, while analyzing thin plates, this finite element model yields a stiff solution. Shear locking became a problem just as in the case of the exact integration in the Timoshenko beam element. One has to employ a large number of elements in order to get satisfactory and convergent result. Researchers[3,6,8] used the experience they had with the Timoshenko beam element and have found that the plate model becomes less stiff by reducing the order of integration scheme to integrate the equations of the transverse shear portion. Two-point Guass integration is used to evaluate the stiffness matrix $[k_b]$ due to bending and a one-point integration is used to evaluate the stiffness matrix $[k_s]$ due to shear strain. Still to get a satisfactory solution, more elements are necessary. The strain-displacement relations were derived separately for the curvature and the shear strain, to facilitate the reduction of the order of integration of the stiffness matrix equation given in Eq. (9.31).

For analysis of thick plates, shear locking does not happen, thus reduced integration is not necessary; however more elements are necessary to get satisfactory results. Another alternative is to employ a nine noded quadrilateral (with the 9-th node at the center of the quadrilateral) and this element has been found to yield very good results, but the number of degrees of freedom is 27 per element. In this element, if the displacement at the center is removed by static condensation technique, it will lead to better results. Considerable research has been done on this topic, and analysts are advised to relevant literature[3,5,6,8] on this to gain confidence in their work.

9.5 ANALYSIS OF SHELL PROBLEMS USING FLAT PLATE ELEMENTS

One type of analysis of shell structures is by assuming that they are composed of flat plate elements that have stiffness in its own plane (in-plane or membrane) and in bending. These flat plate elements can be triangular, rectangular or quadrilateral as the analyst would like to choose. The in-plane stiffness can be idealized by the plane stress condition and the bending stiffness by the Kirchhoff or Mindlin theory. Of course, these theories are good only for small displacements of the shell. Let us consider a triangular flat shell element with three nodes at the corners of the triangle. A minimum of three translations and three rotations per node are necessary, and thus there are 6 d.o.f per node resulting in an 18 d.o.f. shell element. This is the simplest of the shell elements.

For the triangular plane stress element there are 6 d.o.f. and for the triangular plate bending element there are 9 d.o.f. per element, and when these two elements are combined, there are only 15 d.o.f. per element. The reason for the lack of three degrees of freedom is that both plane stress and bending elements do not consider the rotations with respect to the normal to the plane of the plate. There are three such rotations in the normal direction possible, but invariably the plate is very stiff with respect to rotation in the normal direction. It can be assumed that the rotations in the normal direction to be zero. If temporarily, the nodal displacement vector $\{u\}$ is arranged as the combination of in plane displacements and plate bending displacements:

$$\{u\}^t_{plane} = \{u_1 \; v_1 \; u_2 \; v_2 \; u_3 \; v_3\}^t \text{ and}$$

$$\{u\}^t_{bending} = \{w_1 \; \theta_{x1} \; \theta_{y1} \; w_2 \; \theta_{x2} \; \theta_{y2} \; w_3 \; \theta_{x3} \; \theta_{y3}\}^t \tag{9.32}$$

then the element stiffness matrix for the shell element in its own local coordinate system with x,y coordinates representing the plane of the plate, is constructed as

$$\begin{bmatrix} \underset{6\times 6}{k_{plane}} & & \\ & \underset{9\times 9}{k_{bending}} & \\ & & \underset{3\times 3}{0} \end{bmatrix} \begin{Bmatrix} \{u\}_{plane} \\ \{u\}_{bending} \\ \{\theta_z\} \end{Bmatrix} = \begin{Bmatrix} \text{corresponding} \\ \text{force} \\ \text{vector} \end{Bmatrix} \tag{9.33}$$

The above element stiffness matrix first has to be rearranged to represent the nodal displacement vector in the form $u, v, w, \theta_x, \theta_y, \theta_z$ for each node. Then the element stiffness matrix has to be transformed to the global coordinate system using a transformation matrix $[T]$ that transforms the displacement vector in the local coordinate system given by (x, y, z) to a corresponding vector in the global coordinate system $(\bar{x}, \bar{y}, \bar{z},)$. Part of this transformation matrix can be easily developed for a displacement vector at a node with components u, v and w in the local coordinate system by using the direction cosine matrix as given below.

$$\begin{Bmatrix} u \\ v \\ w \end{Bmatrix} = \begin{bmatrix} \lambda_{x\bar{x}} & \lambda_{x\bar{y}} & \lambda_{x\bar{z}} \\ \lambda_{y\bar{x}} & \lambda_{y\bar{y}} & \lambda_{y\bar{z}} \\ \lambda_{z\bar{x}} & \lambda_{z\bar{y}} & \lambda_{z\bar{z}} \end{bmatrix} \begin{Bmatrix} \bar{u} \\ \bar{v} \\ \bar{w} \end{Bmatrix}$$

or

$$\{u\}_{local} = [L]\{\bar{u}\}_{global} \tag{9.34}$$

The global coordinate system is denoted by $\bar{x}, \bar{y}, \bar{z}$ and the corresponding global displacement vector is represented by $\bar{u}, \bar{v}$ and $\bar{w}$. A full transformation matrix consists of 6 such matrices as given below.

$$[T] = \begin{bmatrix} L & 0 & 0 & 0 & 0 & 0 \\ 0 & L & 0 & 0 & 0 & 0 \\ 0 & 0 & L & 0 & 0 & 0 \\ 0 & 0 & 0 & L & 0 & 0 \\ 0 & 0 & 0 & 0 & L & 0 \\ 0 & 0 & 0 & 0 & 0 & L \end{bmatrix} \tag{9.35}$$

The final element global stiffness matrix is obtained as

$$[K]_{global} = [T]^t[k]_{local}\,[T] \tag{9.36}$$

9.6 SUMMARY

A brief introduction to the Kirchoff theory of plates is introduced so that the assumptions in the theory are well understood. Based on this, theories for the development of stiffness matrices for both triangular and rectangular finite elements are shown. For the development of quadrilateral plate elements, Mindlin theory of plates that includes the vertical shear strain is presented. Using a transformation into a square and numerical integration techniques, the stiffness matrix for a quadrilateral plate element is developed. Reduced integration techniques are used to minimize the shear locking effect in the element. To analyze thin linear shell problems, flat plane stress and plate bending elements are combined to form a flat shell finite element. For the analysis of a three-dimensional shell structure, all the individual flat shell elements are transformed to a common set of global axes and combined to form the global stiffness equations.

PROBLEMS

9.1 Calculate the equivalent nodal forces for a triangular plate element for the load cases shown below.

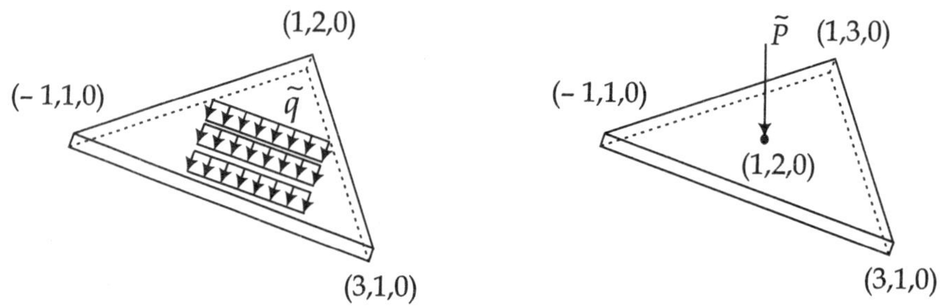

(a) Uniformly Distributed Force

(b) Concentrated Force

Fig. P9.1

9.2 A conformable rectangular plate element with the size of axb is developed using Hermitian polynomials. Derive the consistent forces for a point load at midpoint on a plate.

9.3 A uniform square plate, simply supported carries a uniformly distributed force q/unit square area. The flexural rigidity is D. Using a single element with 12 d.o.f, i.e, 3 d.o.f. per node, calculate the maximum deflection of the plate. Use symmetry in x and y directions.

9.4 A plate bending element has to be developed with two degrees of freedom per node with w and dw/dx only. w varies cubically along the x-axis and linearly along the y-axis. Develop the displacement function w. Discuss its compatibility.

REFERENCES

1. Dym, C.L, and Shames, I.H, "Solid Mechanics: A Variational Approach", McGraw Hill Book Co, New York, 1973.
2. Fung, Y.C, "Foundations of Solid Mechanics", Prentice Hall, Englewood Cliffs, New Jersey, 1965.
3. Hughes, T.J.R, "The Finite Element Method", Prentice Hall, Englewood Cliffs, New Jersey, 1987.
4. Kardestuncer, H., "Finite Element Handbook", McGraw Hill Book Company, New York, 1987.
5. Krishnamoorthy, C.S, "Finite Element Analysis: Theory and Programming", Tata McGraw Hill Publishing Company, New Delhi, 1987.
6. Reddy, J.N, "An Introduction to the Finite Element Method",2nd Ed. McGraw Hill Book Company, New York, 1993.
7. Vallabhan, C.V.Girija, "FEM Class Notes", Texas Tech Univ., Lubbock, 1990.
8. Zienkiewicz, O.C. and Taylor, R.L, "The Finite Element Method", Vols. I & II, McGraw Hill Book Company, London, 1989.

10

CHAPTER

Steady State Flow Problem

10.1 INTRODUCTION

So far the application of finite element methods have been discussed for solving linear elasticity problems. During the development of these methods, researchers have simultaneously made substantial progress on the extension of the method to solve fluid and or heat flow problems in two and three dimensional domains. These problems can be either steady state or transient. Transient problems require time integration schemes, in addition to a solution in a domain at a particular instant. These problems are understood well, if one takes a first step to learn to solve simple steady state problems. Only steady state problems are discussed here, because the relevant equations and the solutions have great similarity to the equilibrium problems in solid mechanics. In fact, the finite element analysis of two dimensional steady state flow problems are easier to comprehend than the finite element analysis of plane stress or strain problems, because the plane elasticity problems with the displacement approach become a vector field problem (with two parameters at each node) as opposed to a scalar field problem (with only one parameter) in a steady state flow condition.

There are two approaches[6,8,9] that are being used in the development of the finite element analysis. One approach is to use a variational functional and minimize it, similar to the minimum total potential energy function in solid mechanics and the other approach is to use the well known 'Galerkin method'. Both approaches are introduced here, even though the latter approach is more general and has gained more popularity among engineers recently. To understand these methods, it is recommended first to go through the classical differential equations for the steady state flow problems and then apply the finite element technique to solve them.

10.2 TWO DIMENSIONAL STEADY STATE SEEPAGE PROBLEM

One of the simplest flow problems that many especially civil engineers are familiar is the steady state fluid flow through a porous media. The same equations are applicable to steady state heat flow, laminar fluid flow, etc., and these problems are being recognized as potential field problems in engineering.[2,3] Flow through a porous soil media under a sheet pile as illustrated in Fig. 10.1 can be also represented by a second order partial differential equation in terms of a potential function, which controls the direction of the flow of the fluid. If h is defined as the potential function, usually called as the hydraulic head, then

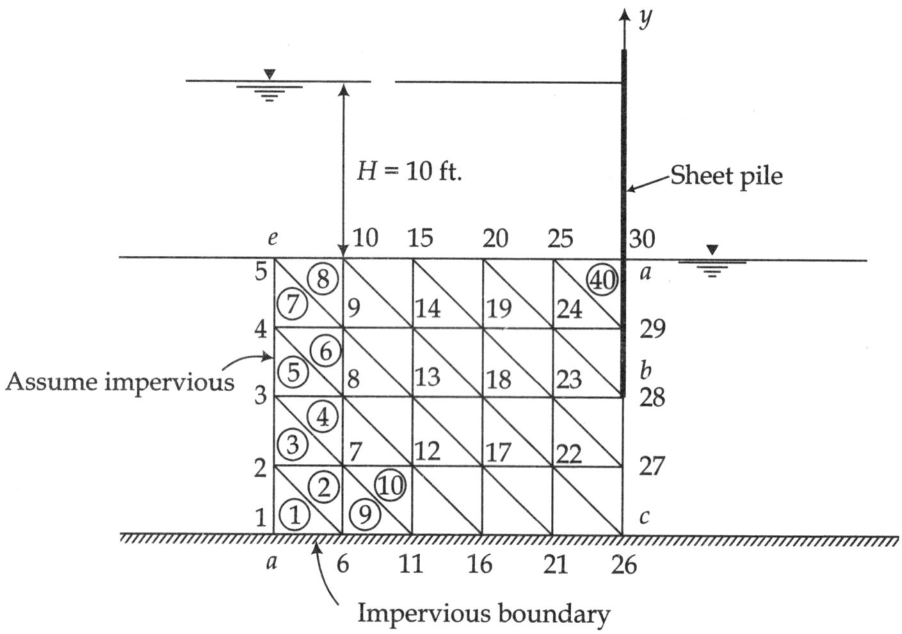

Fig. 10.1 Seepage Under a Sheet Pile (Finite Element Discretization)

$$k_x \frac{\partial^2 h}{\partial x^2} + k_y \frac{\partial^2 h}{\partial y^2} = 0 \text{ in } \Omega \tag{10.1}$$

with specified boundary conditions such as

$$h = \tilde{h}, \text{ prescribed on } S_1$$

and

$$k_x \frac{\partial h}{\partial x} l + k_y \frac{\partial h}{\partial y} m + \tilde{q} \text{ on } S_2 \tag{10.2}$$

Here k_x and k_y are the principal values of the hydraulic conductivity of the soil in the x and y directions respectively and l and m are the direction cosines of the outward normal to the boundary S_2, and $\tilde{q}$ is the rate of flow per unit surface area normal to the boundary. S_1 and S_2

are the two parts of the entire boundary. For outward flow $\tilde{q}$ is positive. The derivation of these equations is usually found in text books on hydraulics, geomechanics, etc. One of the techniques used here is to realize that an abstract functional

$$\Phi = \iint_{\Omega} \frac{1}{2}\left[k_x\left(\frac{\partial h}{\partial x}\right)^2 + k_x\left(\frac{\partial h}{\partial x}\right)^2\right] dx\, dy + \int_{S_2} \tilde{q} h \, dS_2 \tag{10.3}$$

if minimized with respect to the variable h, will result in Eqs. (10.1) and (10.2). These minimization techniques have their roots in the variational calculus just as the equations for the beam are developed in Chapter 1. Having obtained the function Φ which has to be a minimum in the domain as well as in each of the finite elements constituting the entire domain, one can use the same logical approach to derive the system equations for the problem. The relevant four steps can be employed in the same way that was used in formulating the finite element equations for the solid mechanics problems.

10.3 SIMPLE TRIANGULAR FINITE ELEMENT

This steady state flow problem is solved using a simple triangular finite element with three nodes and one degree of freedom per node, resulting in a three degree of freedom element, as shown in Fig. 10.2. Here the finite element is a finite domain of the problem.

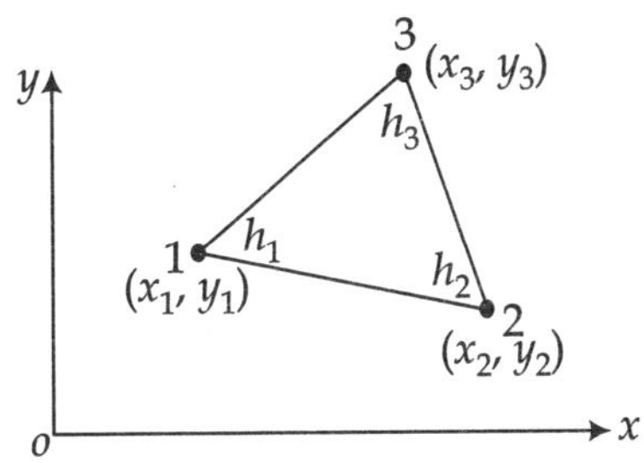

Fig. 10.2 A Simple Triangular Element

Step 1. Assume h Function within the Element

Since there are three degrees of freedom, i.e., three parameters, the following polynomial function for the distribution of the h function within the triangle is assumed.

$$h = \alpha_1 + \alpha_2 x + \alpha_3 y \tag{10.4}$$

Constant head condition and constant hydraulic gradient are required to be contained in the assumed function and they are satisfied by the parameters α_1, α_2 and α_3. Since the head varies linearly within the triangle as well as on the boundary, the interelement compatibility of the h function is automatically satisfied. Following the equations used in the simple 2-D plane stress element, the assumed h function is written using the same shape functions

$$h = \sum_{i=1}^{3} N_i h_i \tag{10.5}$$

where $$N_i = \frac{1}{2A}(a_i + b_i x + c_i y)$$

with notations for a_i, b_i and c_i given in Eq.(4.5) and A is the area of the triangular element.

Step 2. Internal Hydraulic Gradient

Differentiating the h function with respect to the x and y coordinates, the respective hydraulic gradients within the triangle are obtained as,

$$\begin{Bmatrix} \dfrac{\partial h}{\partial x} \\ \dfrac{\partial h}{\partial y} \end{Bmatrix} = \frac{1}{2A}\begin{bmatrix} b_1 & b_2 & b_3 \\ c_1 & c_2 & c_3 \end{bmatrix}\begin{Bmatrix} h_1 \\ h_2 \\ h_3 \end{Bmatrix} = [B]\,\{h\} \tag{10.6}$$

Step 3. Minimization of the Functional

Substituting the above equations in the functional expression in Eq. (10.3),

$$\Phi = \frac{1}{2}\iint_{\Omega} \begin{Bmatrix} \dfrac{\partial h}{\partial x} \\ \dfrac{\partial h}{\partial y} \end{Bmatrix}^t \begin{bmatrix} k_x & 0 \\ 0 & k_y \end{bmatrix} \begin{Bmatrix} \dfrac{\partial h}{\partial x} \\ \dfrac{\partial h}{\partial y} \end{Bmatrix} dx\,dy + \int_{S} \tilde{q} h\, dS \tag{10.7}$$

$$= \frac{1}{2}\iint_{\Omega} \{h\}^t\,[B]^t\,[D][B][h]\; dx\,dy + \int_S \tilde{q}[N]\,\{h\}\, dS$$

where $[D] = \begin{bmatrix} k_x & 0 \\ 0 & k_y \end{bmatrix}$ and S is the boundary of the finite element.

By minimizing the function Φ with respect to the nodal parameters the element system matrix is:

$$[k] = \int_A [B]^t\,[D][B]dx\,dy = [B]^t [D]\,[B]\,A \tag{10.8}$$

$$= \frac{k_x}{4A}\begin{bmatrix} b_1^2 & b_1b_2 & b_1b_3 \\ & b_2^2 & b_2b_3 \\ sym & & b_3^2 \end{bmatrix} + \frac{k_y}{4A}\begin{bmatrix} c_1^2 & c_1c_2 & c_1c_3 \\ & c_2^2 & c_2c_3 \\ sym & & c_3^2 \end{bmatrix}$$

for an arbitrary element.

Step 4. Assembly of the Finite Element Equations to Obtain the Global Equations

Using the element number and nodal number technique, and considering the uniqueness of head at the nodes, the element equations are assembled just as done with the finite elements in solid mechanics. The final equation will be of the form

$$[K]\,\{H\} = \{Q\} \tag{10.9}$$

where H and Q are the global head and flow vectors, and K is the global system matrix. The flow vector Q has to be modified using the specified values of flow on the boundary S_2. The global matrix K and the vector Q has to be modified for the specified values of $\tilde{h}$ on the boundary S_1.

Step 5. Incorporation of the Prescribed Boundary Conditions

The prescribed boundary conditions involving $\tilde{h}$ are incorporated exactly the same way for prescribed displacements as discussed in Chapter 3. After prescribing the boundary conditions, the global matrix becomes nonsingular and solved for the nodal head vector.

Example Problem 10.1 Seepage through Porous Media

An example problem is as shown in Fig. 10.1. For this example problem, all elements can be categorized into two types represented by elements 1 and 2 which are odd and even numbered. (See Fig. 10.1) For simplicity it is assumed that $k_x = k_y = k$, that is the soil is homogeneous and isotropic, and using Eq. (10.8), the respective element stiffness matrices are obtained as:

for odd numbered elements, $$[k] = k\begin{bmatrix} 1 & -0.5 & -0.5 \\ & 0.5 & 0 \\ & & 0.5 \end{bmatrix}$$

for even numbered elements, $$[k] = k\begin{bmatrix} 0.5 & 0 & -0.5 \\ & 0.5 & -0.5 \\ & & 1 \end{bmatrix}$$

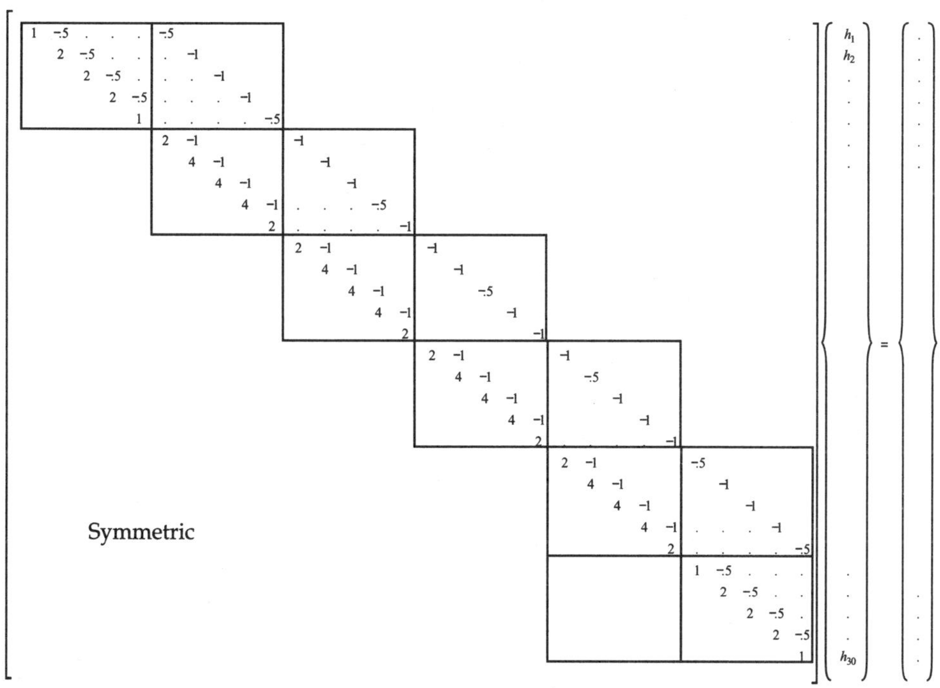

Fig. 10.3 Global System Equations Before Boundary Conditions

Using the techniques developed for assembling the stiffness matrices, the global system equation is developed as shown in Fig. 10.3. It should be noted that the global system matrix is symmetric and banded and hence can be stored as a half-banded matrix. Prescribing the boundary conditions as given below,

on boundary ea, $h_5 = h_{10} = h_{15} = h_{20} = h_{25} = h_{30} = 10$

on boundary cb, $h_{26} = h_{27} = h_{28} = 5$

on boundaries ab, ed and dc, $\tilde{q} = 0$,

will lead to the global system equations. These equations are solved and the results are presented in the Fig. 10.4 This example illustrates how a simple steady state flow problem can be solved by the finite element method.

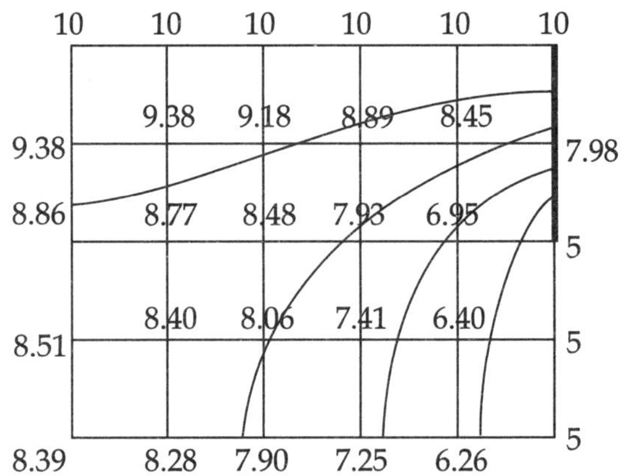

Fig. 10.4 Values of Potential Head at the Nodes (FEM solution)

10.4 APPLICATION OF THE GALERKIN METHOD

The next step is to understand the application of the Galerkin method to solve fluid flow problems. A brief introduction to the Galerkin method was given in Chapter 1. The method belongs to a class of methods called 'weighted residual methods'[1,3,6,8]; however, Galerkin method has its roots in variational calculus as mentioned in Chapter 1. Let the linear differential equation to be solved of the type

$$L(u) - f = 0 \text{ in the domain } \Omega \tag{10.10}$$

with all necessary boundary conditions. As mentioned in Chapter 1, to use the classical Galerkin method, the approximate solution $\bar{u}$ has to be expressed as a linear combination of functions ψ_i, $i = 1,\ldots,n$. that satisfy all the boundary conditions, i.e.,

$$\bar{u} = \sum_{i=1}^{n} c_i \psi_i \text{ for } i = 1, \ldots, n. \tag{10.11}$$

According to the Galerkin method,

$$\int_{\Omega}\left[L(\bar{u}) - f\right]\psi_i \, d\Omega = 0 \quad \text{for } i = 1, \ldots, n \tag{10.12}$$

which would yield a set of n linear equations in c_i, if the operator $L(.)$ is a linear differential operator. Here the functions ψ_i, $i = 1, \ldots, n$ can also be called weighting functions. The technique has been presented in Chapter 1 and is repeated here for convenience. In the past, many scientists and engineers used this technique to solve complex differential equations; but the way the method is applied for developing finite element equations is slightly different. The integral equation shown in Eq. (10.12) is integrated to reduce the order of the differential operator by one order in the case of a second order differential equation. For this reason, the Galerkin method when applied to form finite element equations, is called a "weak form of the Galerkin method".[6,8,9]

One of the best ways to understand the method is to follow a simple example of a weak Galerkin finite element solution of a linear second order differential equation, that might represent a steady state one-dimensional heat flow, flow thorough porous media, the stress analysis of a longitudinal bar, etc. or any simple linear second order differential equation.

10.5 ONE-DIMENSIONAL SECOND ORDER DIFFERENTIAL EQUATION

Let a second order differential equation be as given below:

$$\frac{d^2u}{dx^2} + u = 1 \text{ for } 0 < x < L \tag{10.13}$$

with boundary conditions $u(0) = u_0$ and $u(L) = u_l$. In the finite element method the domain 0 to L is divided into subdomains called 'finite elements'.[4] Let us divide the domain into 4 subdomains as shown in Fig. 10.5(a). Further it is assumed that the function $u(x)$ varies linearly in each subdomain otherwise called 'element'. Recall the use of piece-wise linear functions ψ_i for representing the behavior of a longitudinal bar in Chapter 3, where the function $u(x)$ is approximated as

$$\bar{u}(x) = \sum_{i=1}^{5} u_i \psi_i \tag{10.14}$$

There are 5 weighting functions, one for each node as given in Fig. 10.5(b). One can examine the role of these functions in a global system, and the diagram for this is repeated here (see Fig. 3.9), as the concept is extremely important to the understanding of the Galerkin finite element formulation. Let us take the case of an arbitrary node 3 which is common to two elements 2 and 3 on either side as shown in Fig. 10.5(b). If the function $u(x)$ within each element varies linearly independently, then the function ψ_3 for the node 3, is a combination of the shape functions for the elements 2 and 3 respectively as shown in Fig. 10.5(b).

According to the classical Galerkin method, for $i = 3$, the integral

$$I_3 = \int_0^L \left(\frac{d^2\bar{u}}{dx^2} + \bar{u} - 1 \right) \psi_3 \, dx = 0. \tag{10.15}$$

At this juncture, it is necessary to examine the continuity of the function $\bar{u}(x)$ and its derivatives. Since the shape functions are to represent $\bar{u}(x)$ varying linearly in each domain,

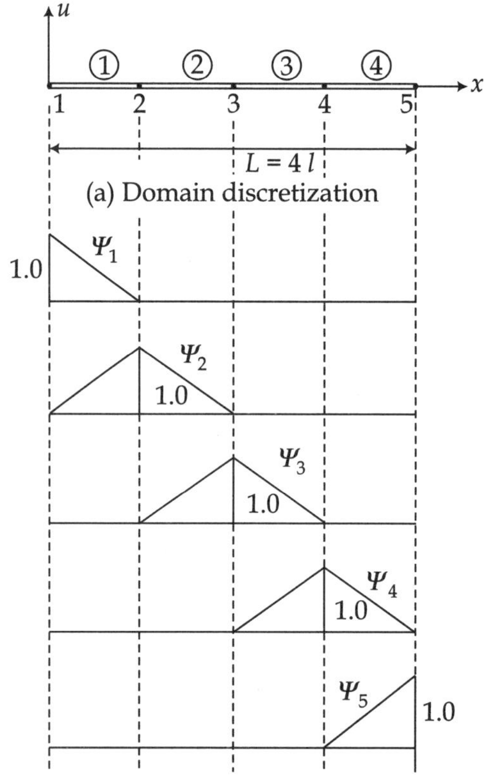

Fig. 10.5

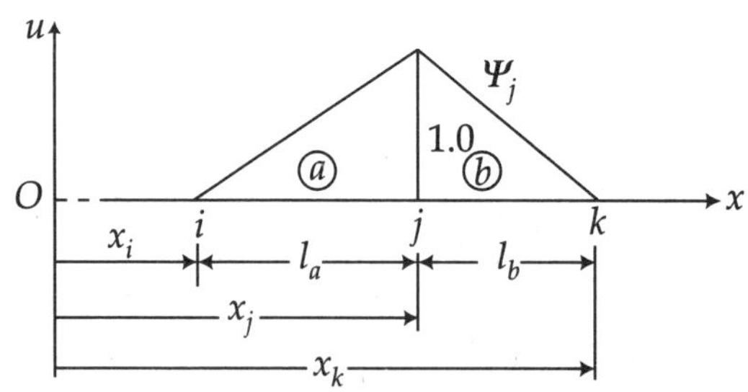

Fig. 10.6 A Sample Basis Function at Node j

with compatibility of its nodal values at the interelement nodes, the function $\bar{u}(x)$ is continuous for the entire domain, but its first derivative is not and the second derivative is zero within each element. Therefore at the nodes, the values of second derivatives are indeterminate. In order that the integral I_3 may have some value, the order of the second derivative in the integral has to be reduced to first derivative using integration by parts. Performing the integration by parts on the second derivative only, the integral I_3 can now be written as

$$I_3 = \frac{d\bar{u}}{dx}\psi_3\Big|_0^L - \int_0^L \frac{d\bar{u}}{dx}\frac{d\psi_3}{dx}dx + \int_0^L (\bar{u}-1)\,\psi_3\,dx = 0 \tag{10.16}$$

Noting that in the term of the above expression, the value of ψ_3 is zero except in elements 2 and 3, substituting the values on the limits of the boundary of the domain, the first term vanishes. Also the limits of the integral I_3 now becomes from x_2 to x_4. Since the basis function ψ_3 is a combination of shape functions for the elements 2 and 3 respectively, the integral I_3 is written as a sum of two integrals, $I_3^{(2)}$ for element 2 and $I_3^{(3)}$ for the element 3,

i.e.,
$$I_3 = I_3^{(2)} + I_3^{(3)} \tag{10.17}$$

The integration within the element 2 and 3 are performed separately as shown below. Here the basis function ψ_j for an arbitrary node j is examined.

10.5.1 Basis Function ψ_j for a Node j

The function ψ_j is a combination of functions for the elements a and b on the left and right side of node j as shown in Fig. 10.6. In other words,

$$\psi_j = \psi_j^a + \psi_j^b \tag{10.18}$$

For the element a on the left side,
$$\psi_j^a = \frac{x - x_i}{l_a}\,u_j \tag{10.19}$$

For the element b on the right side
$$\psi_j^b = \frac{x_k - x}{l_b}\,u_j \tag{10.20}$$

where l_a and l_b are the lengths of elements a and b respectively. It can be very easily verified that these functions taken individually satisfy the properties of the shape functions that have been used before. But the assumed displacement functions,

for the element a on the left side,
$$\bar{u} = \frac{x_j - x}{l_a}u_i + \frac{x - x_i}{l_a}u_j \tag{10.21}$$

for the element b on the right side,
$$\bar{u} = \frac{x_k - x}{l_b}u_j + \frac{x - x_j}{l_b}u_k \tag{10.22}$$

By substituting these equations in Eq. (10.16) and integrating, the system equation corresponding to the node 3 is obtained.

10.5.2 Global Equations

To get numerical values, the domain L is divided equally into 4 elements, i.e., $L = 4\,l$. Applying the weighting functions and integrating for example at node 3,

$$\frac{1}{l}(u_2 - 2u_3 + u_4) + \frac{l}{6}(u_2 + 4u_3 + u_4) - l = 0 \tag{10.23}$$

Repeating this procedure for each node from 1 to 5, the following matrix equation can be developed.

$$\begin{Bmatrix} -\left(\frac{du}{dx}\right)_1 \\ 0 \\ 0 \\ 0 \\ \left(\frac{du}{dx}\right)_5 \end{Bmatrix} - \frac{1}{l}\begin{bmatrix} 1 & -1 & & & \\ -1 & 2 & -1 & & \\ & -1 & 2 & -1 & \\ & & -1 & 2 & -1 \\ & & & -1 & 1 \end{bmatrix}\begin{Bmatrix} u_1 \\ u_2 \\ u_3 \\ u_4 \\ u_5 \end{Bmatrix} + \frac{l}{6}\begin{bmatrix} 2 & 1 & & & \\ 1 & 4 & 1 & & \\ & 1 & 4 & 1 & \\ & & 1 & 4 & 1 \\ & & & 1 & 2 \end{bmatrix}\begin{Bmatrix} u_1 \\ u_2 \\ u_3 \\ u_4 \\ u_5 \end{Bmatrix} - \frac{l}{2}\begin{Bmatrix} 1 \\ 2 \\ 2 \\ 2 \\ 1 \end{Bmatrix} = \begin{Bmatrix} 0 \\ 0 \\ 0 \\ 0 \\ 0 \end{Bmatrix} \tag{10.24}$$

Applying the boundary conditions at the ends, the unknown derivatives at the ends as shown in the above equation can be removed and the equations can be solved for the unknown values of u. This is the weak form of Galerkin or the Galerkin finite element method.

In order to get a numerical solution, following values are given to the domain and the boundary conditions, such as:

$L = 8.0$, and $u_{0|x=0} = 1.0$ and $u_{L|x=8} = 6.0$ the exact solution to the problem is:

$$u = 5.054 \sin x + 1.0. \tag{10.25}$$

Using four elements in the domain, $L = 4\,l = 8.0$ and $u_1 = 1.0$ and $u_5 = 6.0$ substituting $l = 2.0$ and all the boundary conditions, the governing matrix equations are reduced with three unknowns as,

$$\begin{bmatrix} 0.333 & 0.833 & 0 \\ 0.833 & 0.333 & 0.833 \\ 0 & 0.833 & 0.333 \end{bmatrix}\begin{Bmatrix} u_2 \\ u_3 \\ u_4 \end{Bmatrix} = \begin{Bmatrix} 1.167 \\ 2.0 \\ -3.0 \end{Bmatrix}$$

The entire solution including boundary conditions is:

$$\begin{Bmatrix} u_1 \\ u_2 \\ u_3 \\ u_4 \\ u_5 \end{Bmatrix} = \begin{Bmatrix} 1.0 \\ 7.793 \\ -1.717 \\ -4.707 \\ 6.0 \end{Bmatrix}. \tag{10.26a}$$

The result is plotted in Fig. 10.7 along with the exact solution. The problem is repeated for 6 elements with 5 unknowns and the global matrix equations become:

$$\begin{bmatrix} -0.611 & 0.972 & 0 & 0 & 0 \\ 0.972 & -0.611 & 0.972 & 0 & 0 \\ 0 & 0.972 & -0.611 & 0.972 & 0 \\ 0 & 0 & 0.972 & -0.611 & 0.972 \\ 0 & 0 & 0 & 0.972 & -0.611 \end{bmatrix} \begin{Bmatrix} u_2 \\ u_3 \\ u_4 \\ u_5 \\ u_6 \end{Bmatrix} = \begin{Bmatrix} 0.361 \\ 1.333 \\ 1.333 \\ 1.333 \\ -4.5 \end{Bmatrix}$$

The entire solution with 6 elements is:

$$\{u\}^t = \{1.0 \;\; 6.048 \;\; 4.173 \;\; -2.054 \;\; -4.093 \;\; 0.853 \;\; 6.0\} \qquad (10.26b)$$

The same problem is repeated for 8 elements with 7 unknowns and the entire solution is:

$$\{u\}^t = \{1.0 \;\; 5.152 \;\; 5.745 \;\; 2.271 \;\; -2.293 \;\; -4.034 \;\; -1.461 \;\; 3.222 \;\; 6.0\} \qquad (10.26c)$$

All the results are presented along with the exact solution in Fig. 10.7 illustrating the convergence of the solution as the number of elements is increased.

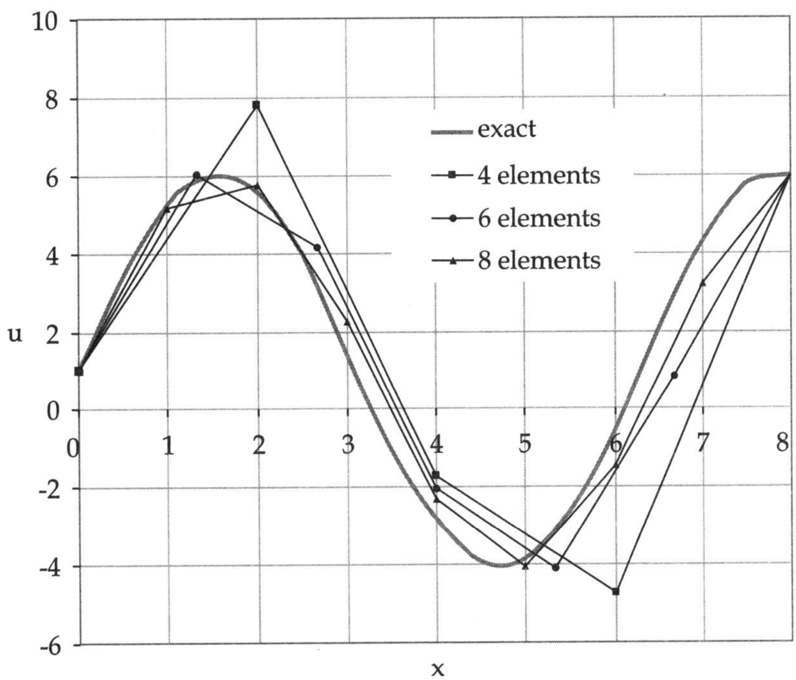

Fig. 10. 7 Finite Element Solution to the Eq. 10.13.

10.6 FINITE ELEMENT GALERKIN METHOD

Section 10.5 was devoted to illustrate the general Galerkin approach where the governing differential equation with discretized domains, was subjected to weighting functions similar to the shape functions in the global sense at each node. By integrating in the whole domain for each node using Eq. (10.12), a global equation was developed for each node. All such equations put together in a matrix form the finite element global equation. After incorporating the given boundary conditions, the global equations were solved, thus getting a solution of the differential equation by the finite element method. This procedure is help beginners to understand the finite element Galerkin method.

In the finite element method, the practice is to have finite elements or subdomains, and equations are developed for each element and assembled together. To illustrate this procedure, system equations are developed specifically for an element 2 from Fig. 10.5. The u-function within this element 2 can be written as:

for $$x_2 \le x \le x_3,$$

$$\bar{u}(x) = N_1 u_1^{(2)} + N_2 u_2^{(2)},$$

where N_1 and N_2 are linear interpolation functions given in Eq. (10.21) or (10.22). The subscripts 1 and 2 for N refers to nodal numbers in a local coordinate system. Applying N_1 as the weighting function for Eq. (10.16) just for the element 2 only,

$$N_1 \frac{d\bar{u}}{dx}\Bigg|_{x_2}^{x_3} - \int_{x_2}^{x_3} \frac{dN_1}{dx}\begin{bmatrix}\frac{dN_1}{dx} & \frac{dN_2}{dx}\end{bmatrix}\begin{Bmatrix}u_1\\u_2\end{Bmatrix} dx + \int_{x_2}^{x_3} N_1 \begin{bmatrix}N_1 & N_2\end{bmatrix}\begin{Bmatrix}u_1\\u_2\end{Bmatrix} dx - \int_{x_2}^{x_3} N_1 dx = 0 \qquad (10.27)$$

Similarly using N_2 as the weighting function, another expression is obtained, and combining these two equations and writing in matrix form,

$$\begin{Bmatrix}-\left(\frac{d\bar{u}}{dx}\right)_{x_2}\\ \left(\frac{d\bar{u}}{dx}\right)_{x_3}\end{Bmatrix} - \int_{x_2}^{x_3}\begin{Bmatrix}\frac{dN_1}{dx}\\ \frac{dN_2}{dx}\end{Bmatrix}\begin{bmatrix}\frac{dN_1}{dx} & \frac{dN_2}{dx}\end{bmatrix}\begin{Bmatrix}u_1\\u_2\end{Bmatrix} dx + \int_{x_2}^{x_3}\begin{Bmatrix}N_1\\N_2\end{Bmatrix}\begin{bmatrix}N_1 & N_2\end{bmatrix}\begin{Bmatrix}u_1\\u_2\end{Bmatrix} dx - \int_{x_2}^{x_3}\begin{Bmatrix}N_1\\N_2\end{Bmatrix} dx = 0 \qquad (10.28)$$

Substituting for $N_1$and N_2, and integrating, the system equation for element 2 is obtained as,

$$\begin{Bmatrix}-\left(\frac{d\bar{u}}{dx}\right)_{x_2}\\ \left(\frac{d\bar{u}}{dx}\right)_{x_3}\end{Bmatrix} - \frac{1}{l}\begin{bmatrix}1 & -1\\ -1 & 1\end{bmatrix}\begin{Bmatrix}u_1\\u_2\end{Bmatrix} + \frac{l}{6}\begin{bmatrix}2 & 1\\ 1 & 2\end{bmatrix} - \frac{l}{2}\begin{Bmatrix}1\\1\end{Bmatrix} = \begin{Bmatrix}0\\0\end{Bmatrix} \qquad (10.29)$$

After getting system equations for the all elements 1 to 4 and superimposing or assembling them, the global equation given in Eq. (10.24) is obtained. Note that the boundary equations on the left hand side of Eq. (10.29) will cancel with each of the neighboring elements except at the boundaries. Here the matrix containing the derivatives of the weighting functions or shape functions is the B matrix as known in the finite element methodology.

10.7 TWO DIMENSIONAL FLOW PROBLEMS BY GALERKIN METHOD

Two dimensional heat flow[2] or flow through porous media is represented by a Laplace Equation in a two dimensional domain, i.e.,

$$\nabla^2\Phi = 0 \text{ in the domain } \Omega \tag{10.30}$$

with prescribed boundary conditions such as

$$\Phi = \tilde{\Phi} \text{ prescribed on } S_1$$

$$\frac{d\Phi}{dn} = \tilde{q} \text{ prescribed on } S_2$$

Assuming that Ψ is one of the admissible functions in the domain of the problem for the Galerkin method, one can write

$$\int_D \Psi \nabla^2 \Phi \, d\Omega = 0 \tag{10.31}$$

Applying the Green's theorem, the order of the derivatives in the original equation is reduced from two to one as it is necessary in the Galerkin finite element method. Thus,

$$\int_\Omega \Psi \nabla^2 \Phi \, d\Omega = -\int_D \left(\frac{\partial \Psi}{\partial x} \cdot \frac{\partial \Phi}{\partial x} + \frac{\partial \Psi}{\partial y} \cdot \frac{\partial \Phi}{\partial y} \right) d\Omega + \int_S \Psi \frac{\partial \Phi}{\partial n} dS = 0 \tag{10.32}$$

At the element level, the value of Φ is interpolated from the nodal parameters, for example in a triangular finite element,

$$\Phi = \sum_{i=1}^{3} N_i \Phi_i \tag{10.33}$$

and the gradient of the Φ function is given in terms of the derivatives of the shape functions as given below.

$$\begin{Bmatrix} \dfrac{\partial \Phi}{\partial x} \\ \dfrac{\partial \Phi}{\partial y} \end{Bmatrix} = \begin{bmatrix} N_{1,x} & N_{2,x} & N_{3,x} \\ N_{1,y} & N_{2,y} & N_{3,y} \end{bmatrix} \begin{Bmatrix} \Phi_1 \\ \Phi_2 \\ \Phi_3 \end{Bmatrix} = [B]\{\Phi\} \tag{10.34}$$

In the Galerkin finite element method, the shape functions are taken as the weighting functions by Ψ, therefore for each triangular element, there are three such functions $\Psi_i = N_i$, $i = 1, 2, 3$. For $i = 1$, and substituting $\Psi_1 = N_1$ in Eq. (10.32) then,

$$-\int_A \begin{Bmatrix} \dfrac{\partial N_1}{\partial x} \\ \dfrac{\partial N_1}{\partial y} \end{Bmatrix}^t \begin{Bmatrix} \dfrac{\partial \Phi}{\partial x} \\ \dfrac{\partial \Phi}{\partial y} \end{Bmatrix} dx\, dy + \int_S N_1 \frac{\partial \Phi}{\partial n} dS = 0 \tag{10.35}$$

$$-\int_A \begin{Bmatrix} \dfrac{\partial N_1}{\partial x} \\ \dfrac{\partial N_1}{\partial y} \end{Bmatrix}^t [B]\{\Phi\}\, dx\, dy + \int_S N_1 \frac{\partial \Phi}{\partial n} dS = 0 \tag{10.36}$$

where A is the domain, S the boundary of the triangle. Similarly, two more equations for the triangle can be obtained for i = 2,3, and the Eq.(10.36) can be augmented as:

$$-\int_A [B]^t [B] \{\Phi\} dx\, dy + \int_S [N]^t \frac{\partial \Phi}{\partial n} dS = \{0\} \tag{10.37}$$

And the system matrix for a simple triangular element will be:

$$[k] = \int_A [B]^t [B] dx\, dy = A\, [B]^t\, [B] \tag{10.38}$$

As all the nodal equations are added from the neighboring finite elements together, the second term in Eq. (10.37) will automatically equal to 0, except on the boundary. The methodology for assembling the system matrices together for all elements and applying the boundary conditions are the same as given before in the earlier example. These resulting equations are the same as obtained by minimizing the functional.

10.8 COMPARISON OF FINITE ELEMENT AND FINITE DIFFERENCE METHODS

Before the evolution of the finite element methods, finite difference methods[3,7] were the primary technique to solve complex differential equations linear and some nonlinear differential equations defining the characteristics of a solid or fluid continuum. The technique essentially consists of dividing the domain into equal subdomains, for one-dimensional problems along the domain axis, for two-dimension problems along x- and y- axes and for three-dimensional problems along the three axes. Problems in cartesian, polar and spherical coordinates were solved using this technique. The development of high speed computers helped to solve many linear and nonlinear problems even time dependant problems, as the technique required to solve a large number of simultaneous equations in order to get satisfactory results. As of now, there are many scientists and researchers still using finite difference methods to solve complex fluid mechanics problems.

Most global differential equations in continuum mechanics are either second order for deformation of an axial bar, plane elasticity, fluid flow and heat flow equations or fourth order as in classical Euler Beam equations, plate equations such as Kirchoff plate or Von Karman plate equations, Airy's stress function equations for plane elasticity[4] etc.

Only a very scanty explanation of the finite difference method is given here to help those who are not familiar with this technique. The basic methodology in the finite difference approach is to represent a derivative as linear combinations of the discrete values of the variables at selected points called nodes, usually at equal distances. There are many ways this can be represented, however the so called central difference equations[3,7] are more accurate for the same number of variables and hence more popular. For example, a second order and fourth order derivatives along x-axis at a particular node i are written as:

$$\left.\frac{d^2u}{dx^2}\right|_i = \frac{u_{i-1} - 2u_i + u_{i+1}}{h^2}$$

$$\left.\frac{d^4u}{dx^4}\right|_i = \frac{u_{i-2} - 4u_{i-1} + 6u_i - 4u_{i+1} + u_{i+2}}{h^4} \tag{10.39}$$

Here h is the distance between the nodes. There are many ways to derive this[3,7] and the equations for other derivatives can also be derived in a similar way. Here a simple second order differential equation already solved by finite element Galerkin method in section 10.5, is solved by finite difference method. The Eq. (10.13), for convenience it is repeated here:

$$\frac{d^2u}{dx^2} + u = 1 \text{ for } 0 < x < L \tag{10.13}$$

Consider the domain is divided into 4 subdomains, with 5 nodes as shown in the Fig. 10.5. At nodes 1 and 5, the values of the function u are prescribed in the problem as boundary conditions, i.e.,

$u_1 = 1.0$ and $u_5 = 6.0$.

Now at an interior node i, the finite difference equivalent of the differential equation is:

$$\frac{1}{h^2}\,(u_{i-1} - 2u_i + u_{i+1}) + u_i = 1. \tag{10.40}$$

Writing these equations for nodes 2, 3 and 4 and substituting values of $u_1$and u_5 and rewriting in matrix form with three unknowns,

$$\begin{bmatrix} h^2-2 & 1 & 0 \\ 1 & h^2-2 & 1 \\ 0 & 1 & h^2-2 \end{bmatrix} \begin{Bmatrix} u_2 \\ u_3 \\ u_4 \end{Bmatrix} = \begin{Bmatrix} h^2-1 \\ h^2 \\ h^2-6 \end{Bmatrix} \tag{10.41}$$

For $L = 8$, $h = 2$, and substituting, the solution vector including the boundary conditions is:

$$\begin{Bmatrix} u_1 \\ u_2 \\ u_3 \\ u_4 \\ u_5 \end{Bmatrix} = \begin{Bmatrix} 1.0 \\ -0.25 \\ 3.5 \\ -2.75 \\ 6.0 \end{Bmatrix}.$$

The problem is repeated with 6 divisions with 5 unknowns and the entire solution is:

$$\{u\}^t = \{1.0 \;\; 9.022 \;\; 2.783 \;\; -6.626 \;\; -2.477 \;\; 7.853 \;\; 6.0\}$$

For purpose of comparison, the same problem is solved with 8 divisions and the result is:

$$\{u\}^t = \{1.0 \;\; 6.0 \;\; 6.0 \;\; 1.0 \;\; -4.0 \;\; -4.0 \;\; 1.0 \;\; 6.0 \;\; 6.0\}$$

The results are plotted for purpose of comparison in Fig. 10.8 with the exact solution. It can be seen that the finite difference solution for this problem with 4 divisions is very poor, because of the first derivative of the u-function changing fast and there are not enough nodes to define them properly. But the solution obtained using the finite element Galerkin method with 4 elements, the results yielded a slightly better solution. If one compares the system equations for both these formulations, the equations for the second derivative are the same, but the u-function representation in the Galerkin method had spread to the neighboring nodes. This is due to the effect of the weighting functions. These discussions are mostly of academic importance, as the number of divisions increased, the finite difference solution improved and with 11 divisions the solution is very close to the exact. To solve real problems, it is customary to take a very large number of unknowns as solution of large matrices is no longer a problem even with laptop computers. The selection of the number of unknowns depends on the problem and the experience of the analyst.

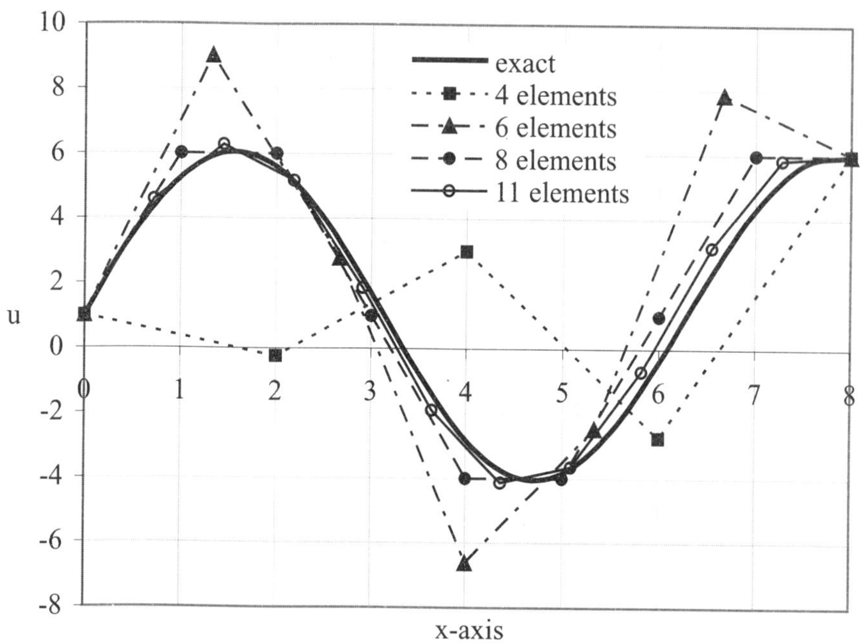

Fig. 10.8 Finite Difference Solution of the Eq. 10.13

It is interesting to note that if a finite difference model is developed for solving the seepage problem cited by Eq. (10.1) and Fig. 10.1 using 5 divisions along x-axis and 4 divisions on the y-axis, the resulting system equations will be exactly the same as that given in Fig. 10.3 for the case of $k_x = k_y = k$ as used in that example. The only difference is that some portion of the final finite difference equations has to be factored to match with the equation given in Eq. (10.3) to make it symmetric and half-banded. The finite difference solution will be exactly the same as those given in Fig. 10.4.

10.8.1 Major Differences between the Two Methods

The main advantage of finite element method is the easy handling of the complex geometry and material properties of a continuum. Where ever the stress gradients are high, smaller elements can be introduced fairly easily. Change in the material properties in different segments of the domain is easily incorporated into the model. All these advantages are absent in the finite difference methods, in other words, it is rather difficult to incorporate these properties in to the system equations. But finite difference equations are easy to generate and when the geometry is simple, the method becomes very handy. They are very useful to solve nonlinear rectangular plate problems with von Karman plate equations where the convergence of the solution at every increment is satisfying the differential equations which represent equilibrium at each node. Still many engineers use the finite difference methods for solving beam on elastic foundation, plates on elastic foundation, pile-soil interaction problems etc. Even now many heat flow, aerodynamic fluid flow, the diffusion problems are solved by finite difference techniques. These are two different methods of solving equations of continuum mechanics, both methods must be studied and continued to be used since each method has its own advantages and disadvantages.

10.9 SUMMARY

Analysis of simple steady state flow problems using the finite element method is presented. Triangular finite elements defining a finite domain are used to form the system equations. A problem of seepage through porous media is presented and solved. Since flow problems are mainly introduced using a weak form of Galerkin finite element method, a topic that is hard to comprehend, here a one-dimensional second order differential equation is solved using Galerkin finite element method. The technique is extended to solve two-dimensional flow problems using finite elements. At the end of the chapter, a short discussion is made comparing the classical finite difference method and the finite element method. An example second order differential equation is solved by both methods and compared.

PROBLEMS

10.1 Derive a system matrix for a rectangular element for solving a steady state seepage or heat flow problems. Use 4 nodal parameters only. Assume isotropic and homogeneous properties of medium.

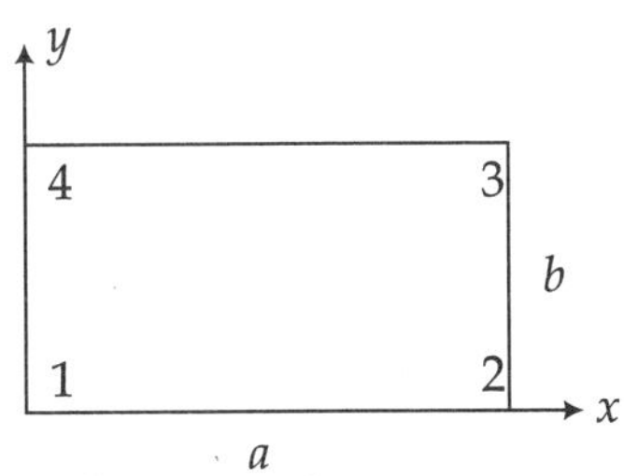

Fig. P10.1

10.2 Solve

$$\frac{d^2v}{dx^2} - 3x = 0,\ 0 < x < 3$$

With $v = 0$ at $x = 0$ and $v = 5$ at $x = 3$

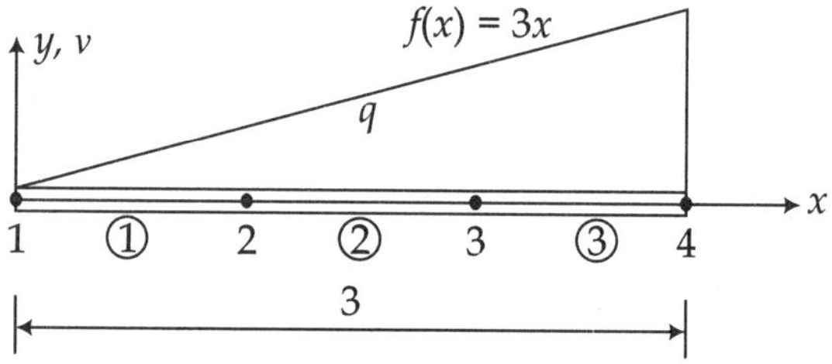

Fig. P10.2

Use 3 finite elements with linear variation of v within the element.

10.3 Derive the finite element model for the differential equation

$$-\frac{d}{dx}\left[(2 - x^2)\frac{du}{dx}\right] = q,\ 0 < x < 1$$

with the boundary conditions $u(0) = 0$ and $\dfrac{du(1)}{dx} = 0$.

10.4 The equation for torsion of a uniform prismatic bar is $\nabla^2\phi = -2G\alpha$, where ϕ is the stress function and α is the angle of twist/unit length. Derive element matrices for a triangular element with

(i) linear variation of ϕ within the element

(ii) quadratic variation of ϕ within the element

Derive the corresponding load vectors also.

10.5 Laplace equation $\nabla^2 u = 0$ has to be solved by using a triangular finite element. By using Galerkin method develop the element stiffness matrix. Solve the Laplace equation for a unit square if $\begin{cases} u = 0 \ @\ x = 0,1 \text{ and } y = 1 \\ u = 1 \ @\ y = 0 \end{cases}$ as shown in the figure below

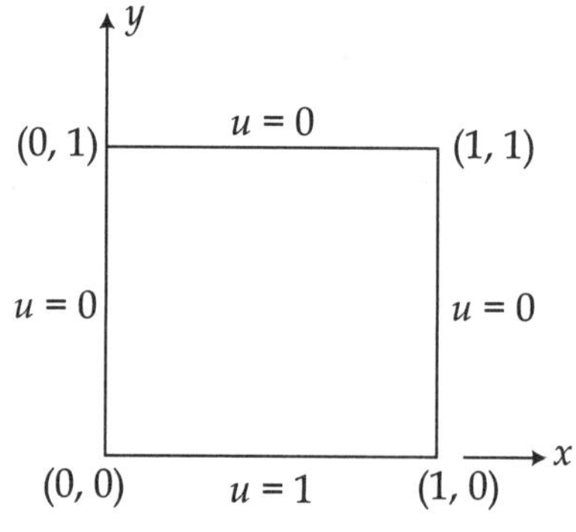

Fig. P10.3

REFERENCES

1. Bathe K.J. and Wilson, E.L, "Numerical Methods in Finite Element Analysis", Prentice Hall, Englewood Cliffs, New Jersey, 1976.
2. Chung, T.J, "Finite Element Analysis in Fluid Dynamics", McGraw Hill Book Company, New York, 1978.
3. Crandall, S, "Engineering Analysis", McGraw-Hill Book Company, New York, 1956.
4. Fung, Y.C., "Foundations of Solid Mechanics", Prentice Hall, Englewood Cliffs, New Jersey, 1965.
5. Grandin, H, "Fundamentals of The Finite Element Method", Macmillan Publishing Company, New York, 1986.
6. Hughes, T.J.R, "The Finite Element Method", Prentice Hall, Englewood Cliffs, New Jersey, 1987.
7. Ketter, R.L. and Prawel, S.P, "Modern Methods of Engineering Computation", McGraw Hill Book Company, New York, 1969.
8. Reddy, J.N, "An Introduction to the Finite Element Method", 2nd Ed. McGraw Hill Book Company,New York, 1993.
9. Zienkiewicz, O.C. and Taylor, R.L, "The Finite Element Method", Vols. I & II, McGraw Hill Book Company, London, 1989.

Appendix - I
Selected References

FINITE ELEMENTS

1. Baker, A.J. and Pepper, D.W., "Finite Elements 1-2-3", McGraw Hill, New York, 1991.
2. Bathe, K.J. and Wilson, E.L., "Numerical Methods in Finite Element Analysis", Prentice-Hall, Englewood Cliffs, New Jersey, 1976.
3. Bathe, K.J., "Finite Element Procedures", Prentice Hall, Englewood Cliffs, New Jersey, 1996.
4. Bazant, Z.P., "Finite Element Analysis of Reinforced Concrete", Published by ASCE, 1982.
5. Bergan, P.G., "Finite Elements in Nonlinear Mechanics", Tapir Publishers, Trondheim, 1977.
6. Brebbia, C.A. and Connor J.J., "Fundamentals of Finite Element Techniques", John Wiley & Sons, New York, 1974.
7. Chandrupatla, T.R. and Belegundu, A.D.,"Introduction to Finite Elements in Engineering", Prentice Hall, New Jersey, 1997.
8. Chung, T.J., "Finite Element Analysis in Fluid Dynamics", McGrawHill Book Company, New York, 1978.
9. Cook, R.D., Malkus, D.S. and Plesha, M.E., "Concepts and Applications of Finite Element Analysis", John Wiley & Sons, New York, 1989.
10. Desai, C.S. and Abel, J.F., "Introduction to Finite Element Method", Van Nostrand Reinhold, New York, 1972.
11. Desai, C.S., "Elementary Finite Element Method", Prentice Hall, Englewood Cliffs, New Jersey, 1979.
12. Gallagher, R.H., "Finite Element Analysis Fundamentals", Prentice Hall, Englewood Cliffs, New Jersey, 1975.
13. Grandin, H., "Fundamentals of The Finite Element Method", Macmillan Publishing Company, New York, 1986.
14. Hughes, T.J.R., 'The Finite Element Method", Prentice Hall, Englewood Cliffs, New Jersey, 1987.
15. Istok, J., "Ground Water Modeling by the Finite Element Method", American Geophysical Union, Washington, 1989.
16. Kardestuncer, H., "Finite Element Handbook", McGraw Hill Book Company, New York, 1987.
17. Kikuchi, N., "Finite Element Methods in Mechanics", Cambridge University Press, Cambridge, London, 1986.
18. Krishnamoorthy, C.S., "Finite Element Analysis: Theory and Programming", Tata McGraw Hill Publishing Company, New Delhi, 1987.
19. Logan, D.L., "A First Course in the Finite Element Method", PWS-Kent Publishing Company, Boston, 1992.
20. Martin, C.H. and Carey, G.F. "Introduction to Finite Element Analysis", McGraw Hill Book Company, New York, 1973.
21. Mitchell, A.R. and Wait R., "The Finite Element Method in Partial Differential Equations", John Wiley & Sons, London, 1977.

22. Oden, J.T., "Finite Elements of Nonlinear Continua", McGraw Hill Book Company, New York, 1972.
23. Rao, S.S, "The Finite Element Method in Engineering", Pergamon Press, Oxford, 1989.
24. Reddy, J.N., "An Introduction to the Finite Element Method", 2nd Ed. McGraw Hill Book Company, New York, 1993.
25. Segerlind, L.J., "Applied Finite Element Analysis", John Wiley & Sons, New York, 1976.
26. Smith, I.M. and Griffiths, D.V., "Programming the Finite Element Method", John Wiley & Sons, New York, 1988.
27. Stasa, F.L., "Applied Finite Element Analysis for Engineers", Holt, Reinhart and Winston, New York, 1985.
28. Weaver, W. and Johnston, P.R., "Finite Elements for Structural Analysis", Prentice Hall, Englewood Cliffs, New Jersey, 1984.
29. Weaver, W.J. and Johnston, P.R. "Structural Dynamics by Finite Elements", Prentice Hall, Englewood Cliffs, New Jersey, 1987.
30. Yang, T.Y., "Finite Element Structural Analysis", Prentice Hall, Englewood Cliffs, New Jersey, 1986.
31. Zienkiewicz, O.C. and Taylor, R.L., "The Finite Element Method", Vols. I & II, McGraw Hill Book Company, London, 1989.

NUMERICAL ANALYSIS

1. Atkinson, K.E., "An Introduction to Numerical Analysis", John Wiley & Sons, New York, 1978.
2. Bathe, K.J. and Wilson, E.L, "Numerical Methods in Finite Element Analysis", Prentice Hall, Englewood Cliffs, New Jersey, 1976.
3. Chapra, S.C. and Canale, R.P., "Numerical Methods for Engineers", McGraw Hill Book Company, New York, 1988.
4. Crandall, S., "Engineering Analysis", McGraw Hill Book Company, New York, 1956.
5. Daniels, R.W., "An Introduction to Numerical Methods and Optimization Techniques", North-Holland, New York, 1978.
6. Faires, J.D. and Burden, R.L., "Numerical Methods", PWS-Kent Publishing Company, Boston, 1993.
7. Ketter, R.L. and Prawel, S.P., "Modern Methods of Engineering Computation", McGraw Hill Book Company, New York, 1969.
8. Pachner, J., "Handbook of Numerical Analysis Applications", McGraw Hill Book Company, New York, 1984.
9. Raltson, A., "A First Course in Numerical Analysis", McGraw Hill Book Company, New York, 1965.
10. Young, D.M. and Gregory R.T., "A Survey of Numerical Mathematics", Addison-Wesley Publishing Company, Reading, Mass, 1973.

MECHANICS OF SOLIDS AND VARIATIONAL PRINCIPLES

1. Boresi, A.P., Sidebottom, O.M., Seely, F.B. and Smith, J.O., "Advanced Mechanics of Materials", John Wiley & Sons, New York, 1978.
2. Chou, P.C. and Pagano, N.J., "Elasticity", D.Van Nostrand Company, Princeton, 1967.
3. Crandall, S.H. and Dahl, N.C. and Lardner, T.J., "An Introduction to the Mechanics of Solids", McGraw Hill Book Company, New York, 1972.

4. Fung, Y.C., "Foundations of Solid Mechanics", Prentice-Hall, Englewood Cliffs, New Jersey, 1965.
5. Little, R.W., "Elasticity", Prentice-Hall, Englewood Cliffs, New Jersey, 1973.
6. Malvern, L.E., "Introduction to the Mechanics of a Continuous Medium", Prentice-Hall, Englewood Cliffs, New Jersey, 1969.
7. Oden, J.T., "Mechanics of Elastic Structures", McGraw Hill Book Company, New York, 1967.
8. Sechler, E.E., "Elasticity in Engineering", John Wiley & Sons, New York, 1961.
9. Szilard, R., "Theory and Analysis of Plates", Prentice Hall, Englewood Cliffs, New Jersey, 1974.
10. Timoshenko, S.P. and Goodier, J.N., "Theory of Elasticity", McGraw Hill Book Company, New York, 1970.
11. Timoshenko, S.P. and Woinowsky-Krieger S., "Theory of Playes and Shells", McGraw Hill Book Company, New York, 1959.
12. Ugural, A.C., "Stresses in Plates and Shells", McGraw Hill Book Company, New York, 1981.
13. Washizu, K., "Variational Methods in Elasticity & Plasticity", Pergamon Press, Oxford, 1982.
14. Wang, C.T., "Applied Elasticity", McGraw Hill Book Company, New York
15. Dym, C.L. and Shames, I.H., "Solid Mechanics: A Variational Approach", McGraw Hill Book Company, 1973.
16. Oden, J.T., "Applied Functional Analysis", Prentice Hall, Englewood Cliffs, New Jersey, 1979.
17. Reddy, J.N., "Energy and Variational Methods in Applied Mechanics", John Wiley & Sons, New York, 1984.

Appendix - II

A.1 SOLUTION OF LINEAR SIMULTANEOUS EQUATIONS

For all linear discrete models, the final equations to be solved end up as a set of linear algebraic equations of the type

$$[A]\{X\} = \{b\} \tag{A.1}$$

where $[A]$ is the coefficient matrix and $\{b\}$ and $\{X\}$ are the known and the unknown vectors of the system respectively. For convenience of writing, the brackets defining the matrices and vectors are removed for the rest of the discussion in this chapter. Since there are as many equations as there are unknowns, the coefficient matrix has to be a square one. In order to have a unique solution for the system, the coefficient matrix has to be nonsingular, i.e., the inverse of the matrix must exist. The solution of linear simultaneous equations is a topic that most students involved in numerical analysis have studied at length. There are numerous methods to solve these equations, some of them are direct and others are indirect (iterative) methods. For finite element methods, majority of the codes use a direct method of solution, even though iterative methods are being scrutinized for solving very large system equations where the equations are sparse. Since direct methods of solution are more popular, only they are discussed here. It will be very superfluous to discuss all the direct methods of solution here, because there are so many of them. Here we need to scrutinize the special properties of the finite element equations and the method of solution must be tailored to take advantage of its inherent properties. For solid mechanics problem, these equations are always symmetric. Besides being symmetric, they can be formulated as a banded matrix by properly numbering the nodes in the system. The coefficient matrix being symmetric and banded, only half of the matrix needs to be stored on the computer. This way one can utilize the maximum core memory capacity of the computer so that equations for very large finite discretizations can be solved. Also unnecessary computations of matrix elements that are already zeros can be avoided to improve the computational efficiency of the code. There are two popular methods of solution; one is called half-band-width solver and the other is called skyline solver. Mathematically, both methods are same, in other words, they use the good old Gauss elimination method, except that, the way the equations are stored and solved are slightly different. In order to understand these techniques, one has to review the operations used in these techniques through simple matrix algebra and pay special attention to the bookkeeping of the elements of the matrix on the computer. The essential feature of the method consists of decomposing the matrix into the product of three matrices, L, D and U which are a lower triangular, diagonal and upper triangular matrices. Incidentally, the upper triangular matrix U is the transpose of the lower triangular matrix L and the diagonals of both L and U are kept equal to one. Perhaps, the best way to understand these techniques is to browse through the classical Gauss elimination procedure.

A.2 GAUSS ELIMINATION PROCEDURE

The simplest way to review this method of solution is to undergo the steps in a numerical example. Let us consider a set of 3 equations, as given below where the coefficient matrix is symmetric.

$$\begin{bmatrix} 5 & -4 & 1 \\ -4 & 6 & -4 \\ 1 & -4 & 7 \end{bmatrix} \begin{Bmatrix} x_1 \\ x_2 \\ x_3 \end{Bmatrix} = \begin{Bmatrix} 1 \\ 1 \\ 1 \end{Bmatrix}$$

If one eliminates all the lower off diagonal elements of the coefficient matrix using linear combination of rows, the equation becomes

$$\begin{bmatrix} 5 & -4 & 1 \\ 0 & 2.8 & -3.2 \\ 0 & 0 & 3.143 \end{bmatrix} \begin{Bmatrix} x_1 \\ x_2 \\ x_3 \end{Bmatrix} = \begin{Bmatrix} 1 \\ 1.8 \\ 2.857 \end{Bmatrix}$$

Now, for the purpose of learning some hidden characteristics, let us make all the diagonal elements to be equal to one, and the matrix equations thus become

$$\begin{bmatrix} 1 & -0.8 & 0.2 \\ 0 & 1 & -1.143 \\ 0 & 0 & 1 \end{bmatrix} \begin{Bmatrix} x_1 \\ x_2 \\ x_3 \end{Bmatrix} = \begin{Bmatrix} 0.2 \\ 0.643 \\ 0.909 \end{Bmatrix}$$

And by performing a simple back substitution procedure, one can solve for the unknown vector as

$$\begin{Bmatrix} x_1 \\ x_2 \\ x_3 \end{Bmatrix} = \begin{Bmatrix} 1.364 \\ 1.682 \\ 0.909 \end{Bmatrix}$$

Now, if we write

$$U = L^t = \begin{bmatrix} 1 & -0.8 & 0.2 \\ 0 & 1 & -1.143 \\ 0 & 0 & 1 \end{bmatrix}, \text{ and } D = \begin{bmatrix} 5 & 0 & 0 \\ 0 & 2.8 & 0 \\ 0 & 0 & 3.143 \end{bmatrix}$$

It can be shown that

$$LDU = \begin{bmatrix} 1 & 0 & 0 \\ -0.8 & 1 & 0 \\ 0.2 & -1.143 & 1 \end{bmatrix} \begin{bmatrix} 5 & 0 & 0 \\ 0 & 2.8 & 0 \\ 0 & 0 & 3.143 \end{bmatrix} \begin{bmatrix} 1 & -0.8 & 0.2 \\ 0 & 1 & -1.143 \\ 0 & 0 & 1 \end{bmatrix}$$

$$= \begin{bmatrix} 5 & -4 & 1 \\ -4 & 6 & -4 \\ 1 & -4 & 7 \end{bmatrix} = A$$

In other words, a symmetric matrix A can be decomposed into the product of L, D and U using Gauss elimination procedure, if the matrix is positive (or negative definite. Here the matrix L is the transpose of the matrix U. Now, if

$$A = LDU \tag{A.2}$$

Then
$$AX = LDU\,X = b.$$

Setting $DU\,X = Y$, $AX = b$ now becomes $LY = b$. L being a lower triangular matrix, the value of Y can be obtained by a forward substitution procedure. Having known Y, now

$$DU\,X = Y \text{ or } UX = D^{-1}Y = Z, \tag{A.3}$$

and the vector X can be determined by back substitution.

The advantage of this procedure is many fold. One big advantage is that if one has to solve many equations of the type

$$AX_i = b_i, \text{ for } i = 1, \ldots, n$$

where n is the number of sets of b vectors, the decomposition of A into LDU has to be done only once. The solution of these vectors X for different b can be obtained independently for each b, by a forward and back substitutions. This technique is extensively used in nonlinear iterative analysis and step-by-step integration procedures where the coefficient matrix remains the same.

A.3 HALF BAND SOLVER

The technique was developed by engineers involved in finite element methods to take advantage of the symmetry and sparseness of the coefficient matrix by storing them in a half-banded form. This was the only way to solve large finite element equations efficiently on the computer. It should be remembered that the procedure essentially consists of storing the matrix in an ingenious way, such that core memory of the computer can be used to accommodate maximum possible size of the finite element discretizations. The matrices are stored in a half banded form as indicated in Fig. 1.7. To understand the technique, let us consider a numerical example as given below:

$$\begin{bmatrix} 5 & -4 & 1 & & & & \\ -4 & 6 & -4 & 1 & & & \\ 1 & -4 & 6 & -4 & 1 & & \\ & 1 & -4 & 6 & -4 & 1 & \\ & & 1 & -4 & 6 & -4 & 1 \\ & & & 1 & -4 & 6 & -4 \\ & & & & 1 & -4 & 5 \end{bmatrix} \begin{Bmatrix} x_1 \\ x_2 \\ x_3 \\ x_4 \\ x_5 \\ x_6 \\ x_7 \end{Bmatrix} = \begin{Bmatrix} 1 \\ 2 \\ 3 \\ 4 \\ 5 \\ 6 \\ 7 \end{Bmatrix}$$

The coefficient matrix is stored in the half-banded form in a new rectangular matrix $\overline{A}$ as shown below.

$$\overline{A} = \begin{bmatrix} 5 & -4 & 1 \\ 6 & -4 & 1 \\ 6 & -4 & 1 \\ 6 & -4 & 1 \\ 6 & -4 & 1 \\ 6 & -4 & 0 \\ 5 & 0 & 0 \end{bmatrix}$$

The matrix A is now decomposed into LDU type as discussed above using Gauss elimination, and the matrices are:

$$D = \begin{bmatrix} 5 \\ 2.8 \\ 2.143 \\ 1.833 \\ 1.655 \\ 1.539 \\ 0.457 \end{bmatrix}, \text{ and } U = \begin{bmatrix} 1 & -0.8 & 0.2 \\ 1 & -1.143 & 0.357 \\ 1 & -1.333 & 0.467 \\ 1 & -1.455 & 0.546 \\ 1 & -1.539 & 0.6044 \\ 1 & -1.600 & 0 \\ 1 & 0 & 0 \end{bmatrix}$$

And these are matrices D and U stored in $\overline{A}$ as

$$\overline{A} = \begin{bmatrix} 5 & -0.8 & 0.2 \\ 2.8 & -1.143 & 0.357 \\ 2.143 & -1.333 & 0.467 \\ 1.833 & -1.455 & 0.546 \\ 1.655 & -1.539 & 0.6044 \\ 1.539 & -1.600 & 0 \\ 0.457 & 0 & 0 \end{bmatrix}$$

This way only the coefficient matrix $\overline{A}$ that was used to store the original matrix is used to store the decomposed matrices D and U. Note that since the first column of U matrix are all ones, the coefficients of the diagonal matrix D are stored over there. Since L is the transpose of U, no additional storage is necessary.

Now $$Y = L^{-1} b = \begin{bmatrix} 1.000 \\ 2.800 \\ 6.000 \\ 11.000 \\ 18.199 \\ 28.000 \\ 40.796 \end{bmatrix} \text{ and } Z = D^{-1} Y = \begin{bmatrix} 0.200 \\ 1.000 \\ 2.800 \\ 6.000 \\ 11.000 \\ 18.200 \\ 89.250 \end{bmatrix} \text{ and}$$

$$X = U^{-1} Z = \begin{bmatrix} 78.75 \\ 147.00 \\ 195.25 \\ 216.00 \\ 204.75 \\ 161.00 \\ 89.25 \end{bmatrix}$$

When the core memory is still not sufficient to accommodate the large finite element equations, the half-banded matrices are chopped into blocks and the decomposition are achieved in blocks that are stored on discs. The details of these methods are not given here, as it becomes rather too cumbersome to explain; however, interested readers are advised to read selected papers on this topic.

A.4 SKYLINE SOLVER

The half-band solver is very efficient if the half-band-width of the matrix is almost a constant. In some instances of finite element discretizations, this may not be possible, and the width or the height of the columns of the coefficient matrix vary substantially. Allowing a large half-band-width to cover all these coefficients will contain too many zeros and can result in a very inefficient storage and computing system. An alternative to the half-band solver is to provide a variable width or column system. If the variable columns are used, the matrix can be represented under an enveloping line that will look like a skyline as shown in the Fig. 1.8. The solution of the system equations using this technique is called skyline solver.

In this technique, the matrix is stored in a one-dimensional array with heights of each of the columns are stored in another one-dimensional array. The matrix is decomposed into *LDU* type matrices, where the elements of the diagonal matrix *D* are stored at the end of each column. The theory of this technique is same as before, however, the bookkeeping of the locations of the elements of the coefficient matrix is too complicated for a simple write-up. Interested readers are again advised to read relevant publications on this technique.

A.5 STATIC CONDENSATION

This technique has been used in the finite element methodology for removing the nodal parameters of nodes which are not necessary in the overall assemblage of the finite element. This way the number of degrees of freedom of the element and the bandwidth can be reduced. The technique is mathematically described at first, but the actual numerical method employed in the computer programs is different and also explained here. Consider a stiffness matrix equation of the order n where the number of degrees to be removed is p and the final size of the matrix is $m = n - p$ as shown in Fig. A.4. First all equations pertaining to the 'p' degrees of freedom are selected and brought to the bottom of the matrix by simple row and column operations and this procedure will not affect the symmetry of the stiffness matrix.

$$\begin{bmatrix} K_{mm} & K_{mp} \\ K_{pm} & K_{pp} \end{bmatrix} \begin{Bmatrix} u_m \\ u_p \end{Bmatrix} = \begin{Bmatrix} \tilde{f}_m \\ \tilde{f}_p \end{Bmatrix} \tag{A.4}$$

In other words,

$$K_{mm}u_m + K_{mp}u_p = \tilde{f}_m \text{ and,} \tag{A.5}$$

$$K_{pm}u_m + K_{pp}u_p = \tilde{f}_p \tag{A.6}$$

Solving for u_p from Eq. A.6,

$$u_p = K_{pp}^{-1}(\tilde{f}_p - K_{pm}u_m) \tag{A.7}$$

Substituting Eq. (A.7) in Eq. (A.5) and rewriting

$$[K_{mm} - K_{mp}K_{pp}^{-1}\, K_{pm}]\{u_m\} = \{\bar{f}_m - K_{mp}K_{pp}^{-1}\bar{f}_p\}$$

i.e.,

$$[K_{mm}^*]\,\{u_m\} = \{f_m^*\} \tag{A.8}$$

The transformed matrix K_{mm}^* is symmetric because matrix K_{mp} is the transpose of the matrix K_{pm}. This procedure where the stiffness matrix is reduced to a lower order of m by removing selected nodal parameters is called static condensation. Normally, the selected nodal displacements parameters that are removed are either prescribed or placed within an element etc.

For purpose of clarity, a numerical example is shown here in Eq. (A.9).

$$\begin{bmatrix} 5 & -4 & 1 & 0 & 0 \\ -4 & 6 & -4 & 1 & 0 \\ 1 & -4 & 6 & -4 & 1 \\ 0 & 1 & -4 & 6 & -4 \\ 0 & 0 & 1 & -4 & 5 \end{bmatrix} \begin{Bmatrix} u_1 \\ u_2 \\ u_3 \\ u_4 \\ u_5 \end{Bmatrix} = \begin{Bmatrix} 1 \\ 2 \\ 3 \\ 4 \\ 5 \end{Bmatrix} \tag{A.9}$$

The above is a matrix equation of order 5, which has to be reduced to an order of 3, by eliminating two selected nodal parameters u_2 and u_4. For convenience, the matrix equation is rearranged with the parameters to be removed at the bottom of the matrix as shown below:

$$\begin{bmatrix} 5 & 1 & 0 & -4 & 0 \\ 1 & 6 & 1 & -4 & -4 \\ 0 & 1 & 5 & 0 & -4 \\ -4 & -4 & 0 & 6 & 1 \\ 0 & -4 & -4 & 1 & 6 \end{bmatrix} \begin{Bmatrix} u_1 \\ u_3 \\ u_5 \\ u_2 \\ u_4 \end{Bmatrix} = \begin{Bmatrix} 1 \\ 3 \\ 5 \\ 2 \\ 4 \end{Bmatrix}$$

It may be noted that the rearranged matrix is also symmetric. Using the notation above, $m = 3$, $p = 2$.

$$\left[K_{pp}^{-1}\right] = \begin{bmatrix} 6 & 1 \\ 1 & 6 \end{bmatrix}^{-1} = \frac{1}{35}\begin{bmatrix} 6 & -1 \\ -1 & 6 \end{bmatrix},$$

$$\left[K_{pp}^{-1}K_{pm}\right] = \frac{1}{35}\begin{bmatrix} 6 & -1 \\ -1 & 6 \end{bmatrix}\begin{bmatrix} -4 & -4 & 0 \\ 0 & -4 & -4 \end{bmatrix} = \frac{1}{35}\begin{bmatrix} -24 & -20 & 4 \\ 4 & -20 & -24 \end{bmatrix}$$

and therefore,

$$[K_{mp}K_{pp}^{-1}K_{pm}] = \frac{1}{35}\begin{bmatrix} -4 & 0 \\ -4 & -4 \\ 0 & -4 \end{bmatrix}\begin{bmatrix} -24 & -20 & 4 \\ 4 & -20 & -24 \end{bmatrix} = \frac{1}{35}\begin{bmatrix} 96 & 80 & -16 \\ 80 & 160 & 80 \\ -16 & 80 & 96 \end{bmatrix}$$

$$= \begin{bmatrix} 2.743 & 2.286 & -0.457 \\ 2.286 & 4.571 & 2.286 \\ -0.457 & 2.286 & 2.743 \end{bmatrix}$$

The condensed matrix given in Eq. (A.8) is,

$$[K_{mm}^*] = \left[K_{mm} - K_{mp}K_{pp}^{-1}K_{pm}\right] = \begin{bmatrix} 2.257 & -1.286 & 0.457 \\ -1.286 & 1.429 & -1.286 \\ 0.457 & -1.286 & 2.257 \end{bmatrix}$$

The associated force vector is:

$$\{f_m^*\} = \left\{\bar{f}_m - K_{mp}K_{pp}^{-1}\bar{f}_p\right\} = \begin{Bmatrix} 1 \\ 3 \\ 5 \end{Bmatrix} - \frac{1}{35}\begin{bmatrix} -4 & 0 \\ -4 & -4 \\ 0 & -4 \end{bmatrix}\begin{bmatrix} 6 & -1 \\ -1 & 6 \end{bmatrix}\begin{Bmatrix} 2 \\ 4 \end{Bmatrix}$$

$$= \begin{Bmatrix} 1.914 \\ 6.429 \\ 7.514 \end{Bmatrix}$$

The same solution can be obtained by simple Gauss elimination procedure by eliminating the fourth and fifth columns except the terms k_{44} and k_{55} which are made equal to 1.0. The result is shown below and the elimination procedure is left as an exercise to the reader.

$$\begin{bmatrix} 2.257 & -1.286 & 0.457 & 0 & 0 \\ -1.286 & 1.429 & -1.286 & 0 & 0 \\ 0.457 & -1.286 & 2.257 & 0 & 0 \\ -0.667 & -0.667 & 0 & 1 & 0.167 \\ 0.114 & -0.571 & -0.686 & 0 & 1 \end{bmatrix} \begin{Bmatrix} u_1 \\ u_3 \\ u_5 \\ u_2 \\ u_4 \end{Bmatrix} = \begin{Bmatrix} 1.914 \\ 6.429 \\ 7.514 \\ 0.333 \\ 0.629 \end{Bmatrix}$$

The top matrix of the order 3 and the corresponding force vector are the same as obtained above using matrix inversion etc.

Appendix - III

LISTING OF A COMPUTER PROGRAM (PLSTR) TO SOLVE PLANE STRESS OR PLANE STRAIN PROBLEMS IN LINEAR ELASTICITY

```
C Program PLSTR
      COMMON   NUM,MBAND,NELM,NBOU,NN,KST
     1        ,NF(60),NBO(60,2),BU(60,2),SM(6,6),T
     2        ,NOD(100,3),X(60),Y(60),A(200,100),F(200)
     3        ,B,C,EBARS
      CHARACTER*4 NAME1, NAME2
      DIMENSION NAME1(3,2),NAME2(3),LM(3)
      DATA NAME1,NAME2 /'PLAN','E-ST','RAIN','PLAN','E-ST','RESS',
     1                  'TRIA','NGUL','AR  '/
      OPEN(5,FILE='PLSTR.DAT',STATUS='OLD')
      OPEN(6,FILE='PLSTR.OUT',STATUS='NEW')
C
C
CCCCCC   DSNAME IS WYL.NP.CVG.FEM.PLSTR
C
C
C        DATA INPUT AND CHECK PRINT.
C
C          KASE   PROBLEM TYPE:
C                  1 : PLANE STRAIN
C                  2 : PLANE STRESS.
C
C
  100 MBAND = 0
C
C--------------------------------------------------------------------I
C  NPROB      PROBLEM ID NUMBER                                      I
C  NELM       NUMBER OF ELEMENTS                                     I
C  NNODE      NUMBER OF NODES                                        I
C  NBOU       NO. OF NODES WHERE DISP. ARE PRESCRIBED                I
```

```
C  NCON       NO. OF NODES WHERE CONC. FORCES ARE PRESCRIBED        I
C  KASE       =1, PLANE STRAIN                                      I
C             =2, PLANE STRESS                                      I
C  ELAS       YOUNGS MODULUS OF ELASTICITY                          I
C  PR         POISSONS RATIO                                        I
C  T          THICKNESS FOR PLANE STRESS & =1 FOR PLANE STRAIN      I
C  X(),Y()    NODAL COORDINATES                                     I
C  NODE( , )  NODAL NUMBERS                                         I
C  NF()       NODE CORR. TO PRESCRIBED DISP.                        I
C  NBO( , )   =1, DISP. IS PRESCRIBED                               I
C             =0, FREE                                              I
C  BU( , )    VALUE OF PRES. DISPLACEMENT                           I
C-------------------------------------------------------------------I
C
      READ(5,*,END=9999) NPROB,NELM,NNODE,NBOU,NCON,KASE
     1              ,ELAS,PR,T
      NUM = 2 * NNODE
      WRITE (6,910) NPROB,NELM,NNODE,NBOU,NCON,ELAS,PR,T
     1        ,(NAME1(K,KASE),K=1,3),(NAME2(L),L=1,3)
C
C
      WRITE (6,930)
      DO 110 I = 1, NNODE
      READ (5,*)J,X(J),Y(J)
      WRITE(6,950)J,X(J),Y(J)
  110 CONTINUE
C
C
      WRITE (6,960)
      DO 120 I = 1, NELM
      READ (5,*) J,NOD(J,1),NOD(J,2),NOD(J,3)
      WRITE(6,970) J,NOD(J,1),NOD(J,2),NOD(J,3)
  120 CONTINUE
C
C
      WRITE (6,980)
      DO 130 I = 1, NBOU
      READ (5,*) NF(I),NBO(I,1),NBO(I,2),BU(I,1),BU(I,2)
```

```
      IF(NBO(I,1).NE.0.AND.NBO(I,2).NE.0) WRITE(6,1000)NF(I),NBO(I,1)
     *,NBO(I,2),BU(I,1),BU(I,2)
      IF(NBO(I,1).EQ.0.AND.NBO(I,2).NE.0) WRITE(6,1007)NF(I),NBO(I,1)
     *,NBO(I,2),BU(I,2)
      IF(NBO(I,1).NE.0.AND.NBO(I,2).EQ.0) WRITE(6,1008)NF(I),NBO(I,1)
     *,NBO(I,2),BU(I,1)
      IF(NBO(I,1).EQ.0.AND.NBO(I,2).EQ.0) WRITE(6,1009)NF(I),NBO(I,1)
     *,NBO(I,2)
  130 CONTINUE
C
C
      DO 140 I = 1, NUM
      F(I) = .0
  140 CONTINUE
      WRITE (6,1010)
      DO 150 I = 1, NCON
      READ (5,*)  K,F(2*K-1),F(2*K)
      WRITE(6,1020) K,F(2*K-1),F(2*K)
  150 CONTINUE
C
C
CCCCCC    END OF DATA INPUT.
C
C
      IF (KASE.EQ.2) GO TO 160
      C = PR / (1.-PR)
      B = (1.-2.*PR) / (2.*(1.-PR))
      EBARS = ELAS * (1. - PR)  / (1. + PR) / (1. - 2.*PR)
      GO TO 180
  160 C = PR
      B = 0.5 * (1.-PR)
      EBARS = ELAS  / (1. - PR * PR)
C
C
CCCCCC   DETERMINE HALF BAND WIDTH
C
C
180     CALL BANDW
```

```
C
CCCCCC   INITIALIZATION
C
C
     DO 280 I = 1, NUM
     DO 280 J = 1, MBAND
     A(I,J) = 0.0
 280 CONTINUE
C
     CALL ASSEMBLY
C
     CALL BOU
     CALL SOL
     KST = 0
     WRITE (6,1050)
     DO 275 NN = 1 , NELM
     CALL PLANE
 275 CONTINUE
     GO TO 100
 9999 STOP
C
C
 900 FORMAT (6I5,3F10.3)
 910 FORMAT
   1 ('1PROBLEM NUMBER ......................................',I5
   2 /' NUMBER OF ELEMENTS ..................................',I5
   3 /' NUMBER OF NODES ....................................',I5
   4 /' NUMBER OF NODAL DISPLACEMENT BOUNDARY CONDITIONS .....',I5
   5 /' NUMBER OF NODAL FORCE BOUNDARY CONDITIONS ............',I5
   6 /' YOUNG"S MODULUS ....................................',F13.3
   7 /' POISSON"S RATIO ....................................',F13.3
   8 /' ELEMENT THICKNESS ..................................',F13.3
   9 /' PROBLEM TYPE .......................................',3A4
   A /' ELEMENT TYPE .......................................',4A4)
 930 FORMAT(/'-COORDINATES OF EACH NODAL POINT'/
   1     /'     NODE        X-           Y-'
   2     /'     NUMBER    DIRECTION    DIRECTION'/)
 940 FORMAT(I5, 2F10.2)
```

```
 950 FORMAT( 6X,I5,5X,F8.2,6X,F8.2)
 960 FORMAT(/'-NODAL NUMBERS OF EACH ELEMENT'/
   1      /'     ELEMENT   RIRST   SECOND   THIRD   FORTH'
   2      /'     NUMBER    NODE     NODE    NODE     NODE'/)
 970 FORMAT(2X,5(4X,I5))
 980 FORMAT(/'-BOUNDARY POINTS AND BOUNDARY CONDITIONS'/
   1 /'  BOUNDARY   BOUNDARY   BOUNDARY   BOUNDARY   BOUNDARY '
   2 /'   POINT    CONDITION  CONDITION  DISPLACE.  DISPLACE.'
   3 /)
 990 FORMAT (3I5, 5E10.3)
1000 FORMAT(4X,I5,2(7X,I5),2(7X,F7.1))
1007 FORMAT(4X,I5,2(7X,I5),7X,'UNKNOWN',7X,F7.1)
1008 FORMAT(4X,I5,2(7X,I5),7X,F7.1,7X,'UNKNOWN')
1009 FORMAT(4X,I5,2(7X,I5),7X,'UNKNOWN',7X,'UNKNOWN')
1010 FORMAT(/'-BOUNDARY POINTS AND BOUNDARY FORCES'/
   1      /'     BOUNDARY    BOUNDARY    BOUNDARY'
   2      /'      POINT       FORCE       FORCE'/)
1020 FORMAT( 6X,I5, 7X, F8.1, 5X,F8.1)
1050 FORMAT(/'-STRAINS AND STRESSES OF EACH ELEMENT'/
   1      /'    ELEMENT    STRAIN      STRAIN      STRAIN   ',
   2      '    STRESS     STRESS     STRESS'
   3      /'    NUMBER       EX          EY          EXY    ',
   4      '      SX         SY         SXY  '/)
      END
      SUBROUTINE BANDW
      COMMON  NUM,MBAND,NELM,NBOU,NN,KST
     1       ,NF(60),NBO(60,2),BU(60,2),SM(6,6),T
     2       ,NOD(100,3),X(60),Y(60),A(200,100),F(200)
     3       ,B,C,EBARS
      MBAND = 0
      DO 190 I = 1 , NELM
      II = MAX0(NOD(I,1),NOD(I,2),NOD(I,3))
      JJ = MIN0(NOD(I,1),NOD(I,2),NOD(I,3))
      MBANDS = II - JJ
      IF ((MBANDS-MBAND) .GT. 0) MBAND = MBANDS
 190  CONTINUE
      MBAND = (MBAND + 1) * 2
      RETURN
```

```
      END
C
      SUBROUTINE ASSEMBLY
      COMMON   NUM,MBAND,NELM,NBOU,NN,KST
     1        ,NF(60),NBO(60,2),BU(60,2),SM(6,6),T
     2        ,NOD(100,3),X(60),Y(60),A(200,100),F(200)
     3        ,B,C,EBARS
      DIMENSION LM(3)
C
CCCCCC   ASSEMBLE THE GLOBAL STIFFNESS MATRIX.
C
C
      DO 270 NN= 1, NELM
      CALL ELSTIFF
      DO 220 I = 1 , 3
      LM(I)= 2 * (NOD(NN,I) - 1 )
  220 CONTINUE
      DO 260 I = 1 , 3
      DO 260 K = 1 , 2
      II = LM(I) + K
      KK = 2 * (I-1) + K
      DO 230 J = 1 , 3
      DO 230 L = 1, 2
      JJ = LM(J) + L - II + 1
      LL = 2 * (J-1) + L
      IF (JJ .LE. 0)GO TO 230
      A(II,JJ) = A(II,JJ) + SM(KK,LL)
  230 CONTINUE
  260 CONTINUE
  270 CONTINUE
C
      RETURN
      END
C
      SUBROUTINE BOU
      COMMON   NUM,MBAND,NELM,NBOU,NN,KST
     1        ,NF(60),NBO(60,2),BU(60,2),SM(6,6),T
     2        ,NOD(100,3),X(60),Y(60),A(200,100),F(200)
```

```
C
C
CCCCCC   MODIFY THE MATRIX FOR BOUNDARY CONDITIONS.
C
C
 500 DO 520 I = 1, NBOU
     N = 2 *(NF(I) -1) + 1
     DO 520 L = 1, 2
     IF ( NBO(I,L)) 507, 517, 507
 507 FF = BU(I,L)
 508 DO 515 M = 2,MBAND
     K = N-M+1
     IF (K) 510, 510, 509
 509 F(K) = F(K) - A(K,M) * FF
     A(K,M) = 0.0
 510 K = N+M-1
     IF ( NUM - K) 515, 512, 512
 512 F(K) = F(K) - A(N,M) * FF
     A(N,M) = 0.0
 515 CONTINUE
     A(N,1) = 1.0
     F(N) = FF
 517 N = N+1
 520 CONTINUE
     RETURN
     END
     SUBROUTINE SOL
     COMMON   NUM,MBAND,NELM,NBOU,NN,KST
    1         ,NF(60),NBO(60,2),BU(60,2),SM(6,6),T

    2         ,NOD(100,3),X(60),Y(60),A(200,100),F(200)
C
C
CCCCCC   SOLVE BANDED MATRIX
C
C
     DO 605 N = 1, NUM
     IF (A(N,1)) 601, 605, 601
```

```
601 F(N) = F(N) / A(N,1)
    DO 604 L =2, MBAND
    IF (A(N,L)) 602, 604, 602
602 AA = A(N,L) / A(N,1)
    I = N+L-1
    J = 0
    DO 603 K = L,MBAND
    J = J+1
603 A(I,J) = A(I,J) - AA * A(N,K)
    F(I) = F(I) - A(N,L) * F(N)
    A(N,L) = AA
604 CONTINUE
605 CONTINUE
C
C
CCCCCC   BACKSUBSTITUTION
C
C
    DO 610 M =2, NUM
    N = NUM + 1 - M
606 DO 608 K = 2, MBAND
    L = N + K - 1
    IF ( L- NUM ) 607, 607, 610
607 F(N) = F(N) - A(N,K) * F(L)
608 CONTINUE
610 CONTINUE
    WRITE (6,900)
    DO 615 I = 1 , NUM , 2
    J = (I+1) / 2
    WRITE (6,910) J , F(I) , F(I+1)
615 CONTINUE
    RETURN
C
C
900 FORMAT(/'-DISPLACEMENTS AT EACH NODAL POINT'/
  1    /'     NODE        X-          Y- '
  2    /'    NUMBER    DIRECTION    DIRECTION'/)
910 FORMAT(7X,I3,7X,E9.3,5X,E9.3)
```

```
END

SUBROUTINE ELSTIFF
COMMON   NUM,MBAND,NELM,NBOU,NN,KST
1        ,NF(60),NBO(60,2),BU(60,2),SM(6,6),T
2        ,NOD(100,3),X(60),Y(60),A(200,100),F(200)
3        ,B,C,EBARS
II = NOD(NN,1)
JJ = NOD(NN,2)
KK = NOD(NN,3)
B1 = Y(JJ) - Y(KK)
B2 = Y(KK)- Y(II)
B3 = Y(II)- Y(JJ)
C1 = X(KK)- X(JJ)
C2 = X(II)- X(KK)
C3 = X(JJ)- X(II)
AREA =.5*(C3*B2-C2*B3)
AREA =ABS(AREA)
EBAR = EBARS * T / 4. / AREA
SM(1,1) = EBAR*(B1*B1+B*C1*C1)
SM(1,2) = EBAR*(C+B)*B1*C1
SM(1,3) = EBAR*(B1*B2+B*C1*C2)
SM(1,4) = EBAR*(C*B1*C2+B*B2*C1)
SM(1,5) = EBAR*(B1*B3+B*C1*C3)
SM(1,6) = EBAR*(C*B1*C3+B*B3*C1)
SM(2,2) = EBAR*(B*B1*B1+C1*C1)
SM(2,3) = EBAR*(C*B2*C1+B*B1*C2)
SM(2,4) = EBAR*(C1*C2+B*B1*B2)
SM(2,5) = EBAR*(C*B3*C1+B*B1*C3)
SM(2,6) = EBAR*(C1*C3+B*B1*B3)
SM(3,3) = EBAR*(B2*B2+B*C2*C2)
SM(3,4) = EBAR*(C+B)*B2*C2
SM(3,5) = EBAR*(B2*B3+B*C2*C3)
SM(3,6) = EBAR*(C*B2*C3+B*B3*C2)
SM(4,4) = EBAR*(B*B2*B2+C2*C2)
SM(4,5) = EBAR*(C*B3*C2+B*B2*C3)
SM(4,6) = EBAR*(C2*C3+B*B2*B3)
SM(5,5) = EBAR*(B3*B3+B*C3*C3)
```

```
      SM(5,6) = EBAR*(C+B)*B3*C3
      SM(6,6) = EBAR*(B*B3*B3+C3*C3)
      DO 10  J = 1,6
      DO 10  K = J,6
      SM(K,J) = SM(J,K)
   10 CONTINUE
      RETURN
      END
C
C
      SUBROUTINE PLANE
      COMMON   NUM,MBAND,NELM,NBOU,NN,KST
     1        ,NF(60),NBO(60,2),BU(60,2),SM(6,6),T
     2        ,NOD(100,3),X(60),Y(60),A(200,100),F(200)
     3        ,B,C,EBARS
      II = NOD(NN,1)
      JJ = NOD(NN,2)
      KK = NOD(NN,3)
      B1 = Y(JJ) - Y(KK)
      B2 = Y(KK)- Y(II)
      B3 = Y(II)- Y(JJ)
      C1 = X(KK)- X(JJ)
      C2 = X(II)- X(KK)
      C3 = X(JJ)- X(II)
      AREA =.5*(C3*B2-C2*B3)
      AREA =ABS(AREA)
      U1 = F(2*II-1)
      U2 = F(2*JJ-1)
      U3 = F(2*KK-1)
      V1 = F(2*II)
      V2 = F(2*JJ)
      V3 = F(2*KK)
      EX =.5*(B1*U1+B2*U2+B3*U3)/AREA
      EY = .5*(C1*V1+C2*V2+C3*V3)/AREA
      EXY = 0.5*(C1*U1+B1*V1+C2*U2+B2*V2+C3*U3+B3*V3)/AREA
      SX = EBARS * (EX + C * EY)
      SY = EBARS * (EY + C * EX)
      SXY = EBARS * B * EXY
```

```
    WRITE(6,900) NN,EX,EY,EXY,SX,SY,SXY
 900 FORMAT( 5X,I5,1X,6(3X,E11.4))
C
C
 100 RETURN
    END
C*
INPUT
   1  08  09  03   3   2  29000.000     .250     2.000
   1     0.00     0.00
   2     2.0      0.0
   3     4.0      0.0
   4     0.0      3.0
   5     2.0      3.0
   6     4.0      3.0
   7     0.0      6.0
   8     2.0      6.0
   9     4.0      6.0
   1   1   2   4
   2   2   5   4
   3   2   3   5
   4   3   6   5
   5   4   5   7
   6   5   8   7
   7   5   6   8
   8   6   9   8
   1   1   1 0 0
   2   0   1 0 0
   3   0   1 0 0
   7     0.00  -2000.00
   8     0.00  -4000.00
   9     0.00  -2000.00
OUTPUT
1PROBLEM NUMBER ........................................  1
 NUMBER OF ELEMENTS ....................................  8
 NUMBER OF NODES .......................................  9
 NUMBER OF NODAL DISPLACEMENT BOUNDARY CONDITIONS .....  3
 NUMBER OF NODAL FORCE BOUNDARY CONDITIONS ............  3
```

YOUNG"S MODULUS 29000.000
POISSON"S RATIO 0.250
ELEMENT THICKNESS 2.000
PROBLEM TYPE ....PLANE-STRESS
ELEMENT TYPE ....TRIANGULAR

-COORDINATES OF EACH NODAL POINT

NODE NUMBER	X- DIRECTION	Y- DIRECTION
1	0.00	0.00
2	2.00	0.00
3	4.00	0.00
4	0.00	3.00
5	2.00	3.00
6	4.00	3.00
7	0.00	6.00
8	2.00	6.00
9	4.00	6.00

-NODAL NUMBERS OF EACH ELEMENT

ELEMENT NUMBER	FIRST NODE	SECOND NODE	THIRD NODE	FORTH NODE
1	1	2	4	
2	2	5	4	
3	2	3	5	
4	3	6	5	
5	4	5	7	
6	5	8	7	
7	5	6	8	
8	6	9	8	

-BOUNDARY POINTS AND BOUNDARY CONDITIONS

BOUNDARY POINT	BOUNDARY CONDITION	BOUNDARY CONDITION	BOUNDARY DISPLACE.	BOUNDARY DISPLACE.
1	1	1	0.0	0.0
2	0	1	UNKNOWN	0.0
3	0	1	UNKNOWN	0.0

-BOUNDARY POINTS AND BOUNDARY FORCES

BOUNDARY POINT	BOUNDARY FORCE	BOUNDARY FORCE
7	0.0	-2000.0
8	0.0	-4000.0
9	0.0	-2000.0

-DISPLACEMENTS AT EACH NODAL POINT

NODE NUMBER	X- DIRECTION	Y- DIRECTION
1	0.000E+00	0.000E+00
2	0.172E-01	0.000E+00
3	0.345E-01	0.000E+00
4	0.389E-07	-.103E+00
5	0.172E-01	-.103E+00
6	0.345E-01	-.103E+00
7	0.743E-07	-.207E+00
8	0.172E-01	-.207E+00
9	0.345E-01	-.207E+00

-STRAINS AND STRESSES OF EACH ELEMENT

ELEMENT NUMBER	STRAIN EX	STRAIN EY	STRAIN EXY	STRESS SX	STRESS SY	STRESS SXY
1	0.8621E-02	-0.3448E-01	0.1298E-07	0.2305E-03	-0.1000E+04	0.1505E-03
2	0.8621E-02	-0.3448E-01	0.4967E-08	0.0000E+00	-0.1000E+04	0.5762E-04
3	0.8621E-02	-0.3448E-01	0.8692E-08	0.2881E-04	-0.1000E+04	0.1008E-03
4	0.8621E-02	-0.3448E-01	0.1242E-08	0.0000E+00	-0.1000E+04	0.1440E-04
5	0.8621E-02	-0.3448E-01	0.8068E-08	0.0000E+00	-0.1000E+04	0.9359E-04
6	0.8621E-02	-0.3448E-01	0.5588E-08	0.2881E-04	-0.1000E+04	0.6482E-04
7	0.8621E-02	-0.3448E-01	0.5588E-08	0.0000E+00	-0.1000E+04	0.6482E-04
8	0.8621E-02	-0.3448E-01	0.1242E-07	0.0000E+00	-0.1000E+04	0.1440E-03

Index